# Handbook of Occupational Safety and Health

*Fifth Edition*

Publié aussi en français sous le titre
*Guide sur la sécurité et la santé au travail*

Published by
Communications and Coordination Directorate
Treasury Board of Canada

© Minister of Supply and Services Canada 1994

Available in Canada through

Associated Bookstores
and other booksellers

or by mail from

Canada Communication Group - Publishing
Ottawa, Canada  K1A 0S9

Catalogue No. BT45-3/1994E
ISBN 0-660-15415-3

# TABLE OF CONTENTS

## SECTION A - FEDERAL LEGISLATION ON OCCUPATIONAL SAFETY AND HEALTH

## SECTION B - TREASURY BOARD MANUAL, PERSONNEL MANAGEMENT COMPONENT, OCCUPATIONAL SAFETY AND HEALTH VOLUME

Chapter 1 - Policies

Chapter 2 - Directives and Standards

Chapter 3 - Fire Protection Services

Chapter 4 - Procedures

**FOREWORD**

The management of workplace safety and health in the Public Service is governed by the "Occupational Safety and Health" (OSH) volume of the *Treasury Board Manual* (TBM). That OSH volume consists of

. the OSH policy and other policies that foster the health, well-being and productivity of employees, and

. related OSH directives (deemed part of collective agreements), standards, procedures, guides and advisory notices.

This *OSH Handbook* contains all of the above information from the TBM OSH volume, as well as the *Canada Labour Code*, Part II and pursuant regulations that apply to the Public Service.

The handbook is intended as a comprehensive reference book to help departments maintain programs to prevent work-related injuries and illnesses and promote employee health and productivity. Public Service unions have been consulted on the directives through the National Joint Council. Because the content of this handbook is subject to ongoing change, users are advised consult the latest version of the original source document when seeking an official interpretation.

> Safety, Health and Employee Services Group
> Staff Relations Division
> Human Resources Policy Branch

## SECTION A - FEDERAL LEGISLATION ON OCCUPATIONAL SAFETY AND HEALTH

### Introduction

Through provisions of the *Financial Administration Act*, the Public Service of Canada is subject to the *Canada Labour Code,* Part II and the pursuant *Canada Occupational Safety and Health Regulations.* Part II of the Code applies to the Public Service in the same manner and to the same extent as it does to any other federally regulated organization, with a few exceptions. For example, references to the "Board" in the Public Service should be read as the Public Service Staff Relations Board as opposed to the Canada Labour Relations Board.

The purpose of the legislation is to prevent accidents and injury to health arising out of, linked with or occurring in the course of employment. Human Resources Development Canada (HRDC, formerly Labour Canada) is responsible for enforcing the legislation.

HRDC safety officers designated under Part II of the Code are authorized to enter Public Service establishments to ensure compliance with the Code. They conduct inspections and investigations of occupational injuries and illnesses, complaints about safety and health hazards in the workplace, and refusals to work. They obtain assurances of voluntary compliance from departments on corrective measures to be taken where unsafe or unhealthy operations or working conditions exist. In the case of serious or persistent infractions, the safety officers can issue directions and may, finally, initiate prosecutions.

Human Resources Development Canada is also responsible for enforcing the *Non-smokers' Health Act*.

### CANADA LABOUR CODE, PART II
### OCCUPATIONAL SAFETY AND HEALTH

#### Interpretation

122. (1) In this Part,

"Board" means the Canada Labour Relations Board continued by section 9;

"collective agreement" has the same meaning as in section 166;

"danger" means any hazard or condition that could reasonably be expected to cause injury or illness to a person exposed thereto before the hazard or condition can be corrected;

"employee" means a person employed by an employer;

"employer" means a person who employs one or more employees and includes an employers' organization and any person who acts on behalf of an employer;

"hazardous substance" includes a controlled product and a chemical, biological or physical agent that, by reason of a property that the agent possesses, is hazardous to the safety or health of a person exposed to it;

"prescribe" means prescribe by regulation of the Governor in Council;

"regional safety officer" means a person designated as a regional safety officer pursuant to subsection 140(1);

"safety and health committee" means a committee established pursuant to section 135;

"safety and health representative" means a person appointed as a safety and health representative pursuant to section 136;

"safety officer" means a person designated as a safety officer pursuant to subsection 140(1) and includes a regional safety officer;

"work place" means any place where an employee is engaged in work for the employee's employer.

(2) In this Part, the expressions "controlled product", "hazard symbol", "Ingredient Disclosure List", "label" and "material safety data sheet" have the same meanings as in the *Hazardous Products Act*.

(3) Except where otherwise provided in this Part, all other words and expressions have the same meanings as in Part I.

## Purpose of Part

122.1 The purpose of this Part is to prevent accidents and injury to health arising out of, linked with or occurring in the course of employment to which this Part applies.

## Application

123.(1) Notwithstanding any other Act of Parliament or any regulations thereunder, this Part applies to and in respect of employment

(a)  on or in connection with the operation of any federal work, undertaking or business other than a work, undertaking or business of a local or private nature in the Yukon Territory or Northwest Territories;
(b)  by a corporation established to perform any function or duty on behalf of the Government of Canada; and
(c)  by a Canadian carrier, as defined in section 2 of the *Telecommunications Act*, that is an agent of Her Majesty in right of a province.

(2)  Except as otherwise expressly provided in any other Act of Parliament, this Part does not apply to or in respect of employment in a portion of the public service of Canada specified from time to time in Schedule I to the *Public Service Staff Relations Act*.

123.1 The Governor in Council may by order exclude, in whole or in part, from the application of this Part or any specified provision thereof employment on or in connection with any work or undertaking that is regulated pursuant to the *Atomic Energy Control Act.*

## Duties of Employers

124. Every employer shall ensure that the safety and health at work of every person employed by the employer is protected.

125. Without restricting the generality of section 124, every employer shall, in respect of every work place controlled by the employer,

(a) ensure that all permanent and temporary buildings and structures meet the prescribed standards;

(b) install guards, guard rails, barricades and fences in accordance with prescribed standards;

(c) investigate, record and report in the manner and to the authorities as prescribed all accidents, occupational diseases and other hazardous occurrences known to the employer;

(d) post at a place accessible to every employee and at every place directed by a safety officer,
  (i) a copy of this Part,
  (ii) a statement of the employer's general policy concerning the safety and health at work of employees, and
  (iii) such other printed material related to safety and health as may be directed by a safety officer or as is prescribed;

(e) keep and maintain in prescribed form and manner prescribed safety and health records;

(f) provide such first-aid facilities and health services as are prescribed;

(g) provide prescribed sanitary and personal facilities;

(h) provide, in accordance with prescribed standards, potable water;

(i) ensure that the vehicles and mobile equipment used by the employees in the course of their employment meet prescribed safety standards;

(j) provide every person granted access to the work place by the employer with such safety materials, equipment, devices and clothing as are prescribed;

(k) ensure that the use, operation and maintenance of
  (i) every boiler and pressure vessel,
  (ii) every escalator, elevator and other device for moving passengers or freight,
  (iii) all equipment for the generation, distribution or use of electricity, and
  (iv) all gas or oil burning equipment or other heat generating equipment is in accordance with prescribed standards;

(l) & (m) [Repealed, R.S., 1985, c. 24 (3rd Supp.), s. 4];

(n) ensure that the levels of ventilation, lighting, temperature, humidity, sound and vibration are in accordance with prescribed standards;

(o) comply with such standards as are prescribed relating to fire safety and emergency measures;

(p) ensure, in the manner prescribed, that employees have safe entry to, exit from and occupancy of the work place;

(q)  provide, in the prescribed manner, each employee with the information, instruction, training and supervision necessary to ensure the safety and health at work of that employee;

(r)  maintain all installed guards, guard-rails, barricades and fences in accordance with prescribed standards;

(s)  ensure that each employee is made aware of every known or foreseeable safety or health hazard in the area where that employee works;

(t)  ensure that the machinery, equipment and tools used by the employees in the course of their employment meet prescribed safety standards and are safe under all conditions of their intended use;

(u)  adopt and implement prescribed safety codes and safety standards;

(v)  ensure that every person granted access to the work place by the employer is familiar with and uses in the prescribed circumstances and manner all prescribed safety materials, equipment, devices and clothing; and

(w)  comply with every oral or written direction given to the employer by a safety officer concerning the safety and health of employees.

125.1 Without restricting the generality of section 124 or limiting the duties of an employer under section 125 but subject to such exceptions as may be prescribed, every employer shall, in respect of every place controlled by the employer,

(a)  ensure that concentrations of hazardous substances in the work place are controlled in accordance with prescribed standards;

(b)  ensure that all hazardous substances in the work place are stored and handled in the manner prescribed;

(c)  ensure that all hazardous substances in the work place, other than controlled products, are identified in the manner prescribed;

(d)  subject to the *Hazardous Materials Information Review Act*, ensure that each controlled product in the work place or each container in the work place in which a controlled product is contained has applied to it a label that discloses prescribed information and has displayed on it, in the manner prescribed, all applicable prescribed hazard symbols; and

(e)  subject to the *Hazardous Materials Information Review Act*, make available, in the manner prescribed, to each of his employees a material safety data sheet, with respect to each controlled product in the work place, that discloses the following information, namely,

   (i)  where the controlled product is a pure substance, the chemical identity of the controlled product and, where the controlled product is not a pure substance, the chemical identity of any ingredient thereof that is a controlled product and the concentration of that ingredient,

   (ii)  where the controlled product contains an ingredient that is included in the Ingredient Disclosure List and the ingredient is in a concentration that is equal to or greater than the concentration specified in the Ingredient Disclosure List for that ingredient, the chemical identity and concentration of that ingredient,

   (iii)  the chemical identity of any ingredient thereof that the employer believes on reasonable grounds may be harmful to an employee and the concentration of that ingredient,

(iv) the chemical identity of any ingredient thereof the toxicological properties of which are not known to the employer and the concentration of that ingredient, and

(v) such other information with respect to the controlled product as may be prescribed.

125.2(1) An employer shall, in respect of any controlled product in a work place controlled by him, provide, as soon as is practicable in the circumstances, any information referred to in paragraph 125.1(e) that is in the employer's possession to any physician or other prescribed medical professional who requests that information for the purpose of making a medical diagnosis of, or rendering medical treatment to, an employee in an emergency.

(2) Any physician or other prescribed medical professional to whom information is provided by an employer pursuant to subsection (1) shall keep confidential any information specified by the employer as being confidential, except for the purpose for which it is provided.

125.3(1) Every employer of employees employed in a coal mine shall

(a) comply with every condition imposed on the employer pursuant to paragraph 137.2(2)(b) or (3)(a);

(b) comply with every provision substituted for a provision of the regulations, in respect of the employer, pursuant to paragraph 137.2(3)(b);

(c) permit inspections and tests to be carried out on behalf of the employees, in any part of the mine and on any machinery or equipment therein, in the prescribed manner and at intervals not greater than the prescribed interval; and

(d) as a condition of carrying out any activity for which the submission of plans and procedures is prescribed, submit to the Coal Mining Safety Commission for approval, in the form and manner and at the time prescribed, plans and procedures relating to that activity and carry out the activity in conformity with plans and procedures as approved.

(2) No employer shall require or permit the use in a coal mine of any mining method, machinery or equipment in respect of which no prescribed safety standards are applicable unless the use thereof has been approved pursuant to paragraph 137.2(2)(a).

(3) Every employer of employees employed in a coal mine shall, at intervals not greater than the prescribed interval, for the purpose of preventing alcohol, articles for use in smoking and drugs, other than drugs exempted by the regulations, from being brought into the mine,

(a) require every person entering an underground portion of the mine who is not employed there to submit to a personal search conducted in the prescribed manner; and

(b) require a proportion, not less than the prescribed proportion, of employees employed in the underground portions of the mine to submit to personal searches conducted in the prescribed manner.

(4) For the purposes of this section and section 137.2, "coal mine" includes any work place above ground that is used in the operation of the mine and is under the control of the employer of employees employed in the mine.

## Duties of Employees

126.(1) While at work, every employee shall

(a) use such safety materials, equipment, devices and clothing as are intended for the employee's protection and furnished to the employee by the employer or as are prescribed;
(b) follow prescribed procedures with respect to the safety and health of employees;
(c) take all reasonable and necessary precautions to ensure the safety and health of the employee, the other employees and any person likely to be affected by the employee's acts or omissions;
(d) comply with all instructions from the employer concerning the safety and health of employees;
(e) cooperate with any person exercising a duty imposed by this Part or any regulations made thereunder;
(f) cooperate with the safety and health committee established for the work place where the employee is employed or, if there is no such committee, with the safety and health representative, if any, appointed for that work place;
(g) report to the employer any thing or circumstance in a work place that is likely to be hazardous to the safety or health of the employee, the other employees or other persons granted access to the work place by the employer;
(h) report in the manner prescribed every accident or other occurrence arising in the course of or in connection with the employee's work that has caused injury to the employee or to any other person; and
(i) comply with every oral or written direction of a safety officer concerning the safety and health of employees.

(2) Nothing in subsection (1) relieves an employer from any duty imposed on the employer under this Part.

## Employment Safety

127.(1) Subject to subsection (2), where an employee is killed or seriously injured in a work place, no person shall, unless authorized to do so by a safety officer, remove or in any way interfere with or disturb any wreckage, article or thing related to the incident except to the extent necessary to

(a) save a life, prevent injury or relieve human suffering in the vicinity;
(b) maintain an essential public service; or
(c) prevent unnecessary damage to or loss of property.

(2) No authorization referred to in subsection (1) is required where an employee is killed or seriously injured by an accident or incident involving

(a) an aircraft, a ship, rolling stock or a commodity pipeline, where the accident or incident is being investigated under the *Aeronautics Act*, the *Canada Shipping Act*, the *Railway Act* or the *Canadian Transportation Accident Investigation and Safety Board Act*; or

(b) a motor vehicle on a public highway.

128.(1) Subject to this section, where an employee while at work has reasonable cause to believe that

(a) the use or operation of a machine or thing constitutes a danger to the employee or to another employee, or

(b) a condition exists in any place that constitutes a danger to the employee,

the employee may refuse to use or operate the machine or thing or to work in that place.

(2) An employee may not pursuant to this section refuse to use or operate a machine or thing or to work in a place where

(a) the refusal puts the life, health or safety of another person directly in danger; or

(b) the danger referred to in subsection (1) is inherent in the employee's work or is a normal condition of employment.

(3) Where an employee on a ship or an aircraft that is in operation has reasonable cause to believe that

(a) the use or operation of a machine or thing on the ship or aircraft constitutes a danger to the employee or to another employee, or

(b) a condition exists in a place on the ship or aircraft that constitutes a danger to the employee,

the employee shall forthwith notify the person in charge of the ship or aircraft of the circumstances of the danger and the person in charge shall, as soon as practicable thereafter, having regard to the safe operation of the ship or aircraft, decide whether or not the employee may discontinue the use or operation of the machine or thing or to work in that place and shall inform the employee accordingly.

(4) An employee who, pursuant to subsection (3), is informed that he may not discontinue the use or operation of a machine or thing or to work in a place shall not, while the ship or aircraft on which the employee is employed is in operation, refuse pursuant to this section to operate the machine or thing or to work in that place.

(5) For the purposes of subsections (3) and (4),

(a) a ship is in operation from the time it casts off from a wharf in any Canadian or foreign port until it is next secured alongside a wharf in Canada; and

(b)     an aircraft is in operation from the time it first moves under its own power for the purpose of taking off from any Canadian or foreign place of departure until it comes to rest at the end of its flight to its first destination in Canada.

(6)  Where an employee refuses to use or operate a machine or thing or to work in a place pursuant to subsection (1), or is prevented from acting in accordance with that subsection pursuant to subsection (4), the employee shall forthwith report the circumstances of the matter to his employer and to

(a)     a member of the safety and health committee, if any, established for the work place affected; or
(b)     the safety and health representative, if any, appointed for the work place affected.

(7)  An employer shall forthwith on receipt of a report under subsection (6) investigate the report in the presence of the employee who made the report and in the presence of

(a)     at least one member of the safety and health committee, if any, to which the report was made under subsection (6) who does not exercise managerial functions;
(b)     the safety and health representative, if any; or
(c)     where no safety and health committee or safety and health representative has been established or appointed for the work place affected, at least one person selected by the employee.

(8)  Where an employer disputes a report made to the employer by an employee pursuant to subsection (6) or takes steps to make the machine or thing or the place in respect of which the report was made safe, and the employee has reasonable cause to believe that

(a)     the use or operation of the machine or thing continues to constitute a danger to the employee or to another employee, or
(b)     a condition continues to exist in the place that constitutes a danger to the employee,

the employee may continue to refuse to use or operate the machine or thing or to work in that place.

129.(1)  Where an employee continues to refuse to use or operate a machine or thing or to work in a place pursuant to subsection 128(8), the employer and the employee shall each forthwith notify a safety officer, and the safety officer shall forthwith, on receipt of either notification, investigate or cause another safety officer to investigate the matter in the presence of the employer and the employee or the employee's representative.

(2)  A safety officer shall, on completion of an investigation made pursuant to subsection (1), decide whether or not

(a)    the use or operation of the machine or thing in respect of which the investigation was made constitutes a danger to any employee, or

(b)    a condition exists in the place in respect of which the investigation was made that constitutes a danger to the employee referred to in subsection (1),

and he shall forthwith notify the employer and the employee of his decision.

(3) Prior to the investigation and decision of a safety officer under this section, the employer may require that the employee concerned remain at a safe location near the place in respect of which the investigation is being made or assign the employee reasonable alternate work, and shall not assign any other employee to use or operate the machine or thing or to work in that place unless that other employee has been advised of the refusal of the employee concerned.

(4) Where a safety officer decides that the use or operation of a machine or thing constitutes a danger to an employee or that a condition exists in a place that constitutes a danger to an employee, the officer shall give such direction under subsection 145(2) as the officer considers appropriate, and an employee may continue to refuse to use or operate the machine or thing or to work in that place until the direction is complied with or until it is varied or rescinded under this Part.

(5) Where a safety officer decides that the use or operation of a machine or thing does not constitute a danger to an employee or that a condition does not exist in a place that constitutes a danger to an employee, an employee is not entitled under section 128 or this section to continue to refuse to use or operate the machine or thing or to work in that place, but the employee may, by notice in writing given within seven days of receiving notice of the decision of a safety officer, require the safety officer to refer his decision to the Board, and thereupon the safety officer shall refer the decision to the Board.

130.(1) Where a decision of a safety officer is referred to the Board pursuant to subsection 129(5), the Board shall, without delay and in a summary way, inquire into the circumstances of the decision and the reasons therefor and may

(a)    confirm the decision; or

(b)    give any direction that it considers appropriate in respect of the machine, thing or place in respect of which the decision was made that a safety officer is required or entitled to give under subsection 145(2).

(2) Where the Board gives a direction under subsection (1), it shall cause to be affixed to or near the machine, thing or place in respect of which the direction is given a notice in the form approved by the Minister, and no person shall remove the notice unless authorized by a safety officer or the Board.

(3) Where the Board directs, pursuant to subsection (1), that a machine, thing or place not be used until its directions are complied with, the employer shall discontinue the use thereof, and no person shall use such machine, thing or

place until the directions are complied with, but nothing in this subsection prevents the doing of anything necessary for the proper compliance therewith.

131. The Minister may, on the joint application of the parties to a collective agreement if the Minister is satisfied that the agreement contains provisions that are at least as effective as those under sections 128 to 130 in protecting the employees to whom the agreement relates from danger to their safety or health, exclude the employees from the application of those sections for the period during which the agreement remains in force.

132. The fact that an employer or employee has complied with or failed to comply with any of the provisions of this Part shall not be construed as affecting any right of an employee to compensation under any statute relating to compensation for employment injury, or as affecting any liability or obligation of any employer or employee under any such statute.

**Employees' Right To Complain**

133.(1) Where an employee alleges that an employer has taken action against the employee in contravention of paragraph 147(a) because the employee has acted in accordance with section 128 or 129, the employee may, subject to subsection (3), make a complaint in writing to the Board of the alleged contravention.

(2) A complaint made pursuant to subsection (1) shall be made to the Board not later than ninety days from the date on which the complainant knew, or in the opinion of the Board ought to have known, of the action or circumstances giving rise to the complaint.

(3) An employee may not make a complaint under this section if the employee has failed to comply with subsection 128(6) or 129(1) in relation to the matter that is the subject-matter of the complaint.

(4) Notwithstanding any law or agreement to the contrary, a complaint referred to in subsection (1) may not be referred by an employee to arbitration.

(5) On receipt of a complaint made under subsection (1), the Board may assist the parties to the complaint to settle the complaint and shall, where it decides not to so assist the parties or the complaint is not settled within a period considered by the Board to be reasonable in the circumstances, hear and determine the complaint.

(6) A complaint made pursuant to subsection (1) in respect of an alleged contravention of paragraph 147(a) by an employer is itself evidence that that contravention actually occurred and, if any party to the complaint proceedings alleges that the contravention did not occur, the burden of proof thereof is on that party.

134. Where, under subsection 133(5), the Board determines that an employer has contravened paragraph 147(a), the Board may, by order, require the

employer to cease contravening that provision and may, where applicable, by order, require the employer to

(a) permit any employee who has been affected by the contravention to return to the duties of the employee's employment;
(b) reinstate any former employee affected by the contravention;
(c) pay to any employee or former employee affected by the contravention compensation not exceeding such sum as, in the opinion of the Board, is equivalent to the remuneration that would, but for the contravention, have been paid by the employer to that employee or former employee; and
(d) rescind any disciplinary action taken in respect of and pay compensation to any employee affected by the contravention, not exceeding such sum as, in the opinion of the Board, is equivalent to any financial or other penalty imposed on the employee by the employer.

## Safety and Health Committees and Representatives

135.(1) Subject to this section, every employer shall, for each work place controlled by the employer at which twenty or more employees are normally employed, establish a safety and health committee consisting of at least two persons one of whom is an employee or, where the committee consists of more than two persons, at least half of whom are employees who

(a) do not exercise managerial functions; and
(b) subject to any regulations made under subsection (11), have been selected by the trade union, if any, representing the employees and by any employees not represented by a trade union.

(2) An employer is not required to establish a safety and health committee under subsection (1) for a work place that is on board a ship in respect of employees whose base is the ship.

(3) Where the Minister is satisfied that the nature of work being done by employees at a work place is relatively free from risks to safety and health, the Minister may, by order, on such terms and conditions as are specified therein, exempt the employer from the requirements of subsection (1) in respect of that work place.

(4) Where, pursuant to a collective agreement or any other agreement between an employer and his employees, a committee of persons has been appointed in respect of a work place controlled by an employer and the committee has, in the opinion of a safety officer, a responsibility for matters relating to safety and health in the work place to such an extent that a safety and health committee established under subsection (1) for that work place would not be necessary,

(a) the safety officer may, by order, exempt the employer from the requirements of subsection (1) in respect of that work place;
(b) the committee of persons that has been appointed for the work place has, in addition to any rights, functions, powers, privileges and obligations under the agreement, the same rights, functions, powers, privileges and obligations as a safety and health committee under this Part; and

(c) the committee of persons so appointed shall, for the purposes of this Part, be deemed to be a safety and health committee established under subsection (1) and all rights and obligations of employers and employees under this Part and the provisions of this Part respecting a safety and health committee apply, with such modifications as the circumstances require, in respect of the committee of persons so appointed.

(5) An employer shall post and keep posted, in a conspicuous place or places where they are likely to come to the attention of the employer's employees, the names and work locations of all the members of the safety and health committee established for the work place controlled by the employer.

(6) A safety and health committee

(a) shall receive, consider and expeditiously dispose of complaints relating to the safety and health of the employees represented by the committee;

(b) shall maintain records pertaining to the disposition of complaints relating to the safety and health of the employees represented by the committee;

(c) shall cooperate with any occupational health service established to serve the work place;

(d) may establish and promote safety and health programs for the education of the employees represented by the committee;

(e) shall participate in all inquiries and investigations pertaining to occupational safety and health including such consultations as may be necessary with persons who are professionally or technically qualified to advise the committee on those matters;

(f) may develop, establish and maintain programs, measures and procedures for the protection or improvement of the safety and health of employees;

(g) shall regularly monitor programs, measures and procedures related to the safety and health of employees;

(h) shall ensure that adequate records are kept on work accidents, injuries and health hazards and shall regularly monitor data relating to those accidents, injuries and hazards;

(i) shall cooperate with safety officers;

(j) may request from an employer such information as the committee considers necessary to identify existing or potential hazards with respect to materials, processes or equipment in the work place; and

(k) shall have full access to all government and employer reports relating to the safety and health of the employees represented by the committee but shall not have access to the medical records of any person except with the consent of that person.

(7) A safety and health committee shall keep accurate records of all matters that come before it pursuant to subsection (6) and shall keep minutes of its meetings and shall make those minutes and records available to a safety officer on the officer's request.

(8) A safety and health committee shall meet during regular working hours at least once each month and, where meetings are urgently required as a result of an emergency or other special circumstance, the committee shall meet as required whether or not during regular working hours.

(9) The members of a safety and health committee are entitled to such time from their work as is necessary to attend meetings or to carry out any of the other functions of a member of the committee, and any time spent by a member while carrying out any of the functions of a member of the committee shall, for the purpose of calculating wages owing to that member, be deemed to have been spent at work.

(10) No member of a safety and health committee is personally liable for anything done or omitted to be done by the member in good faith under the purported authority of this section or any regulations made under this section.

(11) The Governor in Council may make regulations

(a) specifying the qualifications, terms of office and manner of selection of members of a safety and health committee;

(b) specifying the time, place and frequency of regular meetings of a committee;

(c) specifying the method of selecting officers of a committee and their terms of office;

(d) establishing such procedures for the operation of a committee as the Governor in Council considers advisable;

(e) requiring copies of minutes of committee meetings to be provided by and to such persons as the Governor in Council may prescribe; and

(f) requiring a safety and health committee to submit an annual report of its activities to a specified person in the prescribed form within the prescribed time.

(12) Any regulation made pursuant to subsection (11) may be made applicable generally to all safety and health committees, or particularly to one or more committees or classes thereof.

(13) Subject to any regulations made pursuant to subsection (11), a safety and health committee may establish its own rules of procedure in respect of the terms of office, not exceeding two years, of its members, the time, place and frequency of regular meetings of the committee, and such procedures for its operation as it considers advisable.

136.(1) Every employer shall, for each work place controlled by the employer at which five or more employees are normally employed and for which no safety and health committee has been established, appoint the person selected pursuant to subsection (2) as the safety and health representative for that work place.

(2) The employees at a work place referred to in subsection (1) who do not exercise managerial functions shall, or where those employees are represented by a trade union, the trade union shall, in consultation with any employees who are not so represented and subject to any regulations under subsection (7), select from among those employees a person to be appointed as the safety and health representative of that work place and shall advise the employer in writing of the name of the person so selected.

(3) An employer shall post and keep posted, in a conspicuous place or places where they are likely to come to the attention of the employer's employees, the name and work location of the safety and health representative appointed for the work place controlled by the employer.

(4) A safety and health representative

(a) shall receive, consider and expeditiously dispose of complaints relating to the safety and health of the employees represented by the representative;

(b) shall participate in all inquiries and investigations pertaining to occupational safety and health including such consultations as may be necessary with persons who are professionally or technically qualified to advise the representative on those matters;

(c) shall regularly monitor programs, measures and procedures related to the safety and health of employees;

(d) shall ensure that adequate records are kept on work accidents, injuries and health hazards and shall regularly monitor data relating to those accidents, injuries and hazards;

(e) may request from an employer such information as the representative considers necessary to identify existing or potential hazards with respect to materials, processes or equipment in the work place; and

(f) shall have full access to all government and employer reports relating to the safety and health of the employees represented by the representative but shall not have access to the medical records of any person except with the consent of that person.

(5) A safety and health representative is entitled to such time from the representative's work as is necessary to carry out the functions of a representative and any time spent by the representative while carrying out any of those functions shall, for the purpose of calculating wages owing to the representative, be deemed to have been spent at work.

(6) No safety and health representative is personally liable for anything done or omitted to be done by the representative in good faith under the purported authority of this section.

(7) The Governor in Council may make regulations specifying the qualifications, term of office and manner of selection of a safety and health representative.

137. Notwithstanding sections 135 and 136, where an employer controls more than one work place referred to in section 135 or 136 or the size or nature of the operations of the employer or the work place precludes the effective functioning of a single safety and health committee or safety and health representative, as the case may be, for those work places, the employer shall, subject to the approval of or in accordance with the direction of a safety officer, establish or appoint in accordance with section 135 or 136, as the case may require, a safety and health committee or safety and health representative for such of those work places as are specified in the approval or direction.

## Coal Mining Safety Commission

**137.1(1)** There is hereby established a Coal Mining Safety Commission in this section referred to as the "Commission", consisting of not more than five members to be appointed by the Minister to hold office during good behavior.

(2) One member of the Commission shall be designated chairman of the Commission by the Minister and the others shall be equally representative of non-supervisory employees employed in coal mines and of the employers of those employees.

(3) The manner of selection of the members of the Commission other than the chairman and the term of office of the members of the Commission shall be such as may be prescribed.

(4) A quorum of the Commission consists of the chairman, one member representative of employees and one member representative of employers.

(5) No safety officer is eligible to be appointed to the Commission or to be designated for the purposes of subsection 137.2 (1) or (2).

(6) The members of the Commission shall be paid such remuneration as may be fixed by the Governor in Council and, subject to the approval of the Treasury Board, such reasonable travel and living expenses as are incurred by them while carrying out their functions away from their ordinary place of residence.

(7) The Commission may, subject to the approval of the Minister, make by-laws for the conduct of its activities.

(8) The Minister may, at the request of the Commission, make available to the Commission such staff and other assistance as are necessary for the proper conduct of its activities.

(9) The Commission shall, within sixty days following the end of each calendar year, submit a report to the Minister of its activities during the year.

(10) No member of the Commission and no person designated by the Commission pursuant to subsection 137.2 (1) or (2) is personally liable for anything done or omitted to be done in good faith under section 137.2.

**137.2(1)** The Commission or a person designated by the Commission for the purposes of this subsection may approve in writing, with or without modification, plans or procedures submitted in accordance with paragraph 125.3(1)(d).

(2) On the application of an employer, the Commission or a person designated by the Commission for the purposes of this subsection may, where, in the opinion of the Commission or that person, protection of the safety and health of employees would not thereby be diminished,

(a)  approve in writing the use by the employer in coal mines of mining methods, machinery or equipment in respect of which no prescribed safety standards are applicable; or

(b)  approve in writing, notwithstanding anything in this Part, the use by the employer in coal mines, for a specified time and subject to specified conditions, of any mining method, machinery or equipment that does not meet prescribed safety standards applicable in respect of it.

(3)  On the application of an employer, the Commission may, where in its opinion protection of the safety and health of employees would not thereby be diminished, by order,

(a)  exempt the employer from compliance with any provision of the regulations in the operation of coal mines controlled by the employer, subject to any conditions contained in the order; or

(b)  substitute for any provision of the regulations, so far as it applies to coal mines controlled by the employer, another provision having substantially the same purpose and effect.

(4)  The Commission may make recommendations to the Minister for amending or revoking any provision of the regulations applicable to coal mines or for adding any provision thereto.

**Administration**

138.(1)  The Minister may appoint committees of persons to assist or advise the Minister on any matter that the Minister considers advisable concerning occupational safety and health related to employment to which this Part applies.

(2)  The Minister may cause an inquiry to be made into and concerning occupational safety and health in any employment to which this Part applies and may appoint one or more persons to hold the inquiry.

(3)  A person appointed pursuant to subsection (2) has all the powers of a person appointed as a commissioner under Part I of the *Inquiries Act*.

(4)  The Minister may undertake research into the cause of and the means of preventing employment injury and occupational illness and may, where the Minister deems it appropriate, undertake such research in cooperation with any department or agency of the Government of Canada or with any or all provinces or with any organization undertaking similar research.

(5)  The Minister may publish the results of any research undertaken pursuant to subsection (4) and compile, prepare and disseminate data or information bearing on safety or health of employees obtained from that research or otherwise.

(6)  The Minister may undertake programs to reduce or prevent employment injury and occupational illness and may, where the Minister deems it appropriate, undertake those programs in cooperation with any department or

agency of the Government of Canada or with any or all provinces or any organization undertaking similar programs.

## Advisory Council on Occupational Safety and Health

139.(1) There is hereby established a council to be called the Advisory Council on Occupational Safety and Health composed of not more than fourteen members appointed by the Governor in Council to hold office during pleasure.

(2) The membership of the Advisory Council shall consist of persons equally representative of management and labour and not more than two additional persons who are knowledgeable or concerned about occupational safety and health matters.

(3) The Governor in Council shall designate one of the members of the Advisory Council to be chairman and one to be vice-chairman.

(4) The members of the Advisory Council shall be paid such remuneration as may be fixed by the Governor in Council and are entitled, within such limits as may be established by the Treasury Board, to be paid such reasonable travel and living expenses as they incur in the course of their functions under this Act while absent from their ordinary places of residence.

(5) The Advisory Council may, subject to the approval of the Minister, make by-laws for the management of its internal affairs and generally for the conduct of its activities.

(6) The Advisory Council shall

(a) provide advice to the Minister on matters that have been brought to its attention or referred to it concerning occupational safety and health related to employment to which this Part applies;
(b) make recommendations to the Minister concerning the administration of this Part; and
(c) annually report to the Minister on the activities of the Advisory Council during the year for which the report is made.

## Safety Officers and Safety Services

140.(1) The Minister may designate any person as a regional safety officer or as a safety officer for the purposes of this Part.

(2) The Minister may, with the approval of the Governor in Council, enter into an agreement with any province or any provincial body specifying the terms and conditions under which a person employed by that province or provincial body may act as a safety officer for the purposes of this Part and, where such an agreement has been entered into, a person so employed and referred to in the agreement shall be deemed to be designated as a safety officer under subsection (1).

141.(1) A safety officer may, in the performance of the officer's duties and at any reasonable time, enter any work place controlled by an employer and, in respect of any work place, may

(a) conduct examinations, tests, inquiries and inspections or direct the employer to conduct them;
(b) take or remove for analysis, samples of any material or substance or any biological, chemical or physical agent;
(c) be accompanied and assisted by such persons and bring with him such equipment as the safety officer deems necessary to carry out his duties;
(d) take photographs and make sketches;
(e) direct the employer to ensure that any place or thing specified by the safety officer not be disturbed for a reasonable period of time pending an examination, test, inquiry or inspection in relation thereto;
(f) direct the employer to produce documents and information relating to the safety and health of his employees or the safety of the work place and to permit the safety officer to examine and make copies of or extracts from those documents and that information; and
(g) direct the employer to make or provide statements, in such form and manner as the safety officer may specify, respecting working conditions and material and equipment that affect the safety or health of employees.

(2) The Minister shall furnish every safety officer with a certificate of the officer's authority and on entering any work place a safety officer shall, if so required, produce the certificate to the person in charge of that work place.

142. The person in charge of any work place and every person employed at, or in connection with, that work place shall give a safety officer all reasonable assistance to enable the officer to carry out his duties under this Part.

143. No person shall obstruct or hinder, or make a false or misleading statement either orally or in writing to, a safety officer engaged in carrying out his duties under this Part.

144.(1) No safety officer or person who, as a member of a safety and health committee or as a safety and health representative, has assisted the safety officer in carrying out the officer's duties under this Part shall be required to give testimony in any civil suit with regard to information obtained by him in the discharge of his duties except with the written permission of the Minister.

(2) Subject to subsection (2.1), no safety officer who is admitted to any work place pursuant to the powers conferred on a safety officer by section 141 or person accompanying a safety officer therein shall disclose to any person any information obtained by him therein with regard to any secret process or trade secret, except for the purposes of this Part or as required by law.

(2.1) All information that, pursuant to the *Hazardous Materials Information Review Act*, an employer is exempt from disclosing under paragraph 125.1(d) or (e) or under paragraph 13(a) or (b) or 14(a) or (b) of the *Hazardous Products Act* and that is obtained, in a work place controlled by the employer, by a safety officer who is admitted to the work place, pursuant to the powers conferred by section 141 on a safety officer, or by a person accompanying a safety officer therein is privileged and, notwithstanding the *Access to Information Act* or any other Act or law, shall not be disclosed to any other person except for the purpose of this Part.

(3) No person shall, except for the purposes of this Part or for the purposes of a prosecution under this Part, publish or disclose the results of any analysis, examination, testing, inquiry or sampling made or taken by or at the request of a safety officer pursuant to section 141.

(4) No person to whom information obtained pursuant to section 141 is communicated in confidence shall divulge the name of the informant to any person except for the purposes of this Part or is competent or compellable to divulge the name of the informant before any court or other tribunal.

(5) A safety officer is not personally liable for anything done or omitted to be done by the officer in good faith under the authority or purported authority of this Part.

**Special Safety Measures**

145.(1) Where a safety officer is of the opinion that any provision of this Part is being contravened, the officer may direct the employer or employee concerned to terminate the contravention within such time as the officer may specify and the officer shall, if requested by the employer or employee concerned, confirm the direction in writing if the direction was given orally.

(2) Where a safety officer considers that the use or operation of a machine or thing or a condition in any place constitutes a danger to an employee while at work,

(a)    the safety officer shall notify the employer of the danger and issue directions in writing to the employer directing the employer immediately or within such period of time as the safety officer specifies
    (i)    to take measures for guarding the source of danger, or
    (ii)    to protect any person from the danger; and
(b)    the safety officer may, if the officer considers that the danger cannot otherwise be guarded or protected against immediately, issue a direction in writing to the employer directing that the place, machine or thing in respect of which the direction is made shall not be used or operated until the officer's directions are complied with, but nothing in this paragraph prevents the doing of anything necessary for the proper compliance with the direction.

(3) Where a safety officer issues a direction under paragraph (2)(b), the officer shall affix to or near the place, machine or thing in respect of which the direction is made, a notice in such form and containing such information as the Minister may specify, and no person shall remove the notice unless authorized by a safety officer.

(4) Where a safety officer issues a direction under paragraph (2)(b) in respect of any place, machine or thing, the employer shall discontinue the use or operation of the place, machine or thing and no person shall use or operate it until the measures directed by the officer have been taken.

(5) Where a safety officer issues a direction in writing under subsection (1) or (2) or makes a report in writing to an employer on any matter under this Part, the employer shall forthwith

(a)    cause a copy or copies of the direction or report to be posted in such manner as the safety officer may specify; and

(b)    give a copy of the direction or report to the safety and health committee, if any, for the work place affected or the safety and health representative, if any, for that work place.

(6) Where a safety officer issues a direction in writing under subsection (1) or (2) or makes a report referred to in subsection (5) in respect of an investigation made by the officer pursuant to a complaint, the officer shall forthwith give a copy of the direction or report to each person, if any, whose complaint led to the investigation.

146.(1) Any employer, employee or trade union that considers himself or itself aggrieved by any direction issued by a safety officer under this Part may, within fourteen days of the date of the direction, request that the direction be reviewed by a regional safety officer for the region in which the place, machine or thing in respect of which the direction was issued is situated.

(2) The regional safety officer may require that an oral request for a review under subsection (1) be made as well in writing.

(3) The regional safety officer shall in a summary way inquire into the circumstances of the direction to be reviewed and the need therefor and may vary, rescind or confirm the direction and thereupon shall in writing notify the employee, employer or trade union concerned of the decision taken.

(4) A request for a review of a direction under this section shall not operate as a stay of the direction.

(5) Subsection (1) does not apply in respect of a direction of a safety officer that is based on a decision of the officer that has been referred to the Board pursuant to subsection 129(5).

147. No employer shall

(a)    dismiss, suspend, lay off or demote an employee or impose any financial or other penalty on an employee or refuse to pay the employee remuneration in respect of any period of time that the employee would, but for the exercise of his rights under this Part, have worked or take any disciplinary action against or threaten to take any such action against an employee because that employee

     (i)    has testified or is about to testify in any proceeding taken or inquiry held under this Part,

     (ii)   has provided information to a person engaged in the performance of duties under this Part regarding the conditions of work affecting the safety or health of that employee or any of his fellow employees, or

     (iii)  has acted in accordance with this Part or has sought the enforcement of any of the provisions of this Part; or

(b)    fail or neglect to provide

     (i)    a safety and health committee with any information requested by it pursuant to paragraph 135(6)(j), or

     (ii)   a safety and health representative with any information requested by the representative pursuant to paragraph 136(4)(e).

## Offences and Punishment

148.(1) Subject to this section, every person who contravenes any provision of this Part is guilty of an offence and liable on summary conviction to a fine not exceeding fifteen thousand dollars.

(2) Every person who contravenes paragraph 125(w) or 126(1)(i) is guilty of an offence and liable on summary conviction to a fine not exceeding twenty-five thousand dollars.

(3) Every person who contravenes

(a)    paragraph 125(d), (e) or (g) or 126(1)(e), (f) or (h), or

(b)    subsection 135(5), 136(3) or 145(5) is guilty of an offence and liable on summary conviction to a fine not exceeding five thousand dollars.

(3.1) Every person who contravenes paragraph 125.1(c), (d) or (e) or subsection 125.2(1) or (2) or 144(2.1) is guilty of an offence and liable

(a)    on summary conviction, to a fine not exceeding one hundred thousand dollars or to imprisonment for a term not exceeding six months or to both; or

(b)    on conviction on indictment, to a fine not exceeding one million dollars or to imprisonment for a term not exceeding two years or to both.

(4) Every person who contravenes any provision of this Part the direct result of which is the death of or serious injury to an employee is guilty of an offence and liable on summary conviction to a fine not exceeding one hundred thousand dollars.

(5) Every person who wilfully contravenes any provision of this Part knowing that the contravention is likely to cause the death of or serious injury to an employee is guilty of an offence and liable

(a)    on summary conviction, to a fine not exceeding twenty-five thousand dollars; or

(b)    on conviction on indictment, to imprisonment for a term not exceeding two years.

(6) On a prosecution of a person for a contravention of subsection (4) or

(a)   paragraph 125(q), (r), (s), (t), (u), (v) or (w),
(b)   paragraph 126(1)(c), (d), (e), (f), (g), (h) or (i),
(c)   paragraph 147(b),
(d)   subsection 125.2(1), 125.2(2), 127(1), 135(1), 136(1), 144(2), 144(2.1), 144(3), 144(4) or 155(1), or
(e)   section 124, 125.1, 142 or 143,

it is a defence for the person to prove that the person exercised due care and diligence to avoid the contravention.

(7) For the purposes of this section, where regulations are made under subsection 157(1.1) in relation to safety or health matters referred to in a paragraph of sections 125 to 126 by which a standard or other thing is to be prescribed, that standard or other thing shall be deemed to be prescribed within the meaning of that paragraph.

149.(1)  No proceeding in respect of an offence under this Part shall be instituted except with the consent of the Minister.

(2)  Where a corporation commits an offence under this Part, any officer, director or agent of the corporation who directed, authorized, assented to, acquiesced in or participated in the commission of the offence is a party to and guilty of the offence and is liable on conviction to the punishment provided for the offence, whether or not the corporation has been prosecuted or convicted.

(3)  On any prosecution for an offence under this Part, a copy of a direction purporting to have been made under this Part and purporting to have been signed by the person authorized under this Part to make the direction is evidence of the direction without proof of the signature or authority of the person by whom it purports to be signed.

(4)  Proceedings in respect of an offence under this Part may be instituted at any time within but not later than one year after the time when the subject matter of the proceedings arose.

150.  A complaint or information in respect of an offence under this Part may be heard, tried and determined by a magistrate or justice if the accused is resident or carrying on business within the territorial jurisdiction of the magistrate or justice, notwithstanding that the matter of the complaint or information did not arise in that territorial jurisdiction.

151.  In any proceedings in respect of an offence under this Part, an information may include more than one offence committed by the same person and all those offences may be tried concurrently and one conviction for any or all such offences may be made.

152.  The Minister may apply or cause an application to be made to a judge of a superior court or the Federal Court-Trial Division for an order enjoining any person from contravening a provision of this Part, whether or not a prosecution

has been instituted for an offence under this Part, or enjoining any person from continuing any act or default for which the person was convicted of an offence under this Part.

153. The judge of a court to whom an application under section 152 is made may, in his discretion, make the order applied for under that section and the order may be entered and enforced in the same manner as any other order or judgment of that court.

154.(1) Where a person is convicted of an offence under this Part on proceedings by way of summary conviction and the only punishment provided for the offence under this Part is a fine, no imprisonment may be imposed as punishment for the offence or in default of payment of any fine imposed as punishment.

(2) Where a person is convicted of an offence under this Part and the fine that is imposed is not paid when required, the prosecutor may, by filing the conviction, enter as a judgment the amount of the fine and costs, if any, in a superior court of the province in which the trial was held, and the judgment is enforceable against the person in the same manner as if it were a judgment rendered against the person in that court in civil proceedings.

## Providing of Information

155.(1) Where a person is required to provide information for the purposes of this Part, the Minister may require the information to be provided by a notice to that effect served personally or sent by registered mail addressed to the latest known address of the person, and the person shall comply with the notice within such reasonable time as is specified therein.

(2) A certificate purporting to be signed by the Minister or by a person authorized by the Minister,

(a) certifying that a notice was sent by registered mail to the person to whom it was addressed, accompanied by an identified post office certificate of the registration and a true copy of the notice, and
(b) certifying that the information has not been provided as requested in the notice sent by the Minister,

is evidence of the facts set out therein without proof of the signature or official character of the person by whom the certificate purports to be signed.

## Powers of the Canada Labour Relations Board

156.(1) Notwithstanding subsection 14(1), any member of the Board may dispose of any reference or complaint made to the Board under this Part and, in relation to any reference or complaint so made, any member

(a) has all the powers, rights and privileges that are conferred on the Board by this Act other than the power to make regulations under section 15; and

(b)   is subject to all the obligations and limitations that are imposed on the Board by this Act.

(2)  The provisions of Part I respecting orders and decisions of and proceedings before the Board under that Part apply in respect of all orders and decisions of and proceedings before the Board or any member thereof under this Part.

**Regulations**

157.(1)  Subject to this section, the Governor in Council may make regulations

(a)   prescribing anything that by this Part is to be prescribed; and
(b)   respecting such other matters or things as are necessary to carry out the, provisions of this Part.

(1.1)  Where the Governor in Council is of the opinion that a regulation cannot appropriately be made by prescribing a standard or other thing that by a paragraph of sections 125 to 126 is to be prescribed, the Governor in Council may make regulations in relation to the safety and health matters referred to in that paragraph in such manner as the Governor in Council considers appropriate in the circumstances, whether or not the opinion of the Governor in Council is indicated at the time the regulations are made.

Subsections (2) and (2.1) are repealed, Bill C-101 June 23, 1993.

(3)  Regulations of the Governor in Council under subsection (1) or (1.1) in respect of occupational safety and health of employees employed

(a)   on ships, trains or aircraft while in operation shall be made on the recommendation of the Minister and the Minister of Transport;
(b)   on or in connection with exploration or drilling for or the production, conservation, processing or transportation of oil or gas in Canada lands, as defined in the *Canada Oil and Gas Act*, shall be made on the recommendation of the Minister, the Minister of Natural Resources and the Minister of Indian Affairs and Northern Development.

(4)  Regulations made under this section may be made applicable to all employment to which this Part applies, to one or more classes of employment to which this Part applies or to such employment in one or more work places.

(5)  Regulations made under this section incorporating a standard by reference may incorporate the standard as enacted or adopted at a certain date, as amended to a certain date or as amended from time to time.

(6)  Regulations made under this section that prescribe or incorporate a standard but that require the standard to be complied with only to the extent that compliance is practicable or reasonably practicable in circumstances governed by the standard may require the employer to report to a safety officer the reason that full compliance is not practicable or reasonably practicable in particular circumstances.

## REGULATIONS RESPECTING OCCUPATIONAL SAFETY AND HEALTH MADE UNDER PART II OF THE CANADA LABOUR CODE

### PART I

### Short Title

1.1 These Regulations may be cited as the *Canada Occupational Safety and Health Regulations.*

### Interpretation

1.2 In these Regulations,

"Act" means Part II of the *Canada Labour Code; (Loi)*

"ANSI" means the American National Standards Institute; *(ANSI)*

"approved organization" means the St. John Ambulance Association, the Canadian Red Cross Society, the Emergency Care Instruction Services, the Workers' Compensation Board of British Columbia, Criti Care Emergency Medical Services, the Canadian Institute of Safety Search and Rescue, Lecavalier santé-sécurité du travail inc. or any organization authorized by the Commission de la santé et de la sécurité du travail du Québec for the purpose of teaching first aid in the Province of Quebec; *(organisme approuvé)*

"basic first aid certificate" means the certificate issued by an approved organization for successful completion of a one day first aid course; *(certificat de secourisme élémentaire)*

"change room" means a room that is used by employees to change from their street clothes to their work clothes and from their work clothes to their street clothes, and includes a locker room; *(vestiaire)*

"CSA" means the Canadian Standards Association; *(ACNOR)*

"elevating device" means an escalator, elevator or other device for moving passengers or freight; *(appareil élévateur)*

"fire hazard area" means an area that contains or is likely to contain explosive or flammable concentrations of dangerous substances; *(endroit présentant un risque d'incendie)*

"first aid room" means a room used exclusively for first aid or medical purposes; *(salle de premiers soins)*

"high voltage" means a voltage of 751 volts or more between any two conductors or between a conductor and ground; *(haute tension)*

"locked out" means, in respect of any equipment, machine or device, that the equipment, machine or device has been rendered inoperative and cannot be operated or energized without the consent of the person who rendered it inoperative; (*verrouillé*)

"lower explosive limit" means the lower limit of flammability of a chemical agent or a combination of chemical agents at ambient temperature and pressure, expressed

(a) for a gas or vapour, as a percentage in air by volume, and
(b) for dust, as the weight of dust per volume of air; (*limite explosive inférieure*)

"Minister" means the Minister of Human Resources Development; (*Ministre*)

"National Building Code" means the *National Building Code of Canada*, 1985, issued by the Associate Committee on the National Building Code, National Research Council of Canada, dated 1985; (*Code canadien du bâtiment*)

"National Fire Code" means the *National Fire Code of Canada*, 1985, issued by the Associate Committee on the National Fire Code, National Research Council of Canada, dated 1985; (*Code national de prévention des incendies du Canada*)

"oxygen deficient atmosphere" means an atmosphere in which there is less than 18 per cent by volume of oxygen at a pressure of one atmosphere or in which the partial pressure of oxygen is less than 135 mm Hg; (*air à faible teneur en oxygène*)

"personal service room" means a change room, toilet room, shower room, lunch room, living space, sleeping quarters or a combination thereof; (*local réservé aux soins personnels*)

"protection equipment" means safety materials, equipment, devices and clothing; (*équipement de protection* )

"qualified person" means, in respect of a specified duty, a person who, because of his knowledge, training and experience, is qualified to perform that duty safely and properly; (*personne qualifiée*)

"regional office" means, in respect of a work place, the regional office of the Department of Labour for the administrative region of that Department in which the work place is situated; (*bureau régional*)

"toilet room" means a room that contains a toilet or a urinal, but does not include an outdoor privy. (*lieux d'aisances*)

## Prescription

1.3 These Regulations are prescribed for the purposes of sections 125, 125.1, 125.2 and 126 of the Act.

## Application

1.4 These Regulations do not apply in respect of employees employed

(a) on trains while in operation;
(b) on aircraft while in operation;
(c) on ships;
(d) subject to Part II of the *Oil and Gas Occupational Safety and Health Regulations*, on or in connection with exploration or drilling for or the production, conservation, processing or transportation of oil or gas in Canada lands, as defined in the *Canada Oil and Gas Act*; or
(e) on or in connection with a work or undertaking that is excluded from the application of the Act by an order made pursuant to section 123.1 of the Act.

## Records and Reports

1.5 Where an employer is required by section 125 or 125.1 of the Act to keep and maintain a record, report or other document, the employer shall keep and maintain the record, report or other document in such a manner that it is readily available for examination by a safety officer and by the safety and health committee or the safety and health representative, if either exists, for the work place to which it applies.

## Inconsistent Provisions

1.6 In the event of an inconsistency between any standard incorporated by reference in these Regulations and any other provision of these Regulations, the other provision shall prevail to the extent of the inconsistency.

1.7 Notwithstanding any provision in any standard incorporated by reference in these Regulations, a reference to another publication in that standard is a reference to the publication as it read on March 31, 1986.

## PART II - BUILDING SAFETY

### Standards

2.1 The design and construction of every building shall meet the standards set out in Parts 3 to 9 of the National Building Code in so far as is reasonably practicable.

### Doors

2.2(1) Every double action swinging door that is located in an exit, entrance or passageway used for two-way pedestrian traffic shall be designed and fitted in a manner that will permit persons who are approaching from one side of the door to be seen by persons who are on the other side thereof.
(2) The floor of every passageway into which a door or gate extends when open, other than the door of a closet or other small unoccupied storage room, shall be marked in a manner that clearly indicates the area of hazard created by the opening of the door or gate.

(3)  Notwithstanding section 2.1, where an open door or gate extends into a passageway for a distance that will reduce the effective width of the passageway to less than the standard referred to in section 2.1,

(a)   a doorman shall be posted near the open door or gate; or
(b)   a barricade shall be placed across the passageway before the door or gate is opened to prevent persons from using the passageway while the door or gate is open.

## Awnings and Canopies

2.3  Any window awning or canopy or any part of a building that projects over an exterior walkway shall be installed in a manner that permits a clearance of not less than 2.2 m between the walkway surface and the lowest projection of the awning or canopy or projecting part of the building.

## Floor and Wall Openings

2.4(1)  In this section,

"floor opening" means an opening measuring 300 mm or more in its smallest dimension in a floor, platform, pavement or yard; (*ouverture dans le plancher*)

"wall opening" means an opening at least 750 mm high and 300 mm wide in a wall or partition. (*ouverture dans un mur*)

(2)  Where an employee has access to a wall opening from which there is a drop of more than 1.2 m or to a floor opening, guardrails shall be fitted around the wall opening or floor opening or it shall be covered with material capable of supporting all loads that may be imposed on it.

(3)  The material referred to in subsection (2) shall be securely fastened to and supported on structural members.

(4)  Subsection (2) does not apply to the loading and unloading areas of truck, railroad and marine docks.

## Open Top Bins, Hoppers, Vats and Pits

2.5(1)  Where an employee has access to an open top bin, hopper, vat, pit or other open top enclosure from a point directly above the enclosure, the enclosure shall be

(a)   covered with a grating, screen or other covering that will prevent the employee from falling into the enclosure; or
(b)   provided with a walkway that is not less than 500 mm wide and is fitted with guardrails.

(2)  A grating, screen, covering or walkway referred to in subsection (1) shall be so designed, constructed and maintained that it will support a load that is not less than

(a)    the maximum load that may be imposed on it, or

(b)    a live load of 6 kPa,

whichever is the greater.

(3)  Where an employee is working above an open top bin, hopper, vat, pit or other open top enclosure that is not covered with a grating, screen or other covering, the inside wall of the enclosure shall be fitted with a fixed ladder, except where the operations carried on in the enclosure render such a fitting impracticable.

(4)  Every open top bin, hopper, vat, pit or other open top enclosure referred to in subsection (1) whose walls extend less than 1.1 m above an adjacent floor or platform used by an employee shall be

(a)    covered with a grating, screen or other covering;

(b)    fitted with a guardrail; or

(c)    guarded by a person to prevent employees from falling into the enclosure.

### Ladders, Stairways and Ramps

2.6  Where an employee in the course of employment is required to move from one level to another level that is more than 450 mm higher or lower than the first level, the employer shall install a fixed ladder, stairway or ramp between the levels.

2.7  Where one end of a stairway is so close to a traffic route used by vehicles, to a machine or to any other hazard as to be hazardous to the safety of an employee using the stairway, the employer shall

(a)    post a sign at that end of the stairway to warn employees of the hazard; and

(b)    where practicable, install a barricade that will protect employees using the stairway from the hazard.

2.8(1)  Subject to subsection (5), a fixed ladder that is more than 6 m in length shall be fitted with a cage for that portion of its length that is more than 2 m above the base level of the ladder in such a manner that it will catch an employee who loses his grip and falls backwards or sideways off the ladder.

(2)  Subject to subsection (5), a fixed ladder that is more than 9 m in length shall have, at intervals of not more that 6 m, a landing or platform that

(a)    is not less than 0.36 m² in area; and

(b)    is fitted at its outer edges with a guardrail.

(3)  A fixed ladder, cage, landing or platform referred to in subsection (1) or (2) shall be designed and constructed to withstand all loads that may be imposed on it.

(4) A fixed ladder shall be

(a) vertical;
(b) securely held in place at the top and bottom and at intermediate points not more than 3 m apart; and
(c) fitted with
  (i) rungs that are at least 150 mm from the wall and spaced at intervals not exceeding 300 mm, and
  (ii) side rails that extend not less than 900 mm above the landing or platform.

(5) Subsections (1) and (2) do not apply to a fixed ladder that is used with a fall protection system referred to in section 12.10 of Part XII.

## Docks, Ramps and Dock Plates

2.9(1) Every loading and unloading dock and ramp shall be

(a) of sufficient strength to support the maximum load that may be imposed on it;
(b) free of surface irregularities that may interfere with the safe operation of mobile equipment; and
(c) fitted around its sides that are not used for loading or unloading with side rails, curbs or rolled edges of sufficient height and strength to prevent mobile equipment from running over the edge.

(2) Every portable ramp and every dock plate shall be

(a) clearly marked or tagged to indicate the maximum safe load that it is capable of supporting; and
(b) installed so that it cannot slide, move or otherwise be displaced under the load that may be imposed on it.

## Guardrails

2.10(1) Every guardrail shall consist of

(a) a horizontal top rail not less than 900 mm and not more than 1 100 mm above the base of the guardrail;
(b) a horizontal intermediate rail spaced midway between the top rail and the base; and
(c) supporting posts spaced not more than 3 m apart at their centres.

(2) Every guardrail shall be designed to withstand a static load of 890 N applied in any direction at any point on the top rail.

## Toe Boards

2.11 Where there is a hazard that tools or other objects may fall from a platform or other raised area onto an employee,

(a)     a toe board that extends from the floor of the platform or other raised area to a height of not less than 125 mm shall be installed; or

(b)     where the tools or other objects are piled to such a height that a toe board referred to in paragraph (a) does not prevent the tools or other objects from falling, a solid or mesh panel shall be installed from the floor to a height of not less than 450 mm.

## Housekeeping and Maintenance

2.12(1) Every exterior stairway, walkway, ramp and passageway shall be kept free of accumulations of ice and snow.

(2) All dust, dirt, waste and scrap material in every work place in a building shall be removed as often as is necessary to protect the safety and health of employees and shall be disposed of in such a manner that the safety and health of employees is not endangered.

(3) Every travelled surface in a work place shall be

(a)     slip resistant; and
(b)     maintained free from splinters, holes, loose boards and tiles or similar defects.

2.13 Where a floor in a work place is normally wet and employees in the work place do not use non-slip waterproof footwear, the floor shall be covered with a dry false floor or platform or treated with a non-slip material or substance.

2.14 Where a window on any level above the ground floor level of a building is cleaned, the standards set out in CSA Standard Z91-M1980, *Safety Code for Window Cleaning Operations*, the English version of which is dated May, 1980 and the French version of which is dated November, 1983 shall be adopted and implemented.

## Temporary Heat

2.15(1) Subject to subsection (2), where a salamander or other high capacity portable open-flame heating device is used in an enclosed work place, the heating device shall

(a)     be so located, protected and used that there is no hazard of igniting tarpaulins, wood or other combustible materials adjacent to the heating device;
(b)     be used only when there is ventilation provided;
(c)     be so located as to be protected from damage or overturning; and
(d)     not restrict a means of exit.

(2) Where the heating device referred to in subsection (1) does not provide complete combustion of the fuel used in connection with it, it shall be equipped with a securely supported sheet metal pipe that discharges the products of combustion outside the enclosed work place.

## PART III - TEMPORARY STRUCTURES AND EXCAVATIONS

### Application

3.1 This Part applies to portable ladders, temporary ramps and stairs, temporary elevated work bases used by employees and temporary elevated platforms used for materials.

### General

3.2 No employee shall use a temporary structure where it is reasonably practicable to use a permanent structure.

3.3 No employee shall work on a temporary structure in rain, snow, hail or an electrical or wind storm that is likely to be hazardous to the safety or health of the employee, except where the work is required to remove a hazard or to rescue an employee.

3.4 Tools, equipment and materials used on a temporary structure shall be arranged or secured in such a manner that they cannot be knocked off the structure accidentally.

3.5 No employee shall use a temporary structure unless

(a)   he has authority from his employer to use it; and
(b)   he has been trained and instructed in its safe and proper use.

3.6(1)  Prior to a work shift, a qualified person shall make a visual safety inspection of every temporary structure to be used during that shift.

(2)  Where an inspection made in accordance with subsection (1) reveals a defect or condition that adversely affects the structural integrity of a temporary structure, no employee shall use the temporary structure until the defect or condition is remedied.

### Barricades

3.7 Where a vehicle or a pedestrian may come into contact with a temporary structure, a person shall be positioned at the base of the temporary structure or a barricade shall be installed around it to prevent any such contact.

### Guardrails and Toe Boards

3.8(1)  Guardrails and toe boards shall be installed at every open edge of a platform of a temporary structure.

(2)  The guardrails and toe boards referred to in subsection (1) shall meet the standards set out in sections 2.10 and 2.11 of Part II.

### Temporary Stairs, Ramps and Platforms

3.9(1)  Subject to subsection 3.10(3), temporary stairs, ramps and platforms shall be designed, constructed and maintained to support any load that is likely to be imposed on them and to allow safe passage of persons and equipment on them.

(2) Temporary stairs shall have

(a)    uniform steps in the same flight;
(b)    a slope not exceeding 1.2 in 1; and
(c)    a hand-rail that is not less than 900 mm and not more than 1 100 mm above the stair level on open sides including landings.

(3) Temporary ramps and platforms shall be

(a)    securely fastened in place;
(b)    braced if necessary to ensure their stability; and
(c)    provided with cleats or surfaced in a manner that provides a safe footing for employees.

(4) A temporary ramp shall be so constructed that its slope does not exceed

(a)    where the temporary ramp is installed in the stairwell of a building not exceeding two storeys in height, 1 in 1, if cross cleats are provided at regular intervals not exceeding 300 mm; and
(b)    in any other case, 1 in 3.

**Scaffolds**

3.10(1) The erection, use, dismantling or removal of a scaffold shall be carried out by or under the supervision of a qualified person.

(2) The footings and supports of every scaffold shall be capable of carrying, without dangerous settling, all loads that are likely to be imposed on them.

(3) Every scaffold shall be capable of supporting at least four times the load that is likely to be imposed on it.

(4) The platform of every scaffold shall be at least 480 mm wide and securely fastened in place.

**Portable Ladders**

3.11(1) Commercially manufactured portable ladders shall meet the standards set out in CSA Standard CAN3-Z11-M81, *Portable Ladders*, the English version of which is dated September, 1981, as amended to March, 1983 and the French version of which is dated August, 1982, as amended to June, 1983.

(2) Subject to subsection (3), every portable ladder shall, while being used,

(a)    be placed on a firm footing; and
(b)    be secured in such a manner that it cannot be dislodged accidentally from its position.

(3) Where, because of the nature of the location or of the work being done, a portable ladder cannot be securely fastened in place, it shall, while being used, be sloped so that the base of the ladder is not less than one-quarter and not more than one-third of the length of the ladder from a point directly below the top of the ladder and at the same level as the base.

(4) Every portable ladder that provides access from one level to another shall extend at least three rungs above the higher level.

(5) Metal or wire-bound portable ladders shall not be used where there is a hazard that they may come into contact with any live electrical circuit or equipment.

(6) No employee shall work from any of the three top rungs of any single or extension portable ladder or from either of the two top steps of any portable step ladder.

## Excavation

3.12(1) Before the commencement of work on a tunnel, excavation or trench, the employer shall mark the location of all underground pipes, cables and conduits in the area where the work is to be done.

(2) Where an excavation or trench constitutes a hazard to employees, a barricade shall be installed around it.

(3) In a tunnel or in an excavation or trench that is more than 1.4 m deep and whose sides are sloped at an angle of 45° or more to the horizontal

(a)    the walls of the tunnel, excavation or trench, and
(b)    the roof of the tunnel

shall be supported by shoring and bracing that is installed as the tunnel, excavation or trench is being excavated.

(4) Subsection (3) does not apply in respect of a trench where the employer provides a system of shoring composed of steel plates and bracing, welded or bolted together, that can support the walls of the trench from the ground level to the trench bottom and can be moved along as work progresses.

(5) The installation and removal of the shoring and bracing referred to in subsection (3) shall be performed or supervised by a qualified person.

(6) Tools, machinery, timber, excavated materials or other objects shall not be placed within 1 m from the edge of an excavation or trench.

## Safety Nets

3.13(1) Where there is a hazard that tools, equipment or materials may fall onto or from a temporary structure, the employer shall provide a protective structure or a safety net to protect from injury any employee on or below the temporary structure.

(2) The design, construction and installation of a safety net referred to in subsection (1) shall meet the standards set out in ANSI Standard ANSI A10.11-1979, *American National Standard for Safety Nets Used During Construction, Repair and Demolition Operations*, dated August 7, 1979.

## Housekeeping

3.14 Every platform, hand-rail, guardrail and work area on a temporary structure used by an employee shall be kept free of accumulations of ice and snow while the temporary structure is in use.

3.15 The floor of a temporary structure used by an employee shall be kept free of grease, oil or other slippery substance and of any material or object that may cause an employee to trip.

## PART IV - ELEVATING DEVICES

### Application

4.1 This Part does not apply to elevating devices used in the underground workings of mines.

### Standards

4.2(1) Every elevating device and every safety device attached thereto shall

(a)     meet the standards set out in the applicable CSA standard referred to in subsection (2) in so far as is reasonably practicable; and
(b)     be used, operated and maintained in accordance with the standards set out in the applicable CSA standard referred to in subsection (2).

(2) For the purposes of subsection (1), the applicable CSA standard for

(a)     elevators, dumbwaiters, escalators and moving walks is CSA Standard CAN3-B44-M85, *Safety Code for Elevators*, other than clause 9.1.4 thereof, the English version of which is dated November, 1985 and the French version of which is dated March, 1986;
(b)     manlifts is CSA Standard B311-M1979, *Safety Code for Manlifts*, the English version of which is dated October, 1979 and the French version of which is dated July, 1984 and Supplement No. 1-1984 to B311-M1979, the English version of which is dated June, 1984 and the French version of which is dated August, 1984; and
(c)     elevating devices for the handicapped is CSA Standard CAN3-B355-M81, *Safety Code for Elevating Devices for the Handicapped*, the English version of which is dated April, 1981 and the French version of which is dated December, 1981.

## Use and Operation

4.3 No elevating device shall be used or operated with a load in excess of the load that it was designed and installed to move safely.

4.4(1)  Subject to subsection (3), no elevating device shall be used or placed in service while any safety device attached thereto is inoperative.

(2)  Subject to subsection (3), no safety device attached to an elevating device shall be altered, interfered with or rendered inoperative.

(3)  Subsections (1) and (2) do not apply to an elevating device or a safety device that is being inspected, tested, repaired or maintained by a qualified person.

**Inspection and Testing**

4.5  Every elevating device and every safety device attached thereto shall be inspected and tested by a qualified person to determine that the prescribed standards are met

(a)    before the elevating device and the safety device attached thereto are placed in service;
(b)    after an alteration to the elevating device or a safety device attached thereto; and
(c)    once every 12 months.

4.6(1)  A record of each inspection and test made in accordance with section 4.5 shall

(a)    be signed by the person who made the inspection and test;
(b)    include the date of the inspection and test and the identification and location of the elevating device and safety device that were inspected and tested; and
(c)    set out the observations of the person inspecting and testing the elevating device and safety device on the safety of the devices.

(2)  Every record referred to in subsection (1) shall be made by the employer and kept by him in the work place in which the elevating device is located for a period of two years after the date on which it is signed in accordance with paragraph (1)(a).

**Repair and Maintenance**

4.7  Repair and maintenance of elevating devices or safety devices attached thereto shall be performed by a qualified person appointed by the employer.

## PART V - BOILERS AND PRESSURE VESSELS

**Interpretation**

5.1. In this Part,

"boiler code" means CSA Standard B51-M1981, *Code for the Construction and Inspection of Boilers and Pressure Vessels*, the English version of which is dated March, 1981, as amended to May, 1984 and the French version of which is dated September, 1981, as amended to May, 1984; (*code concernant les chaudières*) "maximum allowable working pressure" means the maximum allowable working pressure set out in the record referred to in section 5.17; (*pression de fonctionnement maximale autorisée*)

"maximum temperature" means the maximum temperature set out in the record referred to in section 5.17; (*température maximale*)

"piping system" means an assembly of pipes, pipe fittings, valves, safety devices, pumps, compressors and other fixed equipment that contains a gas, vapour or liquid and is connected to a boiler or pressure vessel. (*réseau de canalisation*)

## Application

5.2  This Part does not apply to

(a)   a heating boiler that has a heating surface of 3 m² or less;
(b)   a pressure vessel that has a capacity of 40 L or less;
(c)   a pressure vessel that is installed for use at a pressure of 100 kPa or less;
(d)   a pressure vessel that has an internal diameter of 150 mm or less;
(e)   a pressure vessel that has an internal diameter of 600 mm or less and that is used for the storage of hot water;
(f)   a pressure vessel that has an internal diameter of 600 mm or less and that is connected to a water-pumping system containing air that is compressed to serve as a cushion;
(g)   an interprovincial pipeline; or
(h)   a refrigeration plant that has a capacity of 18 kW or less of refrigeration.

## Design, Construction, Testing, Inspection and Installation

5.3  Every boiler, pressure vessel and piping system used in a work place shall meet the standards relating to design, construction, testing, inspection and installation set out in clauses 3.8, 3.9, 4.8 to 5.1, 5.3.4 to 6.3, 7.1 and 8.1 of the boiler code, in so far as is reasonably practicable.

5.4  Solid fuel fire-tube boilers operating at a pressure over 103 kPa shall be provided with a fusible plug that meets the standards set out in Appendix A-19 to A-20.8 of Section 1 of the *American Society of Mechanical Engineers Boiler and Pressure Vessel Code*, dated July 1, 1983.

5.5(1)  Every boiler and pressure vessel shall have at least one safety valve or other equivalent fitting to relieve pressure at or below its maximum allowable working pressure.

(2)  Where two or more boilers or pressure vessels are connected to each other and are used at a common operating pressure, they shall each be fitted with one or more safety valves or other equivalent fittings to relieve pressure at or below the maximum allowable working pressure of the boiler or pressure vessel that has the lowest maximum allowable working pressure.

## Low-Water Cut-Off Devices

5.6(1)  Every steam boiler that is not under continuous attendance by a qualified person shall be equipped with a low-water fuel cut-off device that serves no other purpose.

(2)  Subject to subsection (3), where an automatically fired hot-water boiler is installed in a forced circulation system and is not under continuous attendance by a qualified person, the boiler shall be equipped with a low-water fuel cut-off device.

(3)  Where two or more hot-water boilers of the coil or fintube type are installed in one system, a low-water fuel cut-off device is not required on each boiler if

(a)   the low-water fuel cut-off device is installed on the main water outlet header; and
(b)   a flow switch that will cut off the fuel supply to the burner is installed in the outlet piping on each boiler.

(4)  A low-water fuel cut-off device referred to in this section and a flow switch referred to in paragraph (3)(b) shall be installed in such a manner that

(a)   they cannot be rendered inoperative; and
(b)   they can be tested under operating conditions.

## Use, Operation, Repair, Alteration and Maintenance

5.7(1)  In this section, "qualified person" means a person recognized under the laws of the province in which the boiler, pressure vessel or piping system is located as qualified to inspect boilers, pressure vessels or piping systems.

(2)  No person shall use a boiler, pressure vessel or piping system unless it has been inspected by a qualified person in accordance with subsection (3).

(3)  A qualified person shall

(a)   inspect every boiler, pressure vessel and piping system
  (i)    after installation,
  (ii)   after any welding, alteration or repair is carried out on it, and
  (iii)  in accordance with sections 5.12 to 5.14 and 5.16; and
(b)   make a record of each inspection in accordance with section 5.17.

5.8  Every boiler, pressure vessel and piping system in use at a work place shall be operated, maintained and repaired by a qualified person.

5.9  All repairs and welding of boilers, pressure vessels and piping systems shall be carried out in accordance with the standards referred to in clauses 5.1, 6.1 and 7.1 of the boiler code.

5.10  No person shall alter, interfere with or render inoperative any fitting attached to a boiler or pressure vessel except for the purpose of adjusting or testing the fitting.

5.11  The factor of safety for a high-pressure lap-seam riveted boiler shall be increased by at least 0.1 each year after 20 years of use and, if the boiler is relocated at any time, it shall not be operated at a pressure higher than 103 kPa.

**Inspections**

5.12(1)  Subject to subsection 2 and to sections 5.13 and 5.14, every boiler, pressure vessel and piping system in use in a work place shall be inspected by a qualified person as frequently as is necessary to ensure that the boiler, pressure vessel or piping system is safe for its intended use.

(2)  Every boiler in use in a work place shall be inspected

(a)  externally, at least once each year; and
(b)  internally, at least once every two years.

5.13(1)  Every pressure vessel in use in a work place, other than a pressure vessel that is buried, shall be inspected

(a)  externally, at least once each year; and
(b)  subject to subsections (2) and (3), internally, at least once every two years.

(2)  Where a pressure vessel is used to store anhydrous ammonia, the internal inspection referred to in paragraph (1)(b) may be replaced by an internal inspection conducted once every five years if, at the same time, a hydrostatic test at a pressure equal to one and one-half times the maximum allowable working pressure is conducted.

(3)  Air reservoirs used for stationary or portable purposes in the railway industry, instead of being inspected in accordance with subsection (1) and a record completed in accordance with section 5.17 may be inspected, tested and a record made in accordance with the *Air Reservoirs Other Than on Motive Power Equipment Regulations*.

5.14(1)  Subject to subsection (3), Halon 1301 and Halon 1211 containers shall not be recharged without a test of container strength and a complete visual inspection being carried out, if more than five years have elapsed since the date of the last test and inspection.

(2)  Subject to subsection (3), Halon 1301 and Halon 1211 containers that have been continuously in service without discharging may be retained in service for a maximum of 20 years from the date of the last test and inspection at which time they shall be emptied, retested, subjected to a complete visual inspection and re-marked before being placed back in service.

(3) Where a Halon 1301 or Halon 1211 container has been subjected to unusual corrosion, shock or vibration, a visual inspection and a test of container strength shall be carried out.

(4) A Halon 1301 and Halon 1211 container shall be tested by non-destructive test methods such as hydrostatic testing and the containers shall be thoroughly dried before being filled.

## Buried Pressure Vessels

5.16(1) Where a pressure vessel is buried, the installation shall conform to the standards set out in clauses A1.1(a) to (g), (i) to (k) and (n) of Appendix A to the boiler code.

(2) Before backfilling is done over a pressure vessel, notice of the proposed backfilling shall be given to the regional safety officer.

(3) Where test plates are used as an indication of corrosion of a buried pressure vessel, the test plates and, subject to subsection (4), the pressure vessel shall be completely uncovered and inspected by a qualified person at least once every three years.

(4) Where the test plates on an inspection referred to in subsection (3) show no appreciable corrosion, the pressure vessel may be completely uncovered and inspected at intervals exceeding three years if the employer notifies the regional safety officer of the condition of the test plates and of the proposed inspection schedule for the pressure vessel.

(5) Every buried pressure vessel shall be completely uncovered and inspected at least every 15 years.

## Records

5.17(1) A record of each inspection carried out under sections 5.7 and 5.12 to 5.16 shall be completed by the person who carried out the inspection.

(2) Every record referred to in subsection (1)

(a) shall be signed by the person who carried out the inspection; and
(b) shall include
  (i) the date of the inspection,
  (ii) the identification and location of the boiler, pressure vessel or piping system that was inspected,
  (iii) the maximum allowable working pressure and the maximum temperature at which the boiler or pressure vessel may be operated,
  (iv) a declaration as to whether the boiler, pressure vessel or piping system meets the standards prescribed by this Part,
  (v) a declaration as to whether, in the opinion of the person carrying out the inspection, the boiler, pressure vessel or piping system is safe for its intended use, and
  (vi) any other observation that the person considers relevant to the safety of employees.

(3) The employer shall keep every record referred to in subsection (1) for a period of 10 years after the inspection is made at the work place in which the boiler, pressure vessel or piping system is located.

## PART VI - LEVELS OF LIGHTING

### Interpretation

6.1(1) In this Part,

"aerodrome apron" means that part of a land aerodrome intended to accommodate the loading and unloading of passengers and cargo and the refuelling, servicing, maintenance and parking of aircraft; *(aire de trafic)*

"aircraft stand" means that part of an aerodrome apron intended to be used for the parking of aircraft for the purpose of loading or unloading passengers and providing ground services; *(poste de stationnement)*

"primary grain elevator" means a grain elevator the principal use of which is the receiving of grain directly from producers for storage or forwarding; *(installation primaire)*

"task position" means a position at which a visual task is performed; *(poste de travail)*

"VDT" means a visual display terminal. *(TEV)*

(2) For the purposes of this Part, 1 lx is equal to .0929 fc.

### Application

6.2 This Part does not apply in respect of lighting in any underground portion of a coal mine.

### Measurement of Average Levels of Lighting

6.3 For the purposes of sections 6.4 to 6.10, the average level of lighting at a task position or in an area shall be determined

(a) by making four measurements at different places representative of the level of lighting at the task position or, in an area, representative of the level of lighting 1 m above the floor of the area; and
(b) by dividing the aggregate of the results of those measurements by four.

### Lighting — Office Areas

6.4 The average level of lighting at a task position or in an area set out in column I of an item of Schedule I, other than a task position or area referred to in section 6.7 or 6.9, shall not be less than the level set out in column II of that item.

### Lighting — Industrial Areas

6.5 The average level of lighting in an area set out in column I of an item of Schedule II, other than an area referred to in section 6.7 or 6.9, shall not be less than the level set out in column II of that item.

### Lighting — General Areas

6.6 The average level of lighting in an area set out in column I of an item of Schedule III, other than an area referred to in section 6.7 or 6.9, shall not be less than the level set out in column II of that item.

### Lighting — VDT

6.7(1) The average level of lighting at a task position or in an area set out in column I of an item of Schedule IV shall not be more than the level set out in column II of that item.

(2) Reflection glare on a VDT screen shall be reduced to the point where an employee at a task position is able to

(a)    read every portion of any text displayed on the screen; and
(b)    see every portion of the visual display on the screen.

(3) Where VDT work requires the reading of a document, supplementary lighting shall be provided where necessary to give a level of lighting of at least 500 lx on the document.

### Lighting — Aerodrome Aprons and Aircraft Stands

6.8(1) Subject to subsection (2), the average level of lighting at a task position on an aerodrome apron shall not be less than 10 lx.

(2) The average level of lighting at a task position on an aircraft stand shall not be less than 20 lx.

### Lighting — Artefactual Exhibits and Archival Materials

6.9 The average level of lighting in an area in which artefactual exhibits or archival materials are handled or stored shall not be less than 50 lx.

### Emergency Lighting

6.10(1) Emergency lighting shall be provided to illuminate the following areas within buildings:

(a)    exits and corridors;
(b)    principal routes providing access to exits in open floor areas; and
(c)    floor areas where employees normally congregate.

(2) Except in the case of a primary grain elevator in which hand-held lamps are used for emergency lighting, all emergency lighting provided in accordance with subsection (1) shall

(a)   operate automatically in the event that the regular power supply to the building is interrupted;

(b)   provide an average level of lighting of not less than 10 lx; and

(c)   be independent of the regular power source.

(3)  Where a generator is used as a power source for emergency lighting, the inspection, testing and maintenance of the generator shall be in accordance with the requirements referred to in Section 6.7 of the *National Fire Code*, as amended from time to time.

(4)  Where a central storage battery system is used as a power source for emergency lighting or where emergency lighting is provided by a self-contained emergency lighting unit, the battery system or the unit shall be tested

(a)   monthly by hand; and

(b)   annually under simulated power failure or electrical fault conditions.

(5)  Where a battery, other than a hermetically sealed battery, is tested in accordance with paragraph (4)(a), the electrolyte level of the battery shall be checked and, if necessary, adjusted to the proper level.

(6)  Where a self-contained emergency lighting unit is tested in accordance with paragraph (4)(b), all lamps forming part of the unit shall be operated for the time period set out in Sentence 3.2.7.3(2) of the *National Building Code,* as amended from time to time, that is applicable to the class of buildings to which the building in which the unit is installed belongs.

(7)  Every employer shall make a record of the results of each test performed in accordance with subsection (3) or (4) and keep the record for two years after the test.

## Minimum Levels of Lighting

6.11(1)  Subject to subsections (2) to (4), the level of lighting at any place at a task position or in an area that may be measured for the purposes of section 6.3 shall not be less than one third of the level of lighting prescribed by this Part for that task position or area.

(2)  The level of lighting at any place at a task position or in an area set out in column I of item 8 or 9 of Schedule III or column I of item 1 of Schedule IV that may be measured for the purposes of section 6.3 shall not be less than one tenth of the level of lighting prescribed by this Part for that task position or area.

(3)  The level of lighting at any place at a task position referred to in section 6.8 that may be measured for the purposes of section 6.3 shall not be less than one quarter of the level of lighting prescribed by this Part for that task position.

(4)  In a building the construction of which is commenced after October 31, 1990, the level of emergency lighting at any place in an area referred to in subsection 6.10(1) that may be measured for the purposes of section 6.3 shall not be less than 0.25 lx.

| Item | Column I<br>Task position or area | Column II<br>Level in lx |
|------|-----------------------------------|--------------------------|
| 1. | **DESK WORK** | |
| | (a) Task positions at which cartography, designing, drafting, plan-reading or other very difficult visual tasks are performed | 1 000 |
| | (b) Task positions at which business machines are operated or stenography, accounting, typing, filing, clerking, billing, continuous reading or writing or other difficult visual tasks are performed | 500 |
| 2. | **OTHER OFFICE WORK**<br>Conference and interview rooms, file storage areas, switchboard or reception areas or other areas where ordinary visual tasks are performed | 300 |
| 3. | **SERVICE AREAS** | |
| | (a) Stairways and corridors that are | |
| |     (i) used frequently | 100 |
| |     (ii) used infrequently | 50 |
| | (b) Stairways that are used only in emergencies | 30 |

| Item | Column I<br>Area | | Column II<br>Level in lx |
|---|---|---|---|
| 1. | **GARAGES** | | |
| | (a) | Main repair and maintenance areas, other than those referred to in paragraph (b) | 300 |
| | (b) | Main repair and maintenance areas used for repairing and maintaining cranes, bulldozers and other major equipment | 150 |
| | (c) | General work areas adjacent to a main repair and maintenance area referred to in paragraph (b) | 50 |
| | (d) | Fuelling areas | 150 |
| | (e) | Battery rooms | 100 |
| | (f) | Other areas in which there is | |
| | | (i) a high or moderate level of activity | 100 |
| | | (ii) a low level of activity | 50 |
| 2. | **LABORATORIES** | | |
| | (a) | Areas in which instruments are read and where errors in such reading may be hazardous to the safety or health of an employee | 750 |
| | (b) | Areas in which a hazardous substance is handled | 500 |
| | (c) | Areas in which laboratory work requiring close and prolonged attention is performed | 500 |
| | (d) | Areas in which other laboratory work is performed | 300 |
| 3. | **LOADING PLATFORMS, STORAGE ROOMS AND WAREHOUSES** | | |
| | (a) | Active areas in which packages are frequently checked and sorted | 250 |
| | (b) | Areas in which packages are infrequently checked and sorted | 75 |
| | (c) | Docks (indoor and outdoor), piers and other locations where packages or containers are loaded or unloaded | 150 |
| | (d) | Areas in which grain and granular material is loaded or unloaded in bulk | 30 |
| | (e) | Areas in which goods are stored in bulk or where goods in storage are all of one kind | 30 |

| Item | Column I<br>Area | | | Column II<br>Level in lx |
|------|------|------|------|------|
| | (f) | Areas where goods in storage are of different kinds | | 75 |
| | (g) | Any other area | | 10 |
| **4.** | **MACHINE AND WOODWORKING SHOPS** | | | |
| | (a) | Areas in which medium or fine bench or machine work is performed | | 500 |
| | (b) | Areas in which rough bench or machine work is performed | | 300 |
| | (c) | Any other area | | 200 |
| **5.** | **MANUFACTURING AND PROCESSING AREAS** | | | |
| | (a) | Major control rooms or rooms with dial displays | | 500 |
| | (b) | Areas in which a hazardous substance is processed, manufactured or used | | |
| | | (i) | in main work areas | 500 |
| | | (ii) | in surrounding areas | 200 |
| | (c) | Areas in which substances that are not hazardous substances are processed, manufactured or used or where automatically controlled equipment operates | | |
| | | (i) | in main work areas | 100 |
| | | (ii) | in surrounding areas | 50 |
| **6.** | **SERVICE AREAS** | | | |
| | (a) | Stairways and elevating devices that are | | |
| | | (i) | used frequently | 100 |
| | | (ii) | used infrequently | 50 |
| | (b) | Stairways that are used only in emergencies | | 30 |
| | (c) | Corridors and aisles that are used by pedestrians and mobile equipment | | |
| | | (i) | at main intersections | 100 |
| | | (ii) | at other locations | 50 |
| | (d) | Corridors and aisles that are used by mobile equipment only | | 50 |
| | (e) | Corridors and aisles that are used by pedestrians only and are | | |
| | | (i) | used frequently by employees | 50 |
| | | (ii) | used infrequently by employees | 30 |

## SCHEDULE III *(Section 6.6)*

### LEVELS OF LIGHTING — GENERAL AREAS

| Item | Column I<br>Area | | | Column II<br>Level in lx |
|---|---|---|---|---|
| 1. | **BUILDING EXTERIORS** | | | |
| | (a) | Entrances and exits that are | | |
| | | (i) | used frequently | 100 |
| | | (ii) | used infrequently | 50 |
| | (b) | Pedestrian walkways | | |
| | | (i) | at vehicular intersections | 30 |
| | | (ii) | at other locations | 10 |
| | (c) | Areas used by pedestrians and mobile equipment in which there is | | |
| | | (i) | a high or moderate level of activity | 20 |
| | | (ii) | a low level of activity | 10 |
| | (d) | Storage areas in which there is | | |
| | | (i) | a high or moderate level of activity | 30 |
| | | (ii) | a low level of activity | 10 |
| 2. | **FIRST AID ROOMS** | | | |
| | (a) | in treatment and examination area | | 1 000 |
| | (b) | in other areas | | 500 |
| 3. | **FOOD PREPARATION AREAS** | | | 500 |
| 4. | **PERSONAL SERVICE ROOMS** | | | 200 |
| 5. | **BOILER ROOMS** | | | 200 |
| 6. | **ROOMS IN WHICH PRINCIPAL HEATING, VENTILATION OR AIR CONDITIONING EQUIPMENT IS INSTALLED** | | | 50 |
| 7. | **EMERGENCY SHOWER FACILITIES AND EMERGENCY EQUIPMENT LOCATIONS** | | | 50 |
| 8. | **PARKING AREAS** | | | |
| | (a) | Covered | | 50 |
| | (b) | Open | | 10 |
| 9. | **LOBBIES AND ATRIA** | | | 100 |

**LEVELS OF LIGHTING — VDT WORK**

| Item | Column I<br>Task position or area | Column II<br>Level in lx |
|------|-----------------------------------|--------------------------|
| 1. | **VDT WORK** | |
| | (a) Task positions at which data entry and retrieval work are performed intermittently | 500 |
| | (b) Task positions at which data entry work is performed exclusively | 750 |
| | (c) Air traffic controller areas | 100 |
| | (d) Telephone operator areas | 300 |

## PART VII - LEVELS OF SOUND

### Interpretation

7.1 In this Part,

"A-weighted sound pressure level" means a sound pressure level as determined by a measurement system which includes an A-weighting filter that meets the requirements set out in the International Electrotechnical Commission Standard 651 (1979), *Sound Level Meters*, as amended from time to time; *(niveau de pression acoustique pondérée A)*

"dBA" means decibel A-weighted and is a unit of A-weighted sound pressure level; *(dBA)*

"large truck" means a truck with a gross vehicle weight of more than 4 500 kg that is designed primarily for transporting goods and that is operated primarily on public roads; *(poids lourd)*

"noise exposure level ($L_{ex,8}$)" means 10 times the logarithm to the base 10 of the time integral over any 24 hour period of a squared A-weighted sound pressure divided by 8, the reference sound pressure being 20 µPa; *(niveau d'exposition ($L_{ex,8}$))*

"sound level meter" means a device for measuring sound pressure level that meets the performance requirements for a Type 2 instrument as specified in the

International Electrotechnical Commission Standard 651 (1979), *Sound Level Meters*, as amended from time to time; (*sonomètre*)

"sound pressure level" means 20 times the logarithm to the base 10 of the ratio of the root mean square pressure of a sound to the reference sound pressure of 20 µPa, expressed in decibels. (*niveau de pression acoustique*)

## Measurement and Calculation of Exposure

7.2(1) For the purposes of this Part, the exposure of an employee to sound shall be measured using an instrument that

(a)   is recommended for that measurement in clause 4.3 of CSA Standard CAN/CSA-Z107.56-M86, *Procedures for the Measurement of Occupational Noise Exposure*, as amended from time to time; and
(b)   meets the requirements for such an instrument set out in clause 4 of the Standard referred to in paragraph (a).

(2)   The exposure of an employee to sound shall be measured in accordance with clauses 5, 6.4.1, 6.4.4, 6.5.2, 6.5.4, 6.6.2 and 6.6.4 of the Standard referred to in paragraph (1)(a).

(3)   For the purposes of this Part, the measurement and calculation of the noise exposure level ($L_{ex,8}$) to which an employee is exposed shall take into account the exposure of the employee to A-weighted sound pressure levels of 74 dBA and greater.

(4)   The measurement and calculation of the noise exposure level ($L_{ex,8}$) referred to in subsection (3) may also take into account the exposure of the employee to A-weighted sound pressure levels that are less than 74 dBA.

## Hazard Investigation

7.3(1) Where an employee in a work place may be exposed to an A-weighted sound pressure level equal to or greater than 84 dBA for a duration that is likely to endanger the employee's hearing, the employer shall, without delay,

(a)   appoint a qualified person to carry out an investigation of the degree of exposure; and
(b)   notify the safety and health committee or the safety and health representative, if either exists, of the investigation and the name of the person appointed to carry out the investigation.

(2)   Subsection (1) does not apply in respect of an employee engaged in the operation of a large truck.

(3) For the purposes of subsection (1), the measurement of the A-weighted sound pressure level in a work place shall be performed instantaneously, during normal working conditions, using the slow response setting of a sound level meter.

(4) In the investigation referred to in subsection (1), the following matters shall be considered:

(a) the sources of sound in the work place;
(b) the A-weighted sound pressure levels to which the employee is likely to be exposed and the duration of such exposure;
(c) the methods being used to reduce this exposure;
(d) whether the exposure of the employee is likely to exceed the limits prescribed by paragraph 7.4(1)(a); and
(e) whether the employee is likely to be exposed to a noise exposure level ($L_{ex,8}$) equal to or greater than 84 dBA.

(5) On completion of the investigation and after consultation with the safety and health committee or the safety and health representative, if either exists, the person appointed to carry out the investigation shall set out in a written report signed and dated by the person

(a) observations respecting the matters considered in accordance with subsection (4);
(b) recommendations respecting the measures that should be taken in order to comply with sections 7.4 to 7.8; and
(c) recommendations respecting the use of hearing protectors by employees who are exposed to a noise exposure level ($L_{ex,8}$) equal to or greater than 84 dBA and not greater than 87 dBA.

(6) The report shall be kept by the employer at the work place in respect of which it applies for a period of ten years after the date of the report.

(7) Where it is stated in the report that an employee is likely to be exposed to a noise exposure level ($L_{ex,8}$) equal to or greater than 84 dBA, the employer shall, without delay,

(a) post and keep posted a copy of the report in a conspicuous place in the work place in respect of which it applies; and
(b) provide the employee with written information describing the hazards associated with exposure to high levels of sound.

**Limits of Exposure**

7.4(1) No employee in a work place, other than an employee referred to in subsection (2), shall, in any 24 hour period, be exposed to

(a)   an A-weighted sound pressure level set out in column I of Schedule I for a duration of exposure exceeding the applicable duration set out in column II; or

(b)   a noise exposure level ($L_{ex,8}$) that exceeds 87 dBA.

(2)  No employee who operates a large truck shall, in any 24 hour period, be exposed to an A-weighted sound pressure level set out in column I of Schedule II for a duration of exposure exceeding the applicable duration set out in column II.

## Reduction of Sound Exposure

7.5  Insofar as is reasonably practicable, every employer shall, by engineering controls or other physical means other than hearing protectors, reduce the exposure to sound of employees to whom subsection 7.4(1) applies to a level that does not exceed the limits prescribed by that subsection.

## Report to Regional Safety Officer

7.6  Where it is not reasonably practicable, without providing hearing protectors, for an employer to maintain the exposure to sound of an employee to whom subsection 7.4(1) applies at a level that does not exceed the limits prescribed by that subsection, the employer shall, without delay,

(a)   make a report in writing to the regional safety officer setting out the reasons why it is not reasonably practicable to do so; and

(b)   provide a copy of the report to the safety and health committee or the safety and health representative, if either exists.

## Hearing Protection

7.7(1)  When an employer is required to make a report pursuant to section 7.6, the employer shall, as soon as is reasonably practicable, provide every employee whose exposure to sound is likely to exceed the limits prescribed by subsection 7.4(1) with a hearing protector that

(a)   meets the requirements set out in CSA Standard Z94.2-M1984, *Hearing Protectors*, as amended from time to time; and

(b)   prevents the employee using the hearing protector from being exposed to a level of sound that exceeds the limits prescribed by subsection 7.4(1).

## SCHEDULE I *(Subsection 7.4(1))*

### MAXIMUM DURATION OF EXPOSURE TO A-WEIGHTED SOUND PRESSURE LEVELS IN THE WORK PLACE

| Column I | Column II |
|---|---|
| A-weighted sound pressure level (dBA) | Maximum duration of exposure in hours per employee per 24 hour period |
| 87 | 8.0 |
| 88 | 6.4 |
| 89 | 5.0 |
| 90 | 4.0 |
| 91 | 3.2 |
| 92 | 2.5 |
| 93 | 2.0 |
| 94 | 1.6 |
| 95 | 1.3 |
| 96 | 1.0 |
| 97 | 0.80 |
| 98 | 0.64 |
| 99 | 0.50 |
| 100 | 0.40 |
| 101 | 0.32 |
| 102 | 0.25 |
| 103 | 0.20 |
| 104 | 0.16 |
| 105 | 0.13 |
| 106 | 0.10 |
| 107 | 0.080 |
| 108 | 0.064 |
| 109 | 0.050 |
| 110 | 0.040 |
| 111 | 0.032 |
| 112 | 0.025 |
| 113 | 0.020 |
| 114 | 0.016 |
| 115 | 0.013 |
| 116 | 0.010 |
| 117 | 0.008 |
| 118 | 0.006 |
| 119 | 0.005 |
| 120 | 0.004 |

## SCHEDULE II *(Subsection 7.4(2))*

### MAXIMUM PERMITTED DURATION OF EXPOSURE TO A-WEIGHTED SOUND PRESSURE LEVELS IN THE WORKPLACE

| Column I | Column II |
|---|---|
| A-weighted sound pressure level (dBA) | Maximum duration of exposure in hours per employee per 24 hour period |
| 90 | 8.0 |
| 91 | 7.0 |
| 92 | 6.0 |
| 93 | 5.3 |
| 94 | 4.6 |
| 95 | 4.0 |
| 96 | 3.5 |
| 97 | 3.0 |
| 98 | 2.6 |
| 99 | 2.3 |
| 100 | 2.0 |
| 101 | 1.7 |
| 102 | 1.5 |
| 103 | 1.3 |
| 104 | 1.2 |
| 105 | 1.0 |
| 106 | 0.87 |
| 107 | 0.76 |
| 108 | 0.66 |
| 109 | 0.57 |
| 110 | 0.50 |
| 111 | 0.44 |
| 112 | 0.38 |
| 113 | 0.33 |
| 114 | 0.29 |
| 115 | 0.25 |
| Greater than 115 | 0 |

(2)  Where an employer provides a hearing protector to an employee pursuant to subsection (1), the employer shall

(a)   in consultation with the safety and health committee or the safety and health representative, if either exists, formulate a program to train the employee in the fit, care and use of the hearing protector; and
(b)   implement the program.

(3)  Every employer shall ensure that every person, other than an employee, to whom the employer grants access to a work place where the person is likely to be exposed to a level of sound that exceeds the limits set out in subsection 7.4(1) uses a hearing protector that meets the requirements of the standard referred to in paragraph (1)(a).

**Warning Signs**

7.8(1)  At every work place, other than a large truck, where an employee may be exposed to an A-weighted sound pressure level greater than 87 dBA, the employer shall, at conspicuous locations within the work place, post and keep posted signs warning of a potentially hazardous level of sound in the work place.

(2)  For the purposes of subsection (1), the measurement of the A-weighted sound pressure level in a work place shall be performed instantaneously, during normal working conditions, using the slow response setting of a sound level meter.

**PART VIII - ELECTRICAL SAFETY**

**Interpretation**

8.1  In this Part,

"Canadian Electrical Code" means

(a)   CSA Standard C22.1-1990, *Canadian Electrical Code, Part I,* dated January, 1990, and
(b)   CSA Standard C22.3 No. 1-M1979, *Overhead Systems and Underground Systems*, dated April, 1979; *(Code canadien de l'électricité)*

"control device" means a device that will safely disconnect electrical equipment from its source of energy; *(dispositif de commande)*

"electrical equipment" means equipment for the generation, distribution or use of electricity; *(outillage électrique)*

"guarantor" means a person who gives a guarantee of isolation; *(garant)*

"person in charge" means an employee who supervises employees performing work on or a live test of isolated electrical equipment.  *(responsable)*

## Application

8.2  This Part does not apply to the underground workings of mines.

## Standards

8.3(1)  The design, construction and installation of all electrical equipment shall meet the standards set out in the *Canadian Electrical Code, Part I* in so far as is reasonably practicable.

(2)  The operation and maintenance of all electrical equipment shall meet the standards set out in the *Canadian Electrical Code*.

## Safety Procedures

8.4(1)  All testing or work performed on electrical equipment shall be performed by a qualified person or an employee under the direct supervision of a qualified person.

(2)  Where the electrical equipment has a voltage in excess of 5,200 V between any two conductors or in excess of 3,000 V between any conductor and ground,

(a)  the qualified person or the employee referred to in subsection (1) shall use such insulated protection equipment and tools as will protect him from injury during the performance of the work; and

(b)  the employee referred to in subsection (1) shall be instructed and trained in the use of the insulated protection equipment and tools.

8.5(1)  Where electrical equipment is live or may become live, no employee shall work on the equipment unless

(a)  the employer has instructed the employee in procedures that are safe for work on live conductors;

(b)  a safety ground is connected to the equipment; or

(c)  the equipment is isolated.

(2)  Subject to subsections (3) and (4), where an employee is working on or near electrical equipment that is live or may become live, the electrical equipment shall be guarded.

(3)  Subject to subsection (4), where it is not practicable for electrical equipment referred to in subsection (2) to be guarded, the employer shall take measures to protect the employee from injury by insulating the equipment from the employee or the employee from ground.

(4)  Where live electrical equipment is not guarded or insulated in accordance with subsection (2) or (3) or where the employee referred to in subsection (3) is not insulated from ground, no employee shall work so near to any live part of the electrical equipment that is within a voltage range listed in column I of an item of the schedule to this Part that the distance between the body of the

employee or any thing with which the employee is in contact and the live part of the equipment is less than

(a)     the distance set out in column II of that item, where the employee is not a qualified person; or
(b)     the distance set out in column III of that item, where the employee is a qualified person.

(5) No employee shall work near a live part of any electrical equipment referred to in subsection (4) where there is a hazard that an unintentional movement by the employee would bring any part of his body or any thing with which he is in contact closer to that live part than the distance referred to in that subsection.

8.6 No employee shall work on or near high voltage electrical equipment unless he is authorized to do so by his employer.

8.7 A legible sign with the words "Danger — High Voltage" and "Danger — Haute Tension" in letters that are not less than 50 mm in height on a contrasting background shall be posted in a conspicuous place at every approach to live high voltage electrical equipment.

## Safety Watcher

8.8(1) Where an employee is working on or near live electrical equipment and, because of the nature of the work or the condition or location of the work place, it is necessary for the safety of the employee that the work be observed by a person not engaged in the work, the employer shall appoint a safety watcher

(a)     to warn all employees in the work place of the hazard; and
(b)     to ensure that all safety precautions and procedures are complied with.

(2) A safety watcher shall be

(a)     informed of his duties as a safety watcher and of the hazard involved in the work;
(b)     trained and instructed in the procedures to follow in the event of an emergency;
(c)     authorized to stop immediately any part of the work that he considers dangerous; and
(d)     free of any other duties that might interfere with his duties as a safety watcher.

(3) For the purposes of subsection (1), an employer may appoint himself as a safety watcher.

## Coordination of Work

8.9 Where an employee is working on or in connection with electrical equipment, that employee and every other person who is so working, including every safety watcher, shall be fully informed by the employer with respect to the safe coordination of their work.

## Poles and Elevated Structures

8.10(1) Before an employee climbs a pole or elevated structure that is used to support electrical equipment, the employer shall give instructions and training to the employee respecting inspections and tests of the pole or structure to be carried out before the pole or structure is climbed.

(2) Where, as a result of an inspection or test of a pole or elevated structure referred to in subsection (1), it appears to an employee that the pole or structure will be safe for climbing only when temporary supports have been installed, pike-poles alone shall not be used for such supports.

(3) No employee shall work on any pole or elevated structure referred to in subsection (1) unless he has been instructed and trained in the rescue of employees who may be injured in the course of the work.

8.11 Every pole or elevated structure that is embedded in the ground and is used to support electrical equipment shall meet the standards set out in

(a)   CSA Standard CAN3-015-M83, *Wood Utility Poles and Reinforcing Studs*, dated January, 1983; or
(b)   CSA Standard A14-M1979, *Concrete Poles*, the English version of which is dated September, 1979 and the French version of which is dated November, 1987.

## Isolation of Electrical Equipment

8.12(1) Before an employee isolates electrical equipment or changes or terminates the isolation of electrical equipment, the employer shall issue written instructions with respect to the procedures to be followed for the safe performance of that work.

(2) The instructions referred to in subsection (1) shall be signed by the employer and shall specify

(a)   the date and hour when the instructions are issued;
(b)   the date and hour of the commencement and of the termination of the period during which the instructions are to be followed;
(c)   the name of the employee to whom the instructions are issued; and
(d)   where the instructions are in respect of the operation of a control device that affects the isolation of the electrical equipment,
  (i)   the device to which the instructions apply, and
  (ii)   where applicable, the correct sequence of procedures.

(3) A copy of the instructions referred to in subsection (1) shall be shown and explained to the employee.

(4) The instructions referred to in subsection (1) shall be kept readily available for examination by employees for the period referred to in paragraph (2)(b) and thereafter shall be kept by the employer for a period of one year at his place of business nearest to the work place in which the electrical equipment is located.

8.13(1) Subject to subsection (4), no work on or live test of isolated electrical equipment shall be performed unless

(a) isolation of the equipment has been confirmed by test; and
(b) the employer has determined, on the basis of visual observation, that every control device and every locking device necessary to establish and maintain the isolation of the equipment
  (i) is set in the safe position with the disconnecting contacts of control devices safely separated or, in the case of a draw-out type electrical switch gear, is withdrawn to its full extent from the contacts of the electrical switch gear,
  (ii) is locked out, and
  (iii) bears a distinctive tag or sign designed to notify persons that operation of the control device and movement of the locking device are prohibited during the performance of the work or live test.

(2) Where more than one employee is performing any work on or live test of isolated electrical equipment, a separate tag or sign for each such employee shall be attached to each control device and locking device referred to in subsection (1).

(3) The tag or sign referred to in subparagraph (1)(b)(iii) or subsection (2) shall

(a) contain the words "DO NOT OPERATE — *DÉFENSE D'ACTIONNER* " or display a symbol conveying the same meaning;
(b) show the date and hour that the control device and the locking device referred to in paragraph (1)(b) were set in the safe position or were withdrawn to their full extent from the contacts;
(c) show the name of the employee performing the work or live test;
(d) where used in connection with a live test, be distinctively marked as a testing tag or sign;
(e) be removed only by the employee performing the work or live test; and
(f) be used for no purpose other than the purpose referred to in paragraph (1)(b)(iii).

(4) Where, because of the nature of the work in which the electrical equipment is being used, it is not practicable to comply with subsection (1), no work on or live test of electrical equipment shall be performed unless a guarantee of isolation referred to in section 8.14 is given to the person in charge.

**Guarantees of Isolation for Electrical Equipment**

8.14(1) No employee shall give or receive a guarantee of isolation for electrical equipment unless he is authorized in writing by his employer to give or receive a guarantee of isolation.

(2) Not more than one employee shall give a guarantee of isolation for a piece of electrical equipment for the same period of time.

(3) Before an employee performs work on or a live test of isolated electrical equipment, the person in charge shall receive from the guarantor

(a)    a written guarantee of isolation; or

(b)    where it is not practicable for him to receive a written guarantee of isolation, an oral guarantee of isolation.

(4) A written guarantee of isolation referred to in paragraph (3)(a) shall be signed by the guarantor and by the person in charge and shall contain the following information:

(a)    the date and hour when the guarantee of isolation is given to the person in charge;

(b)    the date and hour when the electrical equipment will become isolated;

(c)    the date and hour when the isolation will be terminated, if known;

(d)    the procedures by which isolation is assured;

(e)    the name of the guarantor and the person in charge; and

(f)    a statement as to whether live tests are to be performed.

(5) Where an oral guarantee of isolation referred to in paragraph (3)(b) is given, a written record thereof shall forthwith

(a)    be made by the guarantor; and

(b)    be made and signed by the person in charge.

(6) A written record referred to in subsection (5) shall contain the information referred to in subsection (4).

(7) Every written guarantee of isolation and every written record referred to in subsection (5) shall be

(a)    kept by the person in charge readily available for examination by the employee performing the work or live test until the work or live test is completed;

(b)    given to the employer when the work or live test is completed; and

(c)    kept by the employer for a period of one year after the completion of the work or live test at his place of business nearest to the work place in which the electrical equipment is located.

8.15 Where a written guarantee of isolation or a written record of an oral guarantee of isolation is given to a person in charge and the person in charge is replaced at the work place by another person in charge before the guarantee has terminated, the other person in charge shall sign the written guarantee of isolation or written record of the oral guarantee of isolation.

8.16 Before an employee gives a guarantee of isolation for electrical equipment that obtains all or any portion of its electrical energy from a source that is not under his direct control, the employee shall obtain a guarantee of isolation in respect of the source from the person who is in direct control thereof and is authorized to give the guarantee in respect thereof.

## Live Test

8.17(1)  No employee shall give a guarantee of isolation for the performance of a live test on isolated electrical equipment unless

(a)  any other guarantee of isolation given in respect of the electrical equipment for any part of the period for which the guarantee of isolation is given is terminated;
(b)  every person to whom the other guarantee of isolation referred to in paragraph (a) was given has been informed of its termination; and
(c)  any live test to be performed on the electrical equipment will not be hazardous to the safety or health of the person performing the live test.

(2)  Every person performing a live test shall warn all persons who, during or as a result of the test, are likely to be exposed to a hazard.

## Termination of Guarantee of Isolation

8.18(1)  Every person in charge shall, when work on or a live test of isolated electrical equipment is completed,

(a)  inform the guarantor thereof; and
(b)  make and sign a record in writing containing the date and hour when he so informed the guarantor and the name of the guarantor.

(2)  On receipt of the information referred to in subsection (1), the guarantor shall make and sign a record in writing containing

(a)  the date and hour when the work or live test was completed; and
(b)  the name of the person in charge.

(3)  The records referred to in subsections (1) and (2) shall be kept by the employer for a period of one year after the date of signature thereof at his place of business nearest to the work place in which the electrical equipment is located.

## Safety Grounding

8.19(1)  No employee shall attach a safety ground to electrical equipment unless he has tested the electrical equipment and has established that it is isolated.

(2)  Subsection (1) does not apply in respect of electrical equipment that is grounded by means of a grounding switch that is an integral part of the equipment.

8.20(1)  Subject to subsection (2), no work shall be performed on any electrical equipment in an area in which is located

(a)  a grounding bus,
(b)  a station grounding network,
(c)  a neutral conductor,

(d)   temporary phase grounding, or
(e)   a metal structure

unless the equipment referred to in paragraphs (a) to (e) is connected to a common grounding network.

(2)  Where, after the connections referred to in subsection (1) are made, a safety ground is required to ensure the safety of an employee working on the electrical equipment referred to in that subsection, the safety ground shall be connected to the common grounding network.

8.21  Every conducting part of a safety ground on isolated electrical equipment shall have sufficient current carrying capacity to conduct the maximum current that is likely to be carried on any part of the equipment for such time as is necessary to permit operation of any device that is installed on the electrical equipment so that, in the event of a short circuit or other electrical current overload, the electrical equipment is automatically disconnected from its source of electrical energy.

8.22(1)  For the purposes of subsection (2), a "point of safety grounding" means

(a)   a grounding bus, a station grounding network, a neutral conductor, a metal structure or an aerial ground, or
(b)   one or more metal rods that are not less than 16 mm in diameter and are driven not less than 1 m into undisturbed compact earth at a minimum distance of 4.5 m from the base of the pole, structure, apparatus or other thing to which the electrical equipment is attached or from the area where persons on the ground work and in a direction away from the main work area. (*point de mise à la terre*)

(2)  No safety ground shall be attached to or disconnected from isolated electrical equipment except in accordance with the following requirements:

(a)   the safety ground shall, to the extent that is practicable, be attached to the pole, structure, apparatus or other thing to which the electrical equipment is attached;
(b)   all isolated conductors, neutral conductors and all non-insulated surfaces of the electrical equipment shall be short-circuited, electrically bonded together and attached by a safety ground to a point of safety grounding in a manner that establishes equal voltage on all surfaces that can be touched by persons who work on the electrical equipment;
(c)   the safety ground shall be attached by means of mechanical clamps that are tightened securely and are in direct contact with bare metal;
(d)   the safety ground shall be so secured that none of its parts can make contact accidentally with any live electrical equipment;
(e)   the safety ground shall be attached and disconnected using insulated protection equipment and tools;
(f)   the safety ground shall, before it is attached to isolated electrical equipment, be attached to a point of safety grounding; and

(g)   the safety ground shall, before being disconnected from the point of safety grounding, be removed from the isolated electrical equipment in such a manner that the employee avoids contact with all live conductors.

## Switches and Control Devices

8.23(1)  Every control device shall be so designed and located as to permit quick and safe operation at all times.

(2)  The path of access to every electrical switch, control device or meter shall be free from obstruction.

(3)  Where an electrical switch or other device controlling the supply of electrical energy to electrical equipment is operated only by a person authorized to do so by the employer, the switch or other device shall be fitted with a locking device that only an authorized person can activate.

### SCHEDULE (*Section 8.5(4)*)

### DISTANCES FROM LIVE ELECTRICAL PARTS

| Item | Column I Voltage Range of Part: Part to Ground | Column II Distance in metres | Column III Distance in metres |
|------|-----------------------------------------------|------------------------------|-------------------------------|
| 1. | Over 425 to 12,000 | 3 | 0.9 |
| 2. | Over 12,000 to 22,000 | 3 | 1.2 |
| 3. | Over 22,000 to 50,000 | 3 | 1.5 |
| 4. | Over 50,000 to 90,000 | 4.5 | 1.8 |
| 5. | Over 90,000 to 120,000 | 4.5 | 2.1 |
| 6. | Over 120,000 to 150,000 | 6 | 2.7 |
| 7. | Over 150,000 to 250,000 | 6 | 3.3 |
| 8. | Over 250,000 to 300,000 | 7.5 | 3.9 |
| 9. | Over 300,000 to 350,000 | 7.5 | 4.5 |
| 10. | Over 350,000 to 400,000 | 9 | 5.4 |

## PART IX - SANITATION

### Interpretation

9.1 In this Part,

"ARI" means the Air-Conditioning and Refrigeration Institute of the United States; *(ARI)*

"Canadian Plumbing Code" means the *Canadian Plumbing Code*, 1985; (*Code canadien de la plomberie*)

"field accommodation" means fixed or mobile accommodation that is living, eating or sleeping quarters provided by an employer for the accommodation of employees at a work place; (*logement sur place*)

"mobile accommodation" means field accommodation that may be easily and quickly moved. (*logement mobile*)

### General

9.2(1) Every employer shall maintain each personal service room and food preparation area used by employees in a clean and sanitary condition.

(2) Personal service rooms and food preparation areas shall be so used by employees that the rooms or areas will remain as clean and in such a sanitary condition as is possible.

9.3 All janitorial work that may cause dusty or unsanitary conditions shall be carried out in a manner that will prevent the contamination of the air by dust or other substances injurious to health.

9.4 Each personal service room shall be cleaned at least once every day that it is used.

9.5(1) Every plumbing system that supplies potable water and removes water-borne waste

(a) shall meet the standards set out in the *Canadian Plumbing Code*; and
(b) subject to subsection (2), shall be connected to a municipal sanitation sewer or water main.

(2) Where it is not practicable to comply with paragraph (1)(b), the employer shall provide a waste disposal system that meets the standards set out in ANSI standard ANSI Z4.3-1979, *Minimum Requirements for Nonsewered Waste-Disposal Systems*, dated November 8, 1978.

9.6(1) Each container that is used for solid or liquid waste in the work place shall

(a)   be equipped with a tight-fitting cover;
(b)   be so constructed that it can easily be cleaned and maintained in a sanitary condition;
(c)   be leak-proof; and
(d)   where there may be internal pressure in the container, be so designed that the pressure is relieved by controlled ventilation.

(2) Each container referred to in subsection (1) shall be emptied at least once every day that it is used.

9.7(1) Each enclosed part of a work place, each personal service room and each food preparation area shall be constructed, equipped and maintained in a manner that will prevent the entrance of vermin.

(2) Where vermin have entered any enclosed part of a work place, personal service room or food preparation area, the employer shall immediately take all steps necessary to eliminate the vermin and prevent the re-entry of the vermin.

9.8 No person shall use a personal service room for the purpose of storing equipment unless a closet fitted with a door is provided in that room for that purpose.

9.9 In each personal service room and food preparation area, the temperature, measured one metre above the floor in the centre of the room or area, shall be maintained at a level of not less than 18°C and, where reasonably practicable, not more than 29°C.

9.10(1) In each personal service room and food preparation area, the floors, partitions and walls shall be so constructed that they can be easily washed and maintained in a sanitary condition.

(2) The floor and lower 150 mm of any walls and partitions in any food preparation area or toilet room shall be water-tight and impervious to moisture.

9.11 Where separate personal service rooms are provided for employees of each sex, each room shall be equipped with a door that is self-closing and is clearly marked to indicate the sex of the employees for whom the room is provided.

**Toilet Rooms**

9.12(1) Where it is reasonably practicable, a toilet room shall be provided for employees and, subject to section 9.13, where persons of both sexes are employed at the same work place, a separate toilet room shall be provided for employees of each sex.

(2) Subject to subsections (3) and (4), where a toilet room is provided in accordance with subsection (1), the employer shall provide in that room a number of toilets determined according to the maximum number of employees of each sex who are normally employed by him at any one time at the same work place as follows:

(a)   where the number of such employees does not exceed nine, one toilet;
(b)   where the number of such employees exceeds nine but does not exceed 24, two toilets;
(c)   where the number of such employees exceeds 24 but does not exceed 49, three toilets;
(d)   where the number of such employees exceeds 49 but does not exceed 74, four toilets;
(e)   where the number of such employees exceeds 74 but does not exceed 100, five toilets; and
(f)   where the number of such employees exceeds 100, five toilets and one toilet for every 30 such employees or portion of that number in excess of 100.

(3) Subject to subsection (4), where the class of employment in a work place is the transaction of business or the rendering of professional or personal services, the number of toilets provided by the employer in accordance with subsection (2) may be reduced

(a)   where the number of employees of each sex does not exceed 25, to one toilet;
(b)   where the number of employees of each sex exceeds 25 but does not exceed 50, to two toilets; and
(c)   where the number of employees of each sex exceeds 50, to three toilets and one toilet for every 50 employees or portion of that number in excess of 50.

(4) An employer may substitute urinals for up to two-thirds of the number of toilets required by subsection (2) or (3) to be provided for male employees.

(5) For the purposes of subsections (2) and (3), an employee who is normally away from his work place for more than 75 per cent of his working time and does not normally use the toilet room in the work place shall not be counted.

(6) Where reasonably practicable, toilet rooms and wash basins separate from those used by other employees shall be provided for food handlers.

9.13(1)  Subject to subsection (2), an employer may provide only one toilet for both male and female employees if

(a)    the total number of employees normally employed by him in the work place at any one time does not exceed five; and
(b)    the door of the toilet room is fitted on the inside with a locking device.

(2)  Where the class of employment in a work place is the transaction of business or the rendering of professional or personal services, the employer may provide only one toilet for both male and female employees if

(a)    the total number of employees normally employed by him in the work place at any one time does not exceed 10 or the area of the work place does not exceed 100 m²; and
(b)    the door of the toilet room is fitted on the inside with a locking device.

9.14  Toilet rooms shall be located not more than 60 m from and not more than one storey above or below each work place.

9.15  Every toilet room shall be so designed that

(a)    it is completely enclosed with solid material that is nontransparent from the outside;
(b)    no toilet or urinal is visible when the door of the toilet room is open;
(c)    it has a ceiling height of not less than 2.2 m;
(d)    where the toilet room contains more than one toilet, each toilet is enclosed in a separate compartment fitted with a door and an inside locking device; and
(e)    the walls of each separate toilet compartment are designed and constructed to provide a reasonable amount of privacy for its occupant.

9.16  Toilet paper on a holder or in a dispenser shall be provided

(a)    where there is only one toilet in a toilet room, in that toilet room; and
(b)    in each toilet compartment.

9.17  A covered container for the disposal of sanitary napkins shall be provided in each toilet room provided for the use of female employees.

**Wash Basins**

9.18  Hot water provided for personal washing

(a)    shall be maintained at a temperature of not less than 35°C and not more than 43°C; and
(b)    shall not be heated by mixing with steam.

9.19(1) Subject to sections 9.20 and 9.21, every employer shall provide for each toilet room wash basins supplied with cold water and hot water that meets the requirements of section 9.18 as follows:

(a)   where the room contains one or two toilets or urinals, one wash basin; and

(b)   where the room contains more than two toilets or urinals, one wash basin for every two toilets or urinals.

(2)  Where an outdoor privy is provided by an employer, the employer shall provide wash basins required by subsection (1) as close to the outdoor privy as is reasonably practicable.

9.20  Subject to section 9.21, where a toilet room is provided and the work environment of employees is such that their health is likely to be endangered by a hazardous substance coming into contact with their skin, the employer shall provide a wash room with individual wash basins supplied with cold water and hot water that meets the requirements of section 9.18 as follows:

(a)   where the number of those employees does not exceed five, one wash basin;

(b)   where the number of those employees exceeds five but does not exceed 10, two wash basins;

(c)   where the number of those employees exceeds 10 but does not exceed 15, three wash basins;

(d)   where the number of those employees exceeds 15 but does not exceed 20, four wash basins; and

(e)   where the number of those employees exceeds 20, four wash basins and one additional wash basin for every 15 of those employees or portion of that number in excess of 20.

9.21(1)  An industrial wash trough or circular wash basin of a capacity equivalent to the aggregate of the miminum standard capacities of the wash basins referred to in sections 9.19 and 9.20 may be provided in place of the wash basins.

(2)  An industrial wash trough or circular wash basin referred to in subsection (1) shall be supplied with cold water and hot water that meets the requirements of section 9.18.

9.22  In every personal service room that contains a wash basin, the employer shall provide

(a)   powdered or liquid soap or other cleaning agent in a dispenser at each wash basin or between adjoining wash basins;

(b)     sufficient sanitary hand drying facilities to serve the number of employees using the personal service room; and

(c)     a non-combustible container for the disposal of used towels where towels are provided.

## Showers and Shower Rooms

9.23(1)  A shower room with a door fitted on the inside with a locking device and at least one shower head for every 10 employees or portion of that number shall be provided for employees who regularly perform strenuous physical work in a high temperature or high humidity or whose bodies may be contaminated by a dangerous substance.

(2)  Every shower receptor shall be constructed and arranged in such a way that water cannot leak through the walls or floors.

(3)  No more than six shower heads shall be served by a single shower drain.

(4)  Where two or more shower heads are served by a shower drain, the floor shall be sloped and the drain so located that water from one head cannot flow over the area that serves another head.

(5)  Except for column showers, where a battery of shower heads is installed, the horizontal distance between two adjacent shower heads shall be at least 750 mm.

(6)  Waterproof finish shall be provided to a height of not less than 1.8 m above the floor in shower rooms and shall consist of ceramic, plastic or metal tile, sheet vinyl, tempered hardboard, laminated thermosetting decorative sheets or linoleum.

(7)  Finished flooring in shower rooms shall consist of resilient flooring, felted-synthetic fibre floor coverings, concrete terrazzo, ceramic tile, mastic or other types of flooring providing similar degrees of water resistance.

(8)  Where duck boards are used in showers, they shall not be made of wood.

(9)  Every shower shall be provided with cold water and hot water that meets the requirements of section 9.18.

(10)  Where an employee referred to in subsection (1) takes a shower as a result of his work, a clean towel and soap or other cleaning agent shall be provided to him.

## Potable Water

9.24 Every employer shall provide potable water for drinking, personal washing and food preparation that meets the standards set out in the *Guidelines for Canadian Drinking Water Quality*, 1978 published by authority of the Minister of Health.

9.25 Where it is necessary to transport water for drinking, personal washing or food preparation, only sanitary portable water containers shall be used.

9.26 Where a portable storage container for drinking water is used,

(a) the container shall be securely covered and closed;
(b) the container shall be used only for the purpose of storing potable water;
(c) the container shall not be stored in a toilet room; and
(d) the water shall be drawn from the container by
    (i) a tap,
    (ii) a ladle used only for the purpose of drawing water from the container, or
    (iii) any other means that precludes the contamination of the water.

9.27 Except where drinking water is supplied by a drinking fountain, sanitary single-use drinking cups shall be provided.

9.28 Any ice that is added to drinking water or used for the contact refrigeration of foodstuffs shall

(a) be made from potable water; and
(b) be so stored and handled as to prevent contamination.

9.29 Where drinking water is supplied by a drinking fountain, the fountain shall meet the standards set out in ARI Standard 1010-82, *Standard for Drinking-Fountains and Self-Contained, Mechanically-Refrigerated Drinking-Water Coolers*, dated 1982.

## Field Accommodation

9.30 All field accommodation shall meet the following standards:

(a) it shall be located on well-drained ground;
(b) it shall be so constructed that it can easily be cleaned and disinfected;
(c) the food preparation area and lunch room shall be separated from the sleeping quarters;
(d) where a water plumbing system is provided, the system shall operate under sanitary conditions;

(e)   garbage disposal facilities shall be provided to prevent the accumulation of garbage;

(f)   toilet rooms shall be maintained in a sanitary condition; and

(g)   vermin prevention, heating, ventilation and sanitary sewage systems shall be provided.

9.31(1)  Living quarters provided

(a)   in any fixed accommodation shall comprise
  (i)   for a single occupant, a space of at least 18 $m^3$, and
  (ii)  where there is more than one occupant, 18 $m^3$ plus 12 $m^3$ for each additional occupant; and

(b)   in any mobile accommodation shall comprise
  (i)   for a single occupant, a space of at least 12 $m^3$, and
  (ii)  where there is more than one occupant, 12 $m^3$ plus 8 $m^3$ for each additional occupant.

(2)  The living quarters referred to in subsection (1) shall have no floor dimension that is less than 1.5 m.

(3)  Toilet rooms and locker rooms shall not be counted in the calculation made in accordance with subsection (1).

9.32(1)  All mobile accommodation shall meet the standards set out in CSA Standard Z240.2.1-1979, *Structural Requirements for Mobile Homes*, dated September, 1979, as amended to April, 1984.

(2)  For the purposes of clause 4.12.4 of the standard referred to in subsection (1), there is no other approved method.

9.33  In any field accommodation provided as sleeping quarters for employees

(a)   a separate bed or bunk shall be provided for each employee;

(b)   the beds or bunks shall not be more than double-tiered and shall be so constructed that they can be cleaned and disinfected;

(c)   mattresses, sheets, pillow cases, blankets and bed covers shall be provided for each employee and kept in a clean and sanitary condition;

(d)   clean laundered sheets and pillow cases shall be provided for each employee at least once each week; and

(e)   at least one shelf and a locker fitted with a locking device shall be provided for each employee.

**Preparation, Handling, Storage and Serving of Food**

9.34(1)  Each food handler shall be instructed and trained in food handling practices that prevent the contamination of food.

(2) No person who is suffering from a communicable disease shall work as a food handler.

9.35 Where food is served in a work place, the employer shall adopt and implement Section G of the *Sanitation Code for Canada's Foodservice Industry* published by the Canadian Restaurant and Foodservices Association, dated September, 1984, other than items 2 and 11 thereof.

9.36(1) Where foods stored by an employer for consumption by employees require refrigeration to prevent them from becoming hazardous to health, the food shall be maintained at a temperature of 4°C or lower.

(2) Where foods stored by an employer for consumption by employees require freezing to prevent them from becoming hazardous to health, the foods shall be maintained at a temperature of -11°C or lower.

9.37 All equipment and utensils that come into contact with food shall be

(a)  designed to be easily cleaned;
(b)  smooth, free from cracks, crevices, pitting or unnecessary indentations; and
(c)  cleaned to maintain their surfaces in a sanitary condition.

9.38 No person shall eat, prepare or store food

(a)  in a place where a dangerous substance may contaminate food, dishes or utensils;
(b)  in a personal service room that contains a toilet, urinal or shower; or
(c)  in any other place where food is likely to be contaminated.

## Food Waste and Garbage

9.39(1) No food waste or garbage shall be stored in a food preparation area.

(2) Food waste and garbage shall be handled and removed from a food preparation area or lunch room in accordance with subsections (3) to (5).

(3) Wet food waste and garbage shall be

(a)  disposed of by mechanical grinders or choppers connected to sewage disposal lines; or
(b)  held in leak-proof, non-absorptive, easily-cleaned containers with tight-fitting covers in a separate enclosed area or container until removal for disposal.

(4) Dry food waste and garbage shall be removed or incinerated.

(5) Food waste and garbage containers shall be kept covered and the food waste and garbage removed as frequently as is necessary to prevent unsanitary conditions.

(6) Food waste and garbage containers shall, each time they are emptied, be cleansed and disinfected in an area separate from the food preparation area.

**Lunch Rooms**

9.40 Every lunch room provided by the employer

(a)   shall be separated from any place where a hazardous substance may contaminate food, dishes or utensils;
(b)   shall not be used for any purpose that is incompatible with its use as a lunch room;
(c)   shall not have any dimension of less than 2.3 m;
(d)   shall have a minimum floor area of 9 m²;
(e)   shall have 1.1 m² of floor area for each of the employees who normally use the room at any one time;
(f)   shall be furnished with a sufficient number of tables and chairs to accommodate adequately the number of employees normally using the lunch room at any one time; and
(g)   shall be provided with non-combustible covered receptacles for the disposal of waste food or other waste material.

**Ventilation**

9.41(1)  Each personal service room and food preparation area shall be ventilated to provide at least two changes of air per hour

(a)   by mechanical means, where the room is normally used by ten or more employees at any one time; or
(b)   by mechanical means or natural ventilation through a window or similar opening, where the room is used by fewer than ten employees if
(i)    the window or similar opening is located on an outside wall of the room, and
(ii)   not less than 0.2 m² of unobstructed ventilation is provided for each of the employees who normally use the room at any one time.

(2)  Where an employer provides ventilation by mechanical means in accordance with paragraph (1)(a), the amount of air provided for a type of room set out in column I of an item of the schedule to this Part shall be not less than that set out in column II of that item.

(3)  Where an employer provides for the ventilation of a food preparation area or a lunch room by mechanical means in accordance with paragraph (1)(a), the

rate of change of air shall be not less than nine litres per second for each employee who is normally employed in the food preparation area at any one time or for each employee who uses the lunch room at any one time.

9.42(1) Subject to subsection (2), any exhaust system from a personal service room containing a toilet or a shower shall not be connected with any other exhaust or air supply system.

(2) The exhaust system for a personal service room containing a toilet or shower may be connected with the exhaust duct of another room at the exhaust fan inlet if the system is connected in such a manner that an exchange of air cannot occur between the rooms.

## Clothing Storage

9.43 Clothing storage facilities shall be provided by the employer for the storage of overcoats and outer clothes not worn by employees while they are working.

9.44(1) A change room shall be provided by the employer where

(a) the nature of the work engaged in by an employee makes it necessary for that employee to change from street clothes to work clothes for safety or health reasons; or

(b) an employee is regularly engaged in work in which his work clothing becomes wet or contaminated by a hazardous substance.

(2) Where wet or contaminated work clothing referred to in paragraph (1)(b) is changed, it shall be stored in such a manner that it does not come in contact with clothing that is not wet or contaminated.

(3) No employee shall leave the work place wearing clothing contaminated by a hazardous substance.

(4) Every employer shall supply drying and cleaning facilities for the purpose of drying or cleaning wet or contaminated clothing referred to in paragraph (1)(b).

(5) In each change room,

(a) a floor area of at least 0.4 m² shall be provided for each of the employees who normally use the room at any one time; and

(b) where it is necessary for the employees to change footwear, seats shall be provided in sufficient numbers to accommodate them.

9.45 To the extent that is reasonably practicable, the clothing storage facilities referred to in section 9.43 and the change room referred to in section 9.44 shall be located

(a) near the work place and connected thereto by a completely covered route;
(b) on a direct route to the entrance to the work place;
(c) near a shower room provided pursuant to section 9.23; and
(d) near a toilet room.

**SCHEDULE** *(Subsection 9.41(2))*

**MINIMUM VENTILATION REQUIREMENTS FOR
CHANGE ROOMS, TOILET ROOMS AND SHOWER ROOMS**

| | Column I | Column II |
|---|---|---|
| Item | Type of Room | Ventilation Requirements in litres per second |
| 1. | **CHANGE ROOM** | |
| | (a) For employees with clean work clothes | (a) 5 L/s per m$^2$ of floor area |
| | (b) For employees with wet or sweaty work clothes | (b) 10 L/s per m$^2$ of floor area; 3 L/s exhausted from each locker |
| | (c) For employees who work where work clothes pick up heavy odours | (c) 15 L/s per m$^2$ of floor area; 4 L/s exhausted from each locker |
| 2. | **TOILET ROOM** | 10 L/s per m$^2$ of floor area; at least 10 L/s per toilet compartment; minimum 90 L/s |
| 3. | **SHOWER ROOM** | 10 L/s per m$^2$ of floor area; at least 20 L/s per shower head; minimum 90 L/s |

## PART X - HAZARDOUS SUBSTANCES

### Interpretation

10.01 In this Part,

"hazard information" means, in respect of a hazardous substance, information on the proper and safe storage, handling and use of the hazardous substance, including information relating to its toxicological properties; *(renseignements sur les dangers)*

"product identifier" means, in respect of a hazardous substance, the brand name, code name or code number specified by the supplier or employer or the chemical name, common name, generic name or trade name; *(identificateur du produit)*

"readily available" means present in an appropriate place in a physical copy form that can be handled; *(facilement accessible)*

"supplier" means a person who is a manufacturer, processor or packager of a hazardous substance or a person who, in the course of business, imports or sells a hazardous substance. *(fournisseur)*

## Application

10.1 This Part does not apply to the transportation or handling of dangerous goods as defined in the *Transportation of Dangerous Goods Act*.

## DIVISION I - GENERAL

### Hazard Investigation

10.2(1) Where there is a likelihood that the safety or health of an employee in a work place is or may be endangered by exposure to a hazardous substance, the employer shall, without delay,

(a) appoint a qualified person to carry out an investigation; and
(b) notify the safety and health committee or the safety and health representative, if either exists, of the proposed investigation and of the name of the qualified person appointed to carry out that investigation.

(2) In the investigation referred to in subsection (1), the following criteria shall be taken into consideration:

(a) the chemical, biological and physical properties of the hazardous substance;
(b) the routes of exposure of the hazardous substance;
(c) the effects to health of exposure to the hazardous substance;
(d) the quantity of the hazardous substance handled;
(e) the manner in which the hazardous substance is handled;
(f) the control methods used to eliminate or reduce exposure;

(g)    the value, percentage or level of the hazardous substance to which an employee is likely to be exposed; and

(h)    whether the value, percentage or level referred to in paragraph (g) is likely to

      (i)    exceed that prescribed in section 10.21 or 10.22 or Part VII, or

      (ii)    be less than that prescribed in Part VI.

10.3 On completion of the investigation referred to in subsection 10.2(1) and after consultation with the safety and health committee or the safety and health representative, if either exists, the qualified person shall set out in a written report signed by the qualified person

(a)    his observations respecting the criteria considered in accordance with subsection 10.2(2); and

(b)    his recommendations respecting the manner of compliance with sections 10.5 to 10.25.

10.4 The report referred to in section 10.3 shall be kept by the employer at the work place to which it applies for a period of two years after the date on which the qualified person signed the report.

## Substitution of Substances

10.5(1) A hazardous substance shall not be used for any purpose in a work place where it is reasonably practicable to substitute therefor a substance that is not a hazardous substance.

(2) Where a hazardous substance is required to be used for any purpose in a work place and an equivalent substance that is less hazardous is available to be used for that purpose, the equivalent substance shall be substituted for the hazardous substance where reasonably practicable.

## Ventilation

10.6 Every ventilation system used to control the concentration of an airborne hazardous substance shall be so designed, constructed and installed

(a)    that the concentration of the airborne hazardous substance does not exceed the values and levels prescribed in sections 10.21 and 10.22; and

(b)    that the ventilation system meets the standards set out in

      (i)    Part 6 of the *National Building Code*, or

      (ii)    the publication of the American Conference of Governmental Industrial Hygienists entitled *Industrial Ventilation*, 18th edition, dated 1984.

## Warnings

10.7 Where reasonably practicable, automated warning and detection systems shall be provided by the employer where the seriousness of any exposure to a hazardous substance so requires.

## Storage, Handling and Use

10.8 Every hazardous substance stored, handled or used in a work place shall be stored, handled and used in a manner whereby the hazard related to that substance is reduced to a minimum.

10.9 Where a hazardous substance is stored, handled or used in a work place, any hazard resulting from that storage, handling or use shall be confined to as small an area as practicable.

10.10 Every container for a hazardous substance that is used in a work place shall be so designed and constructed that it protects the employees from any safety or health hazard that is created by the hazardous substance.

10.11 Every container for a hazardous substance that is used or processed in a work place shall, to the extent that is practicable, be limited to the quantity required for use or processing in the work place in one work day.

10.12 Where a hazardous substance is capable of combining with another substance to form an ignitable combination and a hazard of ignition of the combination by static electricity exists in a work place, the employer shall adopt and implement the standards set out in the United States National Fire Prevention Association Inc. publication NFPA 77-1983, *Recommended Practice on Static Electricity*, dated 1983.

## Warning of Hazardous Substances

10.13 Where a hazardous substance is stored, handled or used in a work place, signs shall be posted in conspicuous places warning every person granted access to the work place of the presence of the hazardous substance and of any precautions to be taken to prevent or reduce any hazard or injury to health.

10.14 and 10.15 were revoked in 1988.

## Assembly of Pipes

10.16 Every assembly of pipes, pipe fittings, valves, safety devices, pumps, compressors and other fixed equipment that is used for transferring a hazardous substance from one location to another shall be

(a) labelled to identify the hazardous substance transferred therein; and

(b) fitted with valves and other control and safety devices to ensure its safe operation, maintenance and repair.

## Employee Education

10.17(1) Every employer shall, in consultation with the safety and health committee or the safety and health representative, if either exists, develop and implement an employee education program with respect to hazard prevention and control at the work place.

(2) The employee education program referred to in subsection (1) shall include

(a) the instruction of each employee who handles or is exposed to or is likely to handle or be exposed to a hazardous substance with respect to
  (i) the product identifier of the hazardous substance,
  (ii) all hazard information disclosed by the supplier of the hazardous substance or by the employer on a material safety data sheet or label,
  (iii) all hazard information of which the employer is aware or ought reasonably to be aware,
  (iv) the observations referred to in paragraph 10.3(a),
  (v) the information disclosed on the material safety data sheet referred to in section 10.27 and the purpose and significance of that information, and
  (vi) in respect of controlled products in the work place, the information required to be disclosed on a material safety data sheet and on a label by Division III and the purpose and significance of that information;

(b) the instruction and training of each employee who operates, maintains or repairs an assembly of pipes referred to in section 10.16 with respect to
  (i) every valve and other control and safety device connected to the assembly of pipes, and
  (ii) the procedures to follow for the proper and safe use of the assembly of pipes;

(c) the instruction and training of each employee referred to in paragraphs (a) and (b) with respect to
  (i) the procedures to follow to implement the provisions of sections 10.8, 10.9 and 10.12, and
  (ii) the procedures to follow for the safe storage, handling, use and disposal of hazardous substances, including procedures to be followed in an emergency involving a hazardous substance; and

(d) where the employer makes a machine-readable version of a material safety data sheet available in accordance with subsection 10.33(2), the training of each employee in accessing that material safety data sheet.

(3) Every employer shall, in consultation with the safety and health committee or the safety and health representative, if either exists, review the employee education program referred to in subsection (1) and, if necessary, revise it

(a)  at least once a year;
(b)  whenever there is a change in conditions in respect of the hazardous substances in the work place; and
(c)  whenever new hazard information in respect of a hazardous substance in the work place becomes available to the employer.

10.19  A written record of the employee education program referred to in subsection 10.17(1) shall be kept by the employer

(a)  readily available for examination by the employee; and
(b)  for a period of two years after the employee ceases to be required
　　(i)   to handle or be exposed to the hazardous substance, or
　　(ii)  to operate, maintain or repair the assembly of pipes.

**Medical Examinations**

10.20(1)  Where the report referred to in section 10.3 contains a recommendation for a medical examination, the employer may consult a physician regarding that recommendation.

(2)  Where the employer

(a)  consults a physician pursuant to subsection (1) and the physician confirms the recommendation for a medical examination, or
(b)  does not consult a physician pursuant to subsection (1),

the employer shall not permit an employee to work with the hazardous substance in the work place until a physician, acceptable to the employee, has examined the employee and declared the employee fit for work with the hazardous substance.

(3)  Where an employer consults a physician pursuant to subsection (1), the employer shall keep a copy of the decision of the physician with the report referred to in section 10.3.

(4)  The cost of a medical examination referred to in subsection (2) shall be borne by the employer.

**Control of Hazards**

10.21(1)  No employee shall be exposed to a concentration of

(a)    an airborne chemical agent, other than grain dust, in excess of the value
       for that chemical agent adopted by the American Conference of
       Governmental Industrial Hygienists in its publication entitled *Threshold
       Limit Values and Biological Exposure Indices for 1985-86*; or

(b)    airborne grain dust, respirable and non-respirable, in excess of 10 mg per
       1 m$^3$.

(1.1)  Subsection (1) does not apply in respect of concentrations of carbon
dioxide or respirable dust in an underground portion of a coal mine.

(2)  Where there is a likelihood that the concentration of an airborne chemical
agent may exceed the value referred to in subsection (1), the concentration of
the chemical agent shall be sampled and tested

(a)    in accordance with the standards set out by the American Conference of
       Governmental Industrial Hygienists in its publication entitled *Manual of
       Analytical Methods Recommended For Sampling and Analysis of
       Atmospheric Contaminants*, dated 1958;

(b)    in accordance with the standards set out by the United States National
       Institute for Occupational Safety and Health in the *NIOSH Manual of
       Analytical Methods*, third edition, volumes 1 and 2, dated February, 1984;
       or

(c)    by a method that uses the test procedure set out in the *United States
       Federal Register*, volume 40, number 33, dated February 18, 1975, as
       amended by volume 41, number 53, dated March 17, 1976.

(3)  A record of each test made pursuant to subsection (2) shall be kept by the
employer at his place of business nearest to the work place where the
concentration was sampled for a period of three years after the date of the test.

(4)  A record referred to in subsection (3) shall include

(a)    the date, time and location of the test;
(b)    the hazardous substance for which the test was made;
(c)    the sampling and testing methods used;
(d)    the result obtained; and
(e)    the name and occupation of the person who made the test.

10.22(1)  Subject to subsection (2), the concentration of an airborne chemical
agent or combination of airborne chemical agents in a work place shall be less
than 50 per cent of the lower explosive limit of the chemical agent or
combination of chemical agents.

(2)  Where a source of ignition may ignite the concentration of an airborne
chemical agent or combination of airborne chemical agents in a work place, that

concentration shall not exceed 10 per cent of the lower explosive limit of the chemical agent or combination of chemical agents.

(3) Subsection (2) does not apply in respect of concentrations of methane gas in an underground portion of a coal mine.

10.23(1) Compressed air shall be used in such a manner that the air is not directed forcibly against any person.

(2) Where compressed air is used, its use shall not result in a concentration of a hazardous substance in the atmosphere in excess of the value prescribed in subsection 10.21(1).

(3) To the extent that is reasonably practicable, where compressed air is used, it shall be used only

(a) in a ventilated hood or booth; or
(b) in an area where employees are protected from hazardous substances and flying particles.

**Explosives**

10.24 All blasting using dynamite, blasting caps or other explosives shall be done by a qualified person who, where required under the laws of the province in which the blasting is done, holds a blasting certificate or such other authorization as may be required under those laws.

**Radiation Emitting Devices**

10.25(1) Where a device that is capable of producing and emitting energy in the form of electromagnetic waves or acoustical waves is used in a work place, the employer shall

(a) if the device is listed in the schedule to this Part, make a report in writing to the Radiation Protection Bureau of Health Canada, setting out a description of the device and the location of the work place; and
(b) if the device is referred to in subsection (2), adopt and implement the applicable safety code of the Radiation Protection Bureau of Health Canada as specified in that subsection.

(2) For the purposes of paragraph (1)(b), the applicable safety code is

(a) in respect of radiofrequency and microwave devices in the frequency range 10 MHz-300 GHz, *Safety Code – 6*, dated February, 1979;
(b) in respect of X-ray equipment in medical diagnosis, *Safety Code – 20A*, dated 1981;

**RADIATION EMITTING DEVICES TO BE REPORTED**
**TO THE RADIATION PROTECTION BUREAU**

| Item | Device |
|------|--------|
| 1. | Dental X-Ray Equipment |
| 2. | Baggage Inspection X-Ray Devices |
| 3. | Demonstration-Type Gas Discharge Devices |
| 4. | Photofluorographic X-Ray Equipment |
| 5. | Electron Microscopes |
| 6. | Diagnostic X-Ray Equipment |
| 7. | X-Ray Diffraction Equipment |
| 8. | Cabinet X-Ray Equipment |
| 9. | Therapeutic X-Ray Equipment |
| 10. | Industrial X-Ray Radiography and Fluoroscopy Equipment |
| 11. | Analytical X-Ray Equipment |
| 12. | X-Ray Spectrometers |
| 13. | X-Ray Equipment Used for Irradiation of Materials |
| 14. | Electron Welding Equipment |
| 15. | Electron Processors |
| 16. | High-Tension Vacuum Tubes |
| 17. | Accelerators |
| 18. | X-Ray Gauges |
| 19. | Laser Scanners |
| 20. | Demonstration Lasers |
| 21. | Sunlamps |
| 22. | Ultrasound Therapy Equipment |
| 23. | Industrial Radio-frequency Heaters |
| 24. | Lasers |
| 25. | Ultraviolet Polymerizers |
| 26. | Short-wave Diathermy Devices |
| 27. | Microwave Diathermy Devices |
| 28. | Magnetic Resonance Imaging Devices |
| 29. | Induction Heaters |
| 30. | Radars |
| 31. | Telecommunication Transmitters above 5 W |
| 32. | Diagnostic Ultrasound Equipment |
| 33. | Surgical Ultrasound Equipment |
| 34. | Dental Ultrasound Equipment |
| 35. | Hyperthermia Ultrasound Equipment |
| 36. | Nebulizer Ultrasound Equipment |
| 37. | Non-Portable Ultrasonic Cleaners |
| 38. | Ultrasonic Machining Tools |
| 39. | Ultrasonic Welding Equipment |
| 40. | Airborne Ultrasound Motion Detectors |
| 41. | Airborne Ultrasound Pest Repellers |

(c) in respect of baggage inspection X-ray equipment, *Safety Code — 21*, dated May, 1978;

(d) in respect of dental X-ray equipment, *Safety Code — 22*, dated 1981;

(e) in respect of ultrasound, *Safety Code — 23*, dated 1980 and *Safety Code — 24*, dated 1980; and

(f) in respect of short-wave diathermy, *Safety Code — 25*, dated 1983.

## DIVISION II - HAZARDOUS SUBSTANCES OTHER THAN CONTROLLED PRODUCTS

### Identification

10.26 Every container of a hazardous substance, other than a controlled product, that is stored, handled or used in the work place shall be labelled in a manner that discloses clearly

(a) the name of the substance; and

(b) the hazardous properties of the substance.

10.27 Where a material safety data sheet pertaining to a hazardous substance, other than a controlled product, that is stored, handled or used in a work place may be obtained from the supplier of the hazardous substance, the employer shall

(a) obtain a copy of the material safety data sheet; and

(b) keep a material safety data sheet readily available in the work place for examination by employees.

## DIVISION III - CONTROLLED PRODUCTS

### Interpretation

10.28 In this Division,

"bulk shipment" means a shipment of a controlled product that is contained, without intermediate containment or intermediate packaging, in

(a) a tank with a water capacity of more than 454 L,

(b) a freight container or a portable tank,

(c) a road vehicle, railway vehicle or ship, or

(d) a pipeline; *(expédition en vrac)*

"fugitive emission" means a controlled product in gas, liquid or solid form that escapes from processing equipment, from control emission equipment or from a product; *(émission fugitive)* "hazardous waste" means a controlled product that is intended solely for disposal or is sold for recycling or recovery; (*résidu dangereux*)

"manufactured article" means any article that is formed to a specific shape or design during manufacture, the intended use of which when in that form is dependent in whole or in part on its shape or design, and that, under normal conditions of use, will not release or otherwise cause a person to be exposed to a controlled product; (*article manufacturé*)

"risk phrase" means, in respect of a controlled product, a statement identifying a hazard that may arise from the use of or exposure to the controlled product; (*mention de risque*)

"sale" includes offer for sale, expose for sale and distribute; (*vente*)

"supplier label" means, in respect of a controlled product, a label prepared by a supplier pursuant to the *Hazardous Products Act*; (*étiquette du fournisseur*)

"supplier material safety data sheet" means, in respect of a controlled product, a material safety data sheet prepared by a supplier pursuant to the *Hazardous Products Act*; (*fiche signalétique du fournisseur*)

"work place label" means, in respect of a controlled product, a label prepared by an employer pursuant to this Division; (*étiquette du lieu de travail*)

"work place material safety data sheet" means, in respect of a controlled product, a material safety data sheet prepared by an employer pursuant to subsection 10.32(1) or (2). (*fiche signalétique du lieu de travail*)

## Application

10.29(1)  This Division does not apply in respect of any

(a)    wood or product made of wood;
(b)    tobacco or product made of tobacco; or
(c)    manufactured article.

(2)  This Division, other than section 10.42, does not apply in respect of hazardous waste.

## Material Safety Data Sheets and Labels in Respect of Certain Controlled Products

10.30  Subject to section 10.41, every employer shall adopt and implement the provisions of sections 10.26 and 10.27 in respect of a controlled product and may, in so doing, replace the name of the substance with the product identifier, where the controlled product is a controlled product that

(a) is present in the work place;
(b) was received from a supplier; and
(c) is one of the following:
    (i) an explosive within the meaning of the *Explosives Act*,
    (ii) a cosmetic, device, drug or food within the meaning of the *Food and Drugs Act*,
    (iii) a control product with the meaning of the *Pest Control Products Act*,
    (iv) a prescribed substance within the meaning of the *Atomic Energy Control Act*, and
    (v) a product, material or substance included in Part II of Schedule I to the *Hazardous Products Act* that is packaged as a consumer product.

## Supplier Material Safety Data Sheets

10.31(1) Where a controlled product, other than a controlled product referred to in paragraph 10.30(c), is received by an employer, the employer shall, at the time the controlled product is received in the work place, obtain from the supplier of the controlled product a supplier material safety data sheet, unless the employer has in his possession a supplier material safety data sheet that

(a) is for a controlled product that has the same product identifier;
(b) discloses information that is current at the time that the controlled product is received; and
(c) was prepared and dated not more than three years before the date that the controlled product is received.

(2) Where there is a controlled product in a work place and the supplier material safety data sheet pertaining to the controlled product is three years old or more, the employer shall, where possible, obtain from the supplier an up-to-date supplier material safety data sheet.

(3) Where it is not practicable for an employer to obtain an up-to-date supplier material safety data sheet referred to in subsection (2), the employer shall update the hazard information on the most recent supplier material safety data sheet that the employer has received on the basis of the ingredients disclosed in that supplier material safety data sheet.

(4) Where a controlled product is received in a work place that is a laboratory, the employer is excepted from the requirements of subsection (1) if the controlled product

(a) originates from a laboratory supply house;

(b)   is intended for use in a laboratory;

(c)   is packaged in a container in a quantity of less than 10 kg; and

(d)   is packaged in a container that has applied to it a supplier label.

## Work Place Material Safety Data Sheets

10.32(1)  Subject to section 10.41, where an employer produces a controlled product, other than a fugitive emission, in the work place or imports into Canada a controlled product and brings it into a work place, the employer shall prepare a work place material safety data sheet in respect of the controlled product that discloses the information required to be disclosed by

(a)   subparagraphs 125.1(e)(i) to (v) of the Act; and

(b)   the *Controlled Products Regulations.*

(2)  Subject to section 10.41, where an employer receives a supplier material safety data sheet, the employer may prepare a work place material safety data sheet to be used in the work place in place of the supplier material safety data sheet if

(a)   the work place material safety data sheet discloses at least the information disclosed on the supplier material safety data sheet;

(b)   the information disclosed on the work place material safety data sheet does not disclaim or contradict the information disclosed on the supplier material safety data sheet;

(c)   the supplier material safety data sheet is available for examination by employees in the work place; and

(d)   the work place material safety data sheet discloses that the supplier material safety data sheet is available in the work place.

(3)  Where an employer produces, in a work place that is a laboratory supply house, or imports into Canada and brings it into such a work place, a controlled product that is intended for use in a laboratory, the employer is exempted from the requirements of subsection (1) if the employer

(a)   packages the controlled product in containers in quantities of less than 10 kg per container; and

(b)   subject to section 10.41, discloses on the label of the container of the controlled product the information required to be disclosed by

(i)   subparagraphs 125.1(e)(i) to (v) of the Act, and

(ii)   section 10.38.

(4)  The employer shall update the work place material safety data sheet referred to in subsection (1) or (2) or the label referred to in paragraph (3)(b)

(a) as soon as is practicable in the circumstances but not later than 90 days after new hazard information becomes available to the employer; and

(b) at least once every three years.

(5) Where the information required to be disclosed by this section is not available to the employer or not applicable to the controlled product, the employer shall replace the information by the words "not available" or "not applicable", as the case may be, in the English version and the words "pas disponible" or "sans objet", as the case may be, in the French version of the material safety data sheet.

**Availability of Material Safety Data Sheets**

10.33(1) Subject to subsection (2), every employer, other than an employer referred to in subsection 10.31(4), shall keep readily available for examination by employees and by the safety and health committee or the safety and health representative, if either exists, in any work place in which an employee may handle or be exposed to a controlled product, a copy in English and in French of

(a) in the case of an employer who is an employer referred to in subsection 10.32(1) or (2), the work place material safety data sheet; and

(b) in any other case, the supplier material safety data sheet.

(2) In place of keeping a material safety data sheet in the manner required by subsection (1), an employer may make a computerized version of the material safety data sheet available in English and in French for examination by employees and by the safety and health committee or the safety and health representative, if either exists, by means of a computer terminal if the employer

(a) takes all reasonable steps to keep the terminal in working order;

(b) provides the training referred to in paragraph 10.17(2)(d) to the employees and to the safety and health committee or the safety and health representative, if either exists; and

(c) on the request of an employee, the safety and health committee or the safety and health representative, makes the material safety data sheet readily available to the employee, the safety and health committee or the safety and health representative.

**Labels**

10.34(1) Subject to sections 10.36 to 10.38, each controlled product, other than a controlled product referred to in paragraph 10.30(c), in a work place and each container in which such a controlled product is contained in a work place shall, if the controlled product or the container was received from a supplier,

(a) in the case where the controlled product is in a bulk shipment, be accompanied by a supplier label;

(b) in the case where the employer has undertaken in writing to apply a label to the inner container of the controlled product, have applied to it a supplier label, as soon as possible after the controlled product is received from the supplier; and

(c) in any other case, have applied to it a supplier label.

(2) Subject to sections 10.36 to 10.38 and 10.41, where a controlled product, other than a controlled product referred to in paragraph 10.30(c), is received from a supplier and an employer places the controlled product in the work place in a container, other than the container in which it was received from the supplier, the employer shall apply to the container a supplier label or a work place label that discloses the information referred to in paragraphs 10.35(1)(a) to (c).

(3) Subject to sections 10.40 and 10.41, no person shall remove, deface, modify or alter the supplier label applied to

(a) a controlled product that is in the work place; or

(b) a container of a controlled product that is in the work place.

10.35(1) Subject to sections 10.36 to 10.38, where an employer produces a controlled product in a work place, other than a fugitive emission, or imports into Canada a controlled product and brings it into a work place, and the controlled product is not in a container, the employer shall disclose the following information on a work place label applied to the controlled product or on a sign posted in a conspicuous place in the work place:

(a) the product identifier;

(b) hazard information in respect of the controlled product; and

(c) a statement indicating that a work place material safety data sheet for the controlled product is available in the work place.

(2) Subject to sections 10.36 and 10.38, where an employer produces a controlled product in the work place, other than a fugitive emission, or imports into Canada a controlled product and brings it into the work place, and places the controlled product in a container, the employer shall apply to the container a work place label that discloses the information referred to in paragraphs (1)(a) to (c).

(3) Subsection (2) does not apply in respect of a controlled product that is

(a) intended for export, if the information referred to in paragraphs (1)(a) to (c) is disclosed on a sign posted in a conspicuous place in the work place; or

(b)   packaged in a container for sale in Canada, if the container is or is in the process of being appropriately labelled for that purpose.

## Portable Containers

10.36   Where an employer stores a controlled product in the work place in a container that has applied to it a supplier label or a work place label, a portable container filled from that container does not have to be labelled in accordance with section 10.34 or 10.35 if

(a)   the controlled product is required for immediate use; or
(b)   the following conditions apply in respect of the controlled product:
      (i)    it is under the control of and used exclusively by the employee who filled the portable container,
      (ii)   it is used only during the work shift in which the portable container was filled, and
      (iii)  it is clearly identified by a work place label applied to the portable container that discloses the product identifier.

## Special Cases

10.37   An employer shall, in a conspicuous place near a controlled product, post a sign in respect of the controlled product that discloses the product identifier if the controlled product is

(a)   in a process, reaction or storage vessel;
(b)   in a continuous-run container;
(c)   in a bulk shipment that is not placed in a container at the work place; or
(d)   not in a container and stored in bulk.

## Laboratories

10.38   The label of the container of a controlled product in a laboratory shall disclose

(a)   where the controlled product is used exclusively in the laboratory, the product identifier;
(b)   where the controlled product is a mixture or substance undergoing an analysis, test or evaluation in the laboratory, the product identifier; and
(c)   where the controlled product originates from a laboratory supply house and was received in a container containing a quantity of less than 10 kg, the following information:
      (i)    the product identifier,
      (ii)   where a material safety data sheet is available, a statement to that effect,
      (iii)  risk phrases that are appropriate to the controlled product,

(iv)    precautionary measures to be followed when handling, using or being exposed to the controlled product, and

(v)    where appropriate, first aid measures to be taken in case of exposure to the controlled product.

## Signs

10.39  The information disclosed on a sign referred to in subsection 10.35(1), paragraph 10.35(3)(a), section 10.37 or paragraph 10.42(b) shall be of such a size that it is clearly legible to the employees in the work place.

## Replacing Labels

10.40  Where, in a work place, a label applied to a controlled product or a container of a controlled product becomes illegible or is removed from the controlled product or the container, the employer shall replace the label with a work place label that discloses the following information:

(a)    the product identifier;

(b)    hazard information in respect of the controlled product; and

(c)    a statement indicating that a material safety data sheet for the controlled product is available in the work place.

## Exemptions from Disclosure

10.41(1)  Subject to subsection (2), where an employer has filed a claim for exemption from disclosure of information on a material safety data sheet or on a label pursuant to subsection 11(2) of the *Hazardous Materials Information Review Act*, the employer shall disclose, in place of the information that the employer is exempt from disclosing,

(a)    where there is no final disposition of the proceedings in relation to the claim, the date that the claim for exemption was filed and the registry number assigned to the claim under the *Hazardous Materials Information Review Act*; and

(b)    where the final disposition of the proceedings in relation to the claim is that the claim is valid, a statement that an exemption has been granted and the date on which the exemption was granted.

(2)  Where a claim for exemption is in respect of the chemical name, common name, generic name, trade name or brand name of a controlled product, the employer shall, on the material safety data sheet or label of the controlled product, replace that information with a code name or code number specified by the employer as the product identifier for that controlled product.

## Hazardous Waste

10.42 Where a controlled product in the work place is hazardous waste, the employer shall clearly identify that it is hazardous waste by

(a) applying a label to the hazardous waste or its container; or
(b) posting a sign in a conspicuous place near the hazardous waste or its container.

## Information Required in a Medical Emergency

10.43 For the purposes of subsection 125.2(1) of the Act, a medical professional is a registered nurse registered or licensed under the laws of a province.

## PART XI - CONFINED SPACES

### Interpretation

11.1 In this Part,

"class of confined spaces" means a group of at least two confined spaces that are likely, by reason of their similarity, to present the same hazards to persons entering, exiting or occupying them; (*catégorie d'espaces clos*)

"confined space" means an enclosed or partially enclosed space that

(a) is not designed or intended for human occupancy except for the purpose of performing work,
(b) has restricted means of access and egress, and
(c) may become hazardous to an employee entering it due to
    (i) its design, construction, location or atmosphere,
    (ii) the materials or substances in it, or
    (iii) any other conditions relating to it; (*espace clos*)

"hot work" means any work where flame is used or a source of ignition may be produced. (*travail à chaud*)

### Hazard Assessment

11.2(1) Where it is likely that a person will, in order to perform work for an employer, enter a confined space and an assessment pursuant to this subsection has not been carried out in respect of the confined space, or in respect of the class of confined spaces to which it belongs, the employer shall appoint a qualified person

(a) to carry out an assessment of the physical and chemical hazards to which the person is likely to be exposed in the confined space or the class of confined spaces; and

(b)     to specify the tests that are necessary to determine whether the person would be likely to be exposed to any of the hazards identified pursuant to paragraph (a).

(2)  The qualified person referred to in subsection (1) shall, in a signed and dated report to the employer, record the findings of the assessment carried out pursuant to paragraph (1)(a).

(3)  The employer shall make a copy of any report made pursuant to subsection (2) available to the safety and health committee or the safety and health representative, if either exists.

(4)  Subject to subsection (5), the report made pursuant to subsection (2) shall be reviewed by a qualified person at least once every three years to ensure that its assessment of the hazards with which it is concerned is still accurate.

(5)  If a confined space has not been entered in the three years preceding the time when the report referred to in subsection (4) should have been reviewed and no entry is scheduled, the report need not be reviewed until it becomes likely that a person will, in order to perform work for an employer, enter the confined space.

**Entry Procedures**

11.3  Every employer shall, after considering the report made pursuant to subsection 11.2(2),

(a)     in consultation with the safety and health committee or the safety and health representative, if either exists, establish procedures, with the date on which they are established specified therein, that are to be followed by a person entering, exiting or occupying a confined space assessed pursuant to subsection 11.2(1), or a confined space that belongs to a class of confined spaces assessed pursuant to that subsection, and establish, where reasonably practicable, an entry permit system that provides for
         (i)     specifying, in each case, the length of time for which an entry permit is valid, and
         (ii)    recording
                 (A)    the name of the person entering the confined space, and
                 (B)    the time of entry and the anticipated time of exit;
(b)     specify the protection equipment referred to in Part XII that is to be used by every person who is granted access to the confined space by the employer;
(c)     specify any insulated protection equipment and tools referred to in Part VIII that a person may need in the confined space; and

(d)　specify the protection equipment and emergency equipment to be used by a person who takes part in the rescue of a person from the confined space or in responding to other emergency situations in the confined space.

**Confined Space Entry**

11.4(1) The employer shall, where a person is about to enter a confined space, appoint a qualified person

(a)　to verify, by means of tests, that compliance with the following specifications can be achieved during the period of time that the person will be in the confined space, namely,
  (i)　the concentration of any chemical agent or combination of chemical agents in the confined space to which the person is likely to be exposed will not result in the exposure of the person
    (A)　to a concentration of that chemical agent or combination of chemical agents in excess of the value referred to in paragraph 10.21(1)(a), or
    (B)　to a concentration of that chemical agent or combination of chemical agents in excess of the percentage set out in subsection 10.22(1), or in subsection 10.22(2) under the circumstances described in that subsection,
  (ii)　the concentration of airborne hazardous substances, other than chemical agents, in the confined space is not hazardous to the safety or health of the person, and
  (iii)　the percentage of oxygen in the air in the confined space is not less than 18 per cent by volume and not more than 23 per cent by volume, at normal atmospheric pressure;
(b)　to verify that
  (i)　any liquid in which the person could drown has been removed from the confined space,
  (ii)　any free-flowing solid in which the person may become entrapped has been removed from the confined space,
  (iii)　the entry of any liquid, free-flowing solid or hazardous substance into the confined space has been prevented by a secure means of disconnection or the fitting of blank flanges,
  (iv)　all electrical and mechanical equipment that may present a hazard to the person has been disconnected from its power source, real or residual, and has been locked out, and
  (v)　the opening for entry into and exit from the confined space is sufficient to allow the safe passage of a person using protection equipment; and
(c)　subject to subsection 11.5(1), to verify that the specifications set out in subparagraphs (a)(i) to (iii) are complied with at all times that a person is in the confined space.

(2) The qualified person referred to in subsection (1) shall, in a signed and dated report to the employer, set out the results of the verification carried out in accordance with that subsection, including the test methods, the test results and a list of the test equipment used.

(3) The employer shall

(a) where the report made pursuant to subsection (2) indicates that a person who has entered the confined space has been in danger, send the report to the safety and health committee or the safety and health representative, if either exists; and

(b) in all other cases, make a written copy or a machine-readable version of the report available to the safety and health committee or the safety and health representative, if either exists.

**Emergency Procedures and Equipment**

11.5(1) Where conditions in a confined space or the nature of the work to be performed in a confined space is such that paragraph 11.4(1)(c) cannot be complied with, the employer shall

(a) in consultation with the safety and health committee or the safety and health representative, if either exists, establish emergency procedures to be followed in the event of an accident or other emergency in or near the confined space, which procedures shall specify the date on which they are established and provide for the immediate evacuation of the confined space when
(i) an alarm is activated, or
(ii) there is any significant change in a concentration or percentage referred to in paragraph 11.4(1)(a) that would adversely affect the safety or health of a person in the confined space;

(b) provide the protection equipment referred to in paragraphs 11.3(b), (c) and (d) for each person who is about to enter the confined space;

(c) ensure that a qualified person trained in the entry and emergency procedures established pursuant to paragraph 11.3(a) and paragraph (a) is
(i) in attendance outside the confined space, and
(ii) in communication with the person inside the confined space;

(d) provide the qualified person referred to in paragraph (c) with a suitable alarm device for summoning assistance; and

(e) ensure that two or more persons are in the immediate vicinity of the confined space to assist in the event of an accident or other emergency.

(2) One of the persons referred to in paragraph (1)(e) shall

(a)   be trained in the emergency procedures established pursuant to paragraph (1)(a);

(b)   be the holder of a basic first aid certificate; and

(c)   be provided with the protection equipment and emergency equipment referred to in paragraph 11.3(d).

(3)  The employer shall ensure that every person entering, exiting or occupying a confined space referred to in subsection (1) wears an appropriate safety harness that is securely attached to a lifeline that

(a)   is attached to a secure anchor outside the confined space;

(b)   is controlled by the qualified person referred to in paragraph (1)(c);

(c)   protects the person from the hazard for which it is provided and does not in itself create a hazard; and

(d)   is, where reasonably practicable, equipped with a mechanical lifting device.

## Record of Emergency Procedures and Equipment

11.6(1)  When a person enters a confined space under circumstances such that paragraph 11.4(1)(a) cannot be complied with, the qualified person referred to in paragraph 11.5(1)(c) shall, in a signed and dated report to the employer,

(a)   specify those procedures established pursuant to paragraph 11.5(1)(a) that are to be followed and the protection equipment, insulated protection equipment and tools and the emergency equipment that are to be used; and

(b)   specify any other procedures to be followed and any other equipment that could be needed.

(2)  The report made pursuant to subsection (1) and any procedures specified therein shall be explained by the qualified person to every employee who is about to enter a confined space, and a copy of the report shall be signed and dated by any employee to whom the report and the procedures have been so explained, acknowledging by signature the reading of the report and the explanation thereof.

## Provision and Use of Equipment

11.7(1)  The employer shall

(a)   provide each person who is granted access to a confined space with the protection equipment referred to in paragraph 11.3(b); and

(b)   provide each person who is to undertake rescue operations with the protection equipment and emergency equipment referred to in paragraph 11.3(d).

(2) The employer shall ensure that every person who enters, exits or occupies a confined space follows the procedures established pursuant to paragraph 11.3(a) and uses the protection equipment referred to in paragraphs 11.3(b) and (c).

**Precaution**

11.8 No person shall close off a confined space until a qualified person has verified that no person is inside it.

**Hot Work**

11.9(1) Unless a qualified person has determined that the work can be performed safely, hot work shall not be performed in a confined space that contains

(a)  an explosive or flammable hazardous substance in a concentration in excess of 10 per cent of its lower explosive limit; or
(b)  oxygen in a concentration in excess of 23 per cent.

(2) Where hot work is to be performed in a confined space that contains hazardous concentrations of flammable or explosive materials,

(a)  a qualified person shall patrol the area surrounding the confined space and maintain a fire-protection watch in that area until all fire hazard has passed; and
(b)  fire extinguishers specified as emergency equipment pursuant to paragraph 11.3(d) shall be provided in the area referred to in paragraph (a).

(3) Where an airborne hazardous substance may be produced by hot work in a confined space, no person shall enter or occupy the confined space unless

(a)  section 11.10 is complied with; or
(b)  the person uses a respiratory protective device that meets the requirements of sections 12.1 to 12.3 and 12.7.

**Ventilation Equipment**

11.10(1) Where ventilation equipment is used to maintain the concentration of a chemical agent or combination of chemical agents in a confined space at or below the concentration referred to in subparagraph 11.4(1)(a)(i), or to maintain the percentage of oxygen in the air of a confined space within the limits referred to in subparagraph 11.4(1)(a)(iii), the employer shall not grant access to the confined space to any person unless

(a)  the ventilation equipment is

(i) equipped with an alarm that will, if the equipment fails, be activated automatically and be audible or visible to every person in the confined space, or

(ii) monitored by an employee who is in constant attendance at the equipment and who is in communication with the person or persons in the confined space; and

(b) in the event of failure of the ventilation equipment, sufficient time will be available for the person to escape from the confined space before

(i) the concentration of the chemical agent or combination of chemical agents in the confined space exceeds the concentrations referred to in subparagraph 11.4(1)(a)(i), or

(ii) the percentage of oxygen in the air ceases to remain within the limits referred to in subparagraph 11.4(1)(a)(iii).

(2) If the ventilation equipment fails to operate properly, the employee referred to in subparagraph (1)(a)(ii) shall immediately inform the person or persons in the confined space of the failure of the equipment.

## Training

11.11(1) The employer shall provide every employee who is likely to enter a confined space with instruction and training in

(a) the procedures established pursuant to paragraphs 11.3(a) and 11.5(1)(a); and

(b) the use of the protection equipment referred to in paragraphs 11.3(b), (c) and (d).

(2) The employer shall ensure that no person enters a confined space unless the person is instructed in

(a) the procedures to be followed in accordance with paragraphs 11.3(a) and 11.5(1)(a); and

(b) the use of the protection equipment referred to in paragraphs 11.3(b), (c) and (d).

## Record Keeping

11.12 The employer shall, at the employer's place of business nearest to the work place in which the confined space is located, keep a written copy or a machine-readable version of

(a) any report made pursuant to subsection 11.2(2) and the procedures established pursuant to paragraphs 11.3(a) and 11.5(1)(a) for a period of ten years after the date on which the qualified person signed the report or the procedures were established; and

(b)    any report made pursuant to subsection 11.4(2)

    (i)    for a period of ten years after the date on which the qualified person signed the report where the verification procedures undertaken pursuant to paragraphs 11.4(1)(a) and (c) indicate that the specifications set out in subparagraphs 11.4(1)(a)(i) to (iii) were not complied with, and

    (ii)   in every other case, for a period of two years after the date on which the qualified person signed the report.

## PART XII - SAFETY MATERIALS, EQUIPMENT, DEVICES AND CLOTHING

### General

12.1  Where

(a)    it is not reasonably practicable to eliminate or control a safety or health hazard in a work place within safe limits, and

(b)    the use of protection equipment may prevent or reduce injury from that hazard,

every person granted access to the work place who is exposed to that hazard shall use the protection equipment prescribed by this Part.

12.2  All protection equipment referred to in section 12.1

(a)    shall be designed to protect the person from the hazard for which it is provided; and

(b)    shall not in itself create a hazard.

12.3  All protection equipment provided by the employer shall

(a)    be maintained, inspected and tested by a qualified person; and

(b)    where necessary to prevent a health hazard, be maintained in a clean and sanitary condition by a qualified person.

### Protective Headwear

12.4  Where there is a hazard of head injury in a work place, protective headwear that meets the standards set out in CSA Standard Z94.1-M1977, *Industrial Protective Headwear*, the English version of which is dated April, 1977, as amended to September, 1982 and the French version of which is dated April, 1980, as amended to September, 1982, shall be used.

### Protective Footwear

12.5(1)  Where there is a hazard of a foot injury or electric shock through footwear in a work place, protective footwear that meets the standards set out in CSA Standard Z195-M1984, *Protective Footwear*, the English version of which is dated March, 1984 and the French version of which is dated December, 1984, shall be used.

(2) Where there is a hazard of slipping in a work place, nonslip footwear shall be used.

**Eye and Face Protection**

12.6 Where there is a hazard of injury to the eyes, face, ears or front of the neck of an employee in a work place, the employer shall provide eye or face protectors that meet the standards set out in CSA Standard Z94.3-M1982, *Industrial Eye and Face Protectors*, the English version of which is dated May, 1982 and the French version of which is dated February, 1983.

**Respiratory Protection**

12.7(1) Where there is a hazard of an airborne hazardous substance or an oxygen deficient atmosphere in a work place, the employer shall provide a respiratory protective device that is listed in the *NIOSH Certified Equipment List* as of October 1, 1984, dated February, 1985, published by the National Institute for Occupational Safety and Health.

(2) A respiratory protective device referred to in subsection (1) shall be selected, fitted, cared for, used and maintained in accordance with the standards set out in CSA Standard Z94.4-M1982, *Selection, Care and Use of Respirators*, the English version of which is dated May, 1982, as amended to September, 1984 and the French version of which is dated March, 1983, as amended to September, 1984, excluding clauses 6.1.5, 10.3.3.1.2 and 10.3.3.4.2(c).

(3) Where air is provided for the purpose of a respiratory protective device referred to in subsection (1),

(a)  the air shall meet the standards set out in clauses 5.5.2 to 5.5.11 of CSA Standard CAN3-Z180.1-M85, *Compressed Breathing Air and Systems*, the English version of which is dated December, 1985 and the French version of which is dated November, 1987; and

(b)  the system that supplies air shall be constructed, tested, operated and maintained in accordance with the CSA Standard referred to in paragraph (a).

12.8 Where a steel or aluminium self-contained breathing apparatus cylinder has a dent deeper than 1.5 mm and less than 50 mm in major diameter or shows evidence of deep isolated pitting, cracks or splits, the cylinder shall be removed from service until it has been shown to be safe for use by means of a hydrostatic test at a pressure equal to one and one-half times the maximum allowable working pressure.

## Skin Protection

12.9  Where there is a hazard of injury or disease to or through the skin in a work place, the employer shall provide to every person granted access to the work place

(a)  a shield or screen;
(b)  a cream to protect the skin; or
(c)  an appropriate body covering.

## Fall-Protection Systems

12.10(1)  Where a person, other than an employee who is installing or removing a fall-protection system in accordance with the instructions referred to in subsection (5), works from

(a)  an unguarded structure that is
  (i)  more than 2.4 m above the nearest permanent safe level, or
  (ii)  above any moving parts of machinery or any other surface or thing that could cause injury to an employee upon contact,
(b)  a temporary structure that is more than 6 m above a permanent safe level, or
(c)  a ladder at a height of more than 2.4 m above the nearest permanent safe level where, because of the nature of the work, that person cannot use one hand to hold onto the ladder,

the employer shall provide a fall-protection system.

(2)  The components of a fall-protection system shall meet the following standards:

(a)  CSA Standard Z259.1-1976, *Fall-Arresting Safety Belts and Lanyards for the Construction and Mining Industries*, the English version of which is dated November, 1976, as amended to May, 1979 and the French version of which is dated April, 1980;
(b)  CSA Standard Z259.2-M1979, *Fall-Arresting Devices, Personnel Lowering Devices and Life Lines*, the English version of which is dated November, 1979 and the French version of which is dated October, 1983; and
(c)  CSA Standard Z259.3-M1978, *Lineman's Body Belt and Lineman's Safety Strap*, the English version of which is dated September, 1978, as amended to April, 1981 and the French version of which is dated April, 1980, as amended to April, 1981.

(3) The anchor of a fall-protection system shall be capable of withstanding a force of 17.8 kN.

(4) A fall-protection system that is used to arrest the fall of a person shall prevent that person

(a)  from being subjected to a peak fall arrest force greater than 8 kN; and
(b)  from falling freely for more than 1.2 m.

(5) Where an employee is about to install or remove a fall-protection system, the employer shall

(a)  prepare written instructions for the safe installation or removal of the fall-protection system; and
(b)  keep a copy of the instructions readily available for the information of the employee.

## Protection Against Drowning

12.11(1) Where, in a work place, there is a hazard of drowning, the employer shall provide every person granted access to the work place with

(a)  a life jacket or buoyancy device that meets the standards set out in the Canadian General Standards Board Standard
    (i)   CAN2-65.7-M80, *Life Jackets, Inherently Buoyant Type*, dated April, 1980, or
    (ii)  65-GP-11, *Standard for Personal Flotation Devices*, dated October, 1972; or
(b)  a safety net or a fall-protection system.

(2) Where in a work place, there is a hazard of drowning,

(a)  emergency equipment shall be provided and held in readiness;
(b)  a person who is qualified to operate all the emergency equipment provided shall be available;
(c)  if appropriate, a powered boat shall be provided and held in readiness; and
(d)  written emergency procedures shall be prepared by the employer containing
    (i)   a full description of the procedures to be followed and the responsibilities of all persons granted access to the work place, and
    (ii)  the location of any emergency equipment.

(3) Where a work place is a wharf, dock, pier, quay or other similar structure, a ladder that extends at least two rungs below water level shall be affixed to the face of the structure every 60 m along its length.

## Loose Clothing

12.12  Loose clothing, long hair, dangling accessories, jewellery or other similar items that are likely to be hazardous to the safety or health of an employee in a work place shall not be worn unless they are so tied, covered or otherwise secured as to prevent the hazard.

## Protection Against Moving Vehicles

12.13  Where an employee is regularly exposed to contact with moving vehicles during his work, he shall

(a)  wear a high-visibility vest or other similar clothing, or
(b)  be protected by a barricade

that is readily visible under all conditions of use.

## Records

12.14(1)  A record of all protection equipment provided by the employer shall be kept by him in the work place in which the equipment is located for a period of two years after it ceases to be used.

(2)  The record referred to in subsection (1) shall contain

(a)  a description of the equipment and the date of its acquisition by the employer;
(h)  the date and result of each inspection and test of the equipment;
(c)  the date and nature of any maintenance work performed on the equipment since its acquisition by the employer; and
(d)  the name of the person who performed the inspection, test or maintenance of the equipment.

## Instructions and Training

12.15(1)  Every person granted access to the work place who uses protection equipment shall be instructed by the employer in the use of the equipment.

(2)  Every employee who uses protection equipment shall be instructed and trained in the use, operation and maintenance of the equipment.

(3)  Every person granted access to a work place shall be instructed in respect of the written emergency procedures referred to in paragraph 12.11(2)(d).

(4)  The instructions referred to in subsections (2) and (3) shall be

(a)  set out in writing; and

(b)    kept by the employer readily available for examination by every person granted access to the work place.

## Defective Protection Equipment

12.16   Where an employee finds any defect in protection equipment that may render it unsafe for use, he shall report the defect to his employer as soon as possible.

12.17   An employer shall mark or tag as unsafe and remove from service any protection equipment used by his employees that has a defect that may render it unsafe for use.

## PART XIII - TOOLS AND MACHINERY

### Interpretation

13.1   In this Part,

"explosive actuated fastening tool" means a tool that, by means of an explosive force, propels or discharges a fastener for the purpose of impinging it on, affixing it to or causing it to penetrate another object or material. (*pistolet de scellement à cartouches explosives*)

### Design, Construction, Operation and Use of Tools

13.2   The exterior surface of any tool used by an employee in a fire hazard area shall be made of non-sparking material.

13.3   All portable electric tools used by employees shall meet the standards set out in CSA Standard CAN C22.2 No. 71.1-M89, *Portable Electric Tools*, the English version of which is dated September, 1989 and the French version of which is dated February, 1991.

13.4(1)   Subject to subsection (2), all portable electric tools used by employees shall be grounded.

(2)   Subsection (1) does not apply to tools that

(a)    are powered by a self-contained battery;
(b)    have a protective system of double insulation; or
(c)    are used in a location where reliable grounding cannot be obtained if the tools are supplied from a double insulated portable ground fault circuit interrupter of the class A type that meets the standards set out in CSA Standard C22.2 No. 144-1977, *Ground Fault Circuit Interrupters*, dated March, 1977.

13.5 All portable electric tools used by employees in a fire hazard area shall be marked as appropriate for use or designed for use in the area of that hazard.

13.6 Where an air hose is connected to a portable air-powered tool used by an employee, a restraining device shall be attached

(a) where an employee may be injured by the tool falling, to the tool; and
(b) to all hose connections, in order to prevent injury to an employee in the event of an accidental disconnection of a hose.

13.7(1) All explosive actuated fastening tools used by employees shall meet the standards set out in CSA Standard Z166-1975, *Explosive Actuated Fastening Tools*, dated June, 1975.

(2) No employee shall operate an explosive actuated fastening tool unless authorized to do so by his employer.

(3) Every employee who operates an explosive actuated fastening tool shall operate it in accordance with the CSA Standard referred to in subsection (1).

13.8 All chain saws used by employees shall meet the standards set out in CSA Standard CAN3-Z62.1-M85, *Chain Saws*, dated February, 1985.

**Defective Tools and Machines**

13.9 Where an employee finds any defect in a tool or machine that may render it unsafe for use, he shall report the defect to his employer as soon as possible.

13.10 An employer shall mark or tag as unsafe and remove from service any tool or machine used by his employees that has a defect that may render it unsafe for use.

**Instructions and Training**

13.11 Every employee shall be instructed and trained by a qualified person appointed by his employer in the safe and proper inspection, maintenance and use of all tools and machinery that he is required to use.

13.12(1) Every employer shall maintain a manual of operating instructions for each type of portable electric tool, portable air-powered tool, explosive actuated fastening tool and machine used by his employees.

(2) A manual referred to in subsection (1) shall be kept by the employer readily available for examination by an employee who is required to use the tool or machine to which the manual applies.

## General Requirements for Machine Guards

13.13(1) Every machine that has exposed moving, rotating, electrically charged or hot parts or that processes, transports or handles material that constitutes a hazard to an employee shall be equipped with a machine guard that

(a)  prevents the employee or any part of his body from coming into contact with the parts or material;
(b)  prevents access by the employee to the area of exposure to the hazard during the operation of the machine; or
(c)  makes the machine inoperative if the employee or any part of his clothing is in or near a part of the machine that is likely to cause injury.

(2)  To the extent that is reasonably practicable, a machine guard referred to in subsection (1) shall not be removable.

(3)  A machine guard shall be so constructed, installed and maintained that it meets the requirements of subsection (1).

## Use, Operation, Repair and Maintenance of Machine Guards

13.14  Machine guards shall be operated, maintained and repaired by a qualified person.

13.15  Subject to section 13.16, where a machine guard is installed on a machine, no person shall use or operate the machine unless the machine guard is in its proper position.

13.16(1)  Subject to subsection (2), where it is necessary to remove a machine guard from a machine in order to perform repair or maintenance work on the machine, no person shall perform the repair or maintenance work unless the machine has been locked out in accordance with a written lock out procedure provided by the employer.

(2)  Where it is not reasonably practicable to lock out a machine referred to in subsection (1) in order to perform repair or maintenance work on the machine, the work may be performed if

(a)  the person performing the work follows written instructions provided by the employer that will ensure that any hazard to that person is not significantly greater than it would be if the machine had been locked out; and
(b)  the person performing the work
     (i)   obtains a written authorization from the employer each time the work is performed, and
     (ii)  performs the work under the direct supervision of a qualified person.

13.17 A copy of the instructions referred to in section 13.16 shall be kept readily available by the employer for the information of persons who perform repair and maintenance work on his machines.

## Abrasive Wheels

13.18 Abrasive wheels shall be

(a) used only on machines equipped with machine guards,
(b) mounted between flanges, and
(c) operated in accordance with sections 4 to 6 of CSA Standard B173.5-1979, *Safety Requirements for the Use, Care and Protection of Abrasive Wheels*, dated February, 1979.

13.19 A bench grinder shall be equipped with a work rest or other device that

(a) prevents the work piece from jamming between the abrasive wheel and the wheel guard; and
(b) does not make contact with the abrasive wheel at any time.

## Mechanical Power Transmission Apparatus

13.20 Equipment used in the mechanical transmission of power shall be guarded in accordance with sections 7 to 10 of ANSI Standard ANSI B15.1-1972, *Safety Standard for Mechanical Power Transmission Apparatus*, dated July, 1972.

## Woodworking Machinery

13.21 Woodworking machinery shall be guarded in accordance with clause 3.3 of CSA Standard Z114-M1977, *Safety Code for the Woodworking Industry*, dated March, 1977.

## Punch Presses

13.22 Punch presses shall meet the standards set out in CSA Standard Z142-1976, *Code for the Guarding of Punch Presses at Point of Operation*, dated February, 1976.

## PART XIV - MATERIALS HANDLING

### Interpretation

14.1 In this Part,

"materials handling equipment" means equipment used to transport, lift, move or position materials, goods or things and includes a rail motor car and other mobile equipment but does not include an elevating device; (*appareil de manutention des matériaux*)

"safe working load" means, with respect to materials handling equipment, the maximum load that the materials handling equipment is designed and constructed to handle or support safely; (*charge de travail admissible*)

"signaller" means a person instructed by an employer to direct, by means of visual or auditory signals, the safe movement and operation of materials handling equipment. (*signaleur*)

## Application

14.2 This Part does not apply to or in respect of

(a)   the use and operation of motor vehicles on public roads;
(b)   the use and operation of tackle in the loading or unloading of ships; or
(c)   the underground workings of mines.

## DIVISION I - DESIGN AND CONSTRUCTION

### General

14.3 (1)   Materials handling equipment shall, to the extent that is reasonably practicable, be so designed and constructed that if there is a failure of any part of the materials handling equipment, it will not result in loss of control of the materials handling equipment or create a hazardous condition.

(2)   All glass in doors, windows and other parts of materials handling equipment shall be of a type that will not shatter into sharp or dangerous pieces under impact.

### Protection from Falling Objects

14.4(1)   Where materials handling equipment is used under such circumstances that the operator of the equipment may be struck by a falling object or shifting load, the employer shall equip the materials handling equipment with a protective structure of such a design, construction and strength that it will, under all foreseeable conditions, prevent the penetration of the object or load into the area occupied by the operator.

(2)   A protective structure referred to in subsection (1) shall be

(a)   constructed from non-combustible or fire resistant material; and
(b)   designed to permit quick exit from the materials handling equipment in an emergency.

14.5  Where, during the loading or unloading of materials handling equipment, the load will pass over the operator's position, the operator shall not occupy the materials handling equipment unless it is equipped with a protective structure referred to in section 14.4.

## Protection from Turn Over

14.6  Where mobile equipment is used in circumstances where it may turn over, it shall be fitted with a rollover protection device that meets the standards set out in CSA Standard B352-M1980, *Rollover Protective Structures (ROPS) for Agricultural, Construction, Earthmoving, Forestry, Industrial, and Mining Machines*, the English version of which is dated September, 1980, and the French version of which is dated April, 1991, that will prevent the operator of the mobile equipment from being trapped or crushed under the equipment if it does turn over.

## Fuel Tanks

14.7  Where a fuel tank, compressed gas cylinder or similar container contains a hazardous substance and is mounted on materials handling equipment, it shall be

(a)  so located or protected that under all conditions it is not hazardous to the safety or health of an employee who is required to operate or ride on the materials handling equipment; and
(b)  connected to fuel overflow and vent pipes that are so located that fuel spills and vapours cannot
   (i)   be ignited by hot exhaust pipes or other hot or sparking parts, or
   (ii)  be hazardous to the safety or health of any employee who is required to operate or ride on the materials handling equipment.

## Protection from Elements

14.8(1)  Materials handling equipment that is regularly used outdoors shall be fitted with a roof or other structure that will protect the operator from exposure to any weather condition that is likely to be hazardous to his safety or health.

(2)  Where heat produced by materials handling equipment raises the temperature in the operator's compartment or position to 27°C or more, the compartment or position shall be protected from the heat by an insulated barrier.

## Vibration

14.9  All materials handling equipment shall be so designed and constructed that the operator will not be injured or his control of the materials handling equipment impaired by any vibration, jolting or uneven movement of the materials handling equipment.

## Controls

14.10 The arrangement and design of dial displays and controls and the general layout and design of the operator's compartment or position on all materials handling equipment shall not hinder or prevent the operator from operating the materials handling equipment.

## Fire Extinguishers

14.11(1) Mobile equipment that is used or operated for transporting or handling combustible or flammable substances shall be equipped with a dry chemical fire extinguisher.

(2) The fire extinguisher referred to in subsection (1) shall

(a)   have not less than a 5B rating as defined in the *National Fire Code*;
(b)   meet the standards set out in section 6.2 of the *National Fire Code*; and
(c)   be so located that it is readily accessible to the operator of the mobile equipment while he is in the operating position.

## Means of Entering and Exiting

14.12 All materials handling equipment shall be provided with a step, handhold or other means of entering into and exiting from the compartment or position of the operator and any other place on the equipment that an employee enters in order to service the equipment.

## Lighting

14.13(1) Subject to subsection (2), where mobile equipment is used or operated by an employee in a work place at night or at any time when the level of lighting within the work place is less than one dalx, the mobile equipment shall be

(a)   fitted on the front and rear thereof with warning lights that are visible from a distance of not less than 100 m; and
(b)   provided with lighting that ensures the safe operation of the equipment under all conditions of use.

(2) No mobile equipment shall be operated at night on a route that is used by other vehicles unless it is equipped with such lighting facilities for the equipment as are required under the laws of the province in which the equipment is operated.

## Control Systems

14.14 All mobile equipment shall be fitted with braking, steering and other control systems that

(a) are capable of safely controlling and stopping the movement of the mobile equipment and any hoist, bucket or other part of the mobile equipment; and

(b) respond reliably and quickly to moderate effort on the part of the operator.

14.15 Any mobile equipment that is normally used for transporting employees from place to place in a work place shall be equipped with

(a) a mechanical parking brake; and

(b) a hydraulic or pneumatic braking system.

## Warnings

14.16 Mobile equipment that is operated in an area occupied by employees and that travels at speeds in excess of 8 km per hour or in reverse shall be fitted with a horn or similar audible warning device having a distinctive sound that can be clearly heard above the noise of the equipment and any surrounding noise.

## Seat Belts

14.17 Where mobile equipment is used under conditions where a seat belt or shoulder type strap restraining device is likely to contribute to the safety of the operator or passengers, the mobile equipment shall be fitted with such a belt or device.

## Rear View Mirror

14.18 Where mobile equipment cannot be operated safely in reverse unless it is equipped with an outside rear view mirror, the mobile equipment shall be so equipped.

## Electrical Equipment

14.19 Any materials handling equipment that is electrically powered shall be so designed and constructed that the operator and all other employees are protected from electrical shock or injury by means of protective guards, screens or panels secured by bolts, screws or other equally reliable fasteners.

## Automatic Equipment

14.20 Where materials handling equipment that is controlled or operated by a remote or automatic system may make physical contact with an employee, it shall be prevented from doing so by the provision of an emergency stop system or barricades.

## Conveyors

14.21 The design, construction, installation, operation and maintenance of each conveyor, cableway or other similar materials handling equipment shall meet the standards set out in ANSI Standard ANSI B20.1-1976, *Safety Standards for Conveyors and Related Equipment*, dated 1976.

## DIVISION II - MAINTENANCE, OPERATION AND USE

### Inspection, Testing and Maintenance

14.22(1) Before materials handling equipment is operated for the first time in a work place, the employer shall set out in writing instructions for the inspection, testing and maintenance of that materials handling equipment.

(2) The instructions referred to in subsection (1) shall specify the nature and frequency of inspections, tests and maintenance.

(3) A qualified person shall

(a)   comply with the instructions referred to in subsection (1); and
(b)   make and sign a report of each inspection, test or maintenance work performed by him.

(4) The report referred to in paragraph (3)(b) shall

(a)   include the date of the inspection, test or maintenance performed by the qualified person;
(b)   identify the materials handling equipment that was inspected, tested or maintained; and
(c)   set out the safety observations of the qualified person inspecting, testing or maintaining the materials handling equipment.

(5) The employer shall keep at the work place at which the materials handling equipment is located a copy of

(a)   the instructions referred to in subsection (1), and
(b)   the report referred to in paragraph (3)(b)

for a period of one year after the instructions are set out in writing or the report is signed.

## Operator Training

14.23(1) Every operator of materials handling equipment shall be instructed and trained by the employer in the procedures to be followed for

(a)    the inspection of the materials handling equipment;
(b)    the fuelling of the materials handling equipment, where applicable; and
(c)    the safe and proper use of the equipment.

(2) Every employer shall keep a record of any instruction or training given to an operator of materials handling equipment for as long as the operator remains in his employ.

## Operation

14.24 No employer shall require an employee to operate materials handling equipment unless the employee

(a)    is capable of operating the equipment safely; and
(b)    possesses any operator's licence that may be required under the laws of the province in which the equipment is operated.

14.25(1) No person shall operate materials handling equipment unless

(a)    he has a clear and unobstructed view of the area in which the equipment is being operated and, in the case of mobile equipment, of the course to be travelled by the mobile equipment; or
(b)    where the person is an employee, the person is authorized by the employer to do so and is directed by a signaller.

(2) No materials handling equipment shall be used on a ramp with a slope greater than the maximum slope recommended by the manufacturer of the equipment.

14.26(1) Every employer shall establish a code of signals for the purposes of paragraph 14.25(1)(b) and shall

(a)    instruct every signaller and operator of materials handling equipment employed by him in the use of the code; and
(b)    keep a copy of the code in a place where it is readily available for examination by the signallers and operators.

(2) No signaller shall perform duties other than signalling while any materials handling equipment under his direction is in motion.

14.27(1)  Subject to subsection (2), where it is not practicable for a signaller to use visual signals, a telephone, radio or other signalling device shall be provided by the employer for the use of the signaller.

(2)  No radio transmitting equipment shall be used in any work place for the transmission of signals when such use may activate electric blasting equipment in that place.

### Repairs

14.28(1)  Subject to subsection (2), any repair, modification or replacement of a part of any materials handling equipment shall not decrease the safety factor of the materials handling equipment or part.

(2)  If a part of less strength or quality than the original part is used in the repair, modification or replacement of a part of any materials handling equipment, the use of the materials handling equipment shall be restricted by the employer to such loading and use as will ensure the retention of the original safety factor of the equipment or part.

### Transporting and Positioning Employees

14.29(1)  Materials handling equipment shall not be used for transporting an employee unless the equipment is specifically designed for that purpose.

(2)  Materials handling equipment shall not be used for positioning an employee unless the equipment is equipped with a platform, bucket or basket designed for that purpose.

### Loading, Unloading and Maintenance While in Motion

14.30  No materials, goods or things shall be picked up from or placed on any mobile equipment while the equipment is in motion unless the equipment is specifically designed for that purpose.

14.31  Except in the case of an emergency, no employee shall get on or off any mobile equipment while it is in motion.

14.32(1)  Subject to subsection (2), no repair, maintenance or cleaning work shall be performed on any materials handling equipment while the materials handling equipment is being operated.

(2)  Fixed parts of materials handling equipment may be repaired, maintained or cleaned while the materials handling equipment is being operated if they are so isolated or protected that the operation of the materials handling equipment does not affect the safety of the employee performing the repair, maintenance or cleaning work.

## Positioning the Load

14.33  Where mobile equipment is travelling with a raised or suspended load, the operator of the equipment shall ensure that the load is carried as close to the ground or floor level as the situation permits and in no case shall the load be carried at a point above

(a)  the centre of gravity of the loaded mobile equipment; or
(b)  the point at which the loaded mobile equipment becomes unstable.

## Tools

14.34  Where tools, tool boxes or spare parts are carried on materials handling equipment, they shall be securely stored.

## Housekeeping

14.35  The floor, cab and other occupied parts of materials handling equipment shall be kept free of any grease, oil, materials, tools or equipment that may cause an employee to slip or trip.

## Parking

14.36  No mobile equipment shall be parked in a corridor, aisle, doorway or other place where it may interfere with the safe movement of persons, materials, goods or things.

## Materials Handling Area

14.37(1)  In this section, "materials handling area" means an area within which

(a)  mobile equipment, or
(b)  other materials handling equipment with wide swinging booms or other similar parts

may create a hazard to any person.

(2)  The main approaches to any materials handling area shall be posted with warning signs or shall be under the control of a signaller while operations are in progress.

(3)  No personnel shall enter a materials handling area while operations are in progress unless that person

(a)  is a safety officer;
(b)  is an employee whose presence in the materials handling area is essential to the conduct, supervision or safety of the operations; or

(c)   is a person who has been instructed by the employer to be in the
      materials handling area while operations are in progress.

(4) If any person other than a person referred to in subsection (3) enters a
materials handling area while operations are in progress, the employer shall
cause the operations in that area to be immediately discontinued and not
resumed until that person has left the area.

**Overhead and Underground Hazards**

14.38(1)  Subject to subsection (2), no materials handling equipment shall be
operated in an area in which it may contact an electrical cable, pipeline or other
overhead or underground hazard known to the employer, unless the operator
has been

(a)   warned of the presence of the hazard;
(b)   informed of the location of the hazard; and
(c)   informed of the safety clearance that must be maintained with respect to
      the hazard in order to avoid accidental contact with it.

(2)  Where an employer is unable to determine with reasonable certainty the
location of the hazard or the safety clearance referred to in subsection (1),
every electrical cable shall be de-energized and every pipeline containing a
hazardous substance shall be shut down and drained before any operation
involving the use of materials handling equipment commences within the area
of possible contact with the hazard.

**Rear Dumping**

14.39  Where rear dumping mobile equipment is used to discharge a load at
the edge of a sudden drop in grade level that may cause the mobile equipment
to tip,

(a)   a bumping block shall be used, or
(b)   a signaller shall give directions to the operator of the equipment
      to prevent the mobile equipment from being backed over the edge.

**Fuelling**

14.40  Where materials handling equipment is fuelled in a work place, the
fuelling shall be done in accordance with the instructions given by the employer
pursuant to section 14.23 in a place where the vapours from the fuel are readily
dissipated.

**Ropes, Slings and Chains**

14.41 The employer shall, with respect to the use and maintenance of any rope or sling or any attachment or fitting thereon used by an employee, adopt and implement the recommendations set out in chapter 5 of the *Accident Prevention Manual for Industrial Operations, Engineering and Technology*, 8th Edition, published by the National Safety Council of the United States, dated 1980.

14.42 The employer shall, with respect to the use and maintenance of any chain used by an employee, adopt and implement the code of practice set out in CSA Standard B75-1947, *Code of Practice for the Use and Care of Chains*, dated May, 1947.

**Safe Working Loads**

14.43(1) No materials handling equipment shall be used or operated with a load that is in excess of its safe working load.

(2) The safe working load of materials handling equipment shall be clearly marked on the equipment or on a label securely attached to a permanent part of the equipment in a position where the mark or label can be easily read by the operator of the equipment.

**Aisles and Corridors**

14.44(1) Where in a work place an aisle, corridor or other course of travel

(a) is a principal traffic route for pedestrians and mobile equipment, and
(b) exceeds 15 m in length,

the employer shall provide a clearly marked walkway not less than 750 mm wide along one side of the aisle, corridor or other course of travel for the use of pedestrians only.

(2) Subsection (1) does not apply where a signaller or traffic lights are provided for the purpose of controlling traffic and protecting pedestrians.

(3) Where an aisle, corridor or other course of travel that is a principal traffic route intersects with another route, warning signs marked with the words "DANGEROUS INTERSECTION — CROISEMENT DANGEREUX", in letters not less than 50 mm in height on a contrasting background, shall be posted along the approaches to the intersection.

(4)  At blind corners, mirrors shall be installed that permit a mobile equipment operator to see a pedestrian, vehicle or mobile equipment approaching the blind corner.

## Clearances

14.45(1)  Subject to subsection (3), in any passageway that is regularly travelled by mobile equipment, the overhead and side clearances shall be

(a)  in the case of an overhead clearance, at least 150 mm above
   (i)  that part of the mobile equipment or its load that is the highest when the mobile equipment is in its highest normal operating position at the point of clearance, and
   (ii)  the top of the head of an employee riding on the mobile equipment when the employee is occupying his highest normal position at the point of clearance; and
(b)  in the case of a side clearance, adequate to permit the mobile equipment and its load to be manoeuvred safely by an operator, but in no case less than 150 mm on each side measured from the furthest projecting part of the equipment or its load, when the equipment is being operated in a normal manner.

(2)  Where an overhead clearance measured in accordance with subparagraph (1)(a)(i) or (ii) is less than 300 mm

(a)  the top of the doorway or object that restricts the clearance shall be marked with a distinguishing colour or mark; and
(b)  the height of the passageway in metres shall be shown near the top of the passageway in letters that are not less than 50 mm in height and are on a contrasting background.

(3)  Subparagraph (1)(a)(i) and subsection (2) do not apply to

(a)  mobile equipment whose course of travel is controlled by fixed rails or guides;
(b)  that portion of the route of any mobile equipment that is inside a railway car, truck or trailer truck and the warehouse doorway leading directly thereto; or
(c)  a load the nature of which precludes compliance with that subparagraph or subsection if precautions are taken to prevent contact with objects that may restrict the movement of the equipment.

## DIVISION III - MANUAL HANDLING OF MATERIALS

14.46  Where, because of the weight, size, shape, toxicity or other characteristic of materials, goods or things, the manual handling of the materials, goods or things may be hazardous to the safety or health of an employee, the employer shall issue instructions that the materials, goods or things shall, where reasonably practicable, not be handled manually.

14.47  Where an employee is required to lift or carry loads in excess of 10 kg manually, the employee shall be instructed and trained by the employer

(a)     in a safe method of lifting and carrying the loads; and
(b)     in a work procedure appropriate to the conditions of the work place and the employee's physical condition.

14.48  Where an employee is required to lift or carry loads in excess of 45 kg manually, the instructions given to the employee in accordance with section 14.47 shall be

(a)     set out in writing;
(b)     readily available to the employee to whom they apply; and
(c)     kept by the employer for a period of two years after they cease to apply.

## DIVISION IV - STORAGE OF MATERIALS

14.49(1)  All materials, goods and things shall be stored and placed in such a manner that the maximum safe load-carrying capacity of the floor or other supporting structures is not exceeded.

(2)  No materials, goods or things shall be stored or placed in a manner that may

(a)     reduce the distribution of light;
(b)     obstruct or encroach upon passageways, traffic lanes or exits;
(c)     impede the safe operation of materials handling equipment;
(d)     obstruct the ready access to or the use and operation of fire fighting equipment;
(e)     interfere with the operation of fixed fire protection equipment; or
(f)     be hazardous to the safety or health of any employee.

## PART XV - HAZARDOUS OCCURRENCE INVESTIGATION, RECORDING AND REPORTING

### Interpretation

15.1  In this Part,

"disabling injury" means an employment injury or an occupational disease that

(a)     prevents an employee from reporting for work or from effectively performing all the duties connected with the employee's regular work on any day subsequent to the day on which the injury or disease occurred, whether or not that subsequent day is a working day for that employee,

(b) results in the loss by an employee of a body member or part thereof or in the complete loss of the usefulness of a body member or part thereof, or

(c) results in the permanent impairment of a body function of an employee; (*blessure invalidante*)

"district office" means, in respect of a work place, the district office of Human Resources Development Canada that is

(a) closest to the work place, and

(b) in the administrative region of that Department in which the work place is situated; (*bureau de district*)

"minor injury" means an employment injury or an occupational disease for which medical treatment is provided and excludes a disabling injury. (*blessure légère*)

## Application

15.2(1) Subject to subsection (2), this Part does not apply in respect of employees employed in a coal mine or in an underground portion of any other type of mine.

(2) Section 15.10 applies in respect of employees employed in a coal mine.

## Reports by Employee

15.3 Where an employee becomes aware of an accident or other occurrence arising in the course of or in connection with the employee's work that has caused or is likely to cause injury to that employee or to any other person, the employee shall, without delay, report the accident or other occurrence to his employer, orally or in writing.

## Investigations

15.4(1) Where an employer becomes aware of an accident, occupational disease or other hazardous occurrence affecting any of his employees in the course of employment, the employer shall, without delay,

(a) appoint a qualified person to carry out an investigation of the hazardous occurrence;

(b) notify the safety and health committee or the safety and health representative, if either exists, of the hazardous occurrence and of the name of the person appointed to investigate it; and

(c) take necessary measures to prevent a recurrence of the hazardous occurrence.

(2) Where the hazardous occurrence referred to in subsection (1) is an accident involving a motor vehicle on a public road that is investigated by a police authority, the investigation referred to in paragraph (1)(a) shall be carried out by obtaining from the appropriate police authority a copy of its report respecting the accident.

(3) As soon as possible after receipt of the report referred to in subsection (2), the employer shall provide a copy thereof to the safety and health committee or the safety and health representative, if either exists.

## Telephone or Telex Reports

15.5 The employer shall report to a safety officer, by telephone or telex, the date, time, location and nature of any accident, occupational disease or other hazardous occurrence referred to in section 15.4 that had one of the following results as soon as possible but not later than 24 hours after becoming aware of that result, namely,

(a)  the death of an employee;
(b)  a disabling injury to two or more employees;
(c)  the loss by an employee of a body member or part thereof or the complete loss of the usefulness of a body member or a part thereof;
(d)  the permanent impairment of a body function of an employee;
(e)  an explosion;
(f)  damage to a boiler or pressure vessel that results in fire or the rupture of the boiler or pressure vessel; or
(g)  any damage to an elevating device that renders it unserviceable or a free fall of an elevating device.

## Records

15.6(1) The employer shall, within 72 hours after a hazardous occurrence referred to in paragraph 15.5(f) or (g), record in writing

(a)  a description of the hazardous occurrence and the date, time and location of the occurrence;
(b)  the causes of the occurrence; and
(c)  the corrective measures taken or the reason for not taking corrective measures.

(2) The employer shall, without delay, submit a copy of the record referred to in subsection (1) to the safety and health committee or the safety and health representative, if either exists.

## Minor Injury Records

15.7(1)  Every employer shall keep a record of each minor injury of which the employer is aware that affects any employee in the course of employment.

(2)  A record kept pursuant to subsection (1) shall contain

(a)  the date, time and location of the occurrence that resulted in the minor injury;
(b)  the name of the employee affected;
(c)  a brief description of the minor injury; and
(d)  the causes of the minor injury.

## Written Reports

15.8(1)  The employer shall make a report in writing, without delay, in the form set out in Schedule I to this Part setting out the information required by that form, including the results of the investigation referred to in paragraph 15.4(1)(a), where that investigation discloses that the hazardous occurrence resulted in any one of the following circumstances:

(a)  a disabling injury to an employee;
(b)  an electric shock, toxic atmosphere or oxygen deficient atmosphere that caused an employee to lose consciousness;
(c)  the implementation of rescue, revival or other similar emergency procedures; or
(d)  a fire or an explosion.

(2)  The employer shall submit a copy of the report referred to in subsection (1)

(a)  without delay, to the safety and health committee or the safety and health representative, if either exists; and
(b)  within 14 days after the hazardous occurrence, to a safety officer at the regional office or district office.

15.9  Where an accident referred to in subsection 15.4(2) results in a circumstance referred to in subsection 15.8(1), the employer shall, within 14 days after the receipt of the police report of the accident, submit a copy of that report to a safety officer at the regional office or district office.

# SCHEDULE I *(section 15.8)*

| TYPE OF OCCURRENCE/GENRE DE SITUATION | |
|---|---|
| ☐ Explosion / Explosion | ☐ Loss of Consciousness / Évanouissement |
| ☐ Disabling Injury / Blessure invalidante | ☐ Emergency Procedure / Procédures d'urgence |
| ☐ Other / Autre _____ Specify/Préciser | |

Department File No./N° de dossier du ministère

Regional Office or District Office / Bureau régional ou bureau de district

Employer ID No./Numéro d'identification de l'employeur

| Employer Name and Mailing Address/Nom et adresse postale de l'employeur | Postal Code / Code postal |
|---|---|
| | Telephone Number/Numéro de téléphone |

| Site of Hazardous Occurrence/Lieu de la situation hasardeuse | Date and Time of Hazardous Occurrence/Date et heure de la situation hasardeuse |
|---|---|
| | Weather/Conditions météorologiques |
| Witnesses/Témoins | Supervisor's Name/Nom du surveillant |

Description of what happened /Description des circonstances

Brief description and estimated cost of property damage/Description sommaire et coût estimatif des dommages matériels

| Injured Employee's Name (if applicable)/Nom de l'employé blessé (s'il y a lieu) | Age/Âge | Occupation/Profession |
|---|---|---|
| | | Years of experience in occupation/ Nombre d'années d'expérience dans la profession |
| Description of injury/Description de la blessure | Sex/Sexe | Direct cause of injury/Cause directe de la blessure |

Was training in accident prevention given to injured employee in relation to duties performed at the time of the hazardous occurrence?
L'employé blessé a-t-il reçu un entraînement en prévention des accidents relativement aux fonctions qu'il exerçait au moment de la situation hasardeuse?

☐ Yes/Oui  ☐ No/Non  Specify/Préciser

Direct causes of Hazardous Occurrence/Causes directes de la situation hasardeuse

Corrective measures and date employer will implement/ Mesures correctives qui seront appliquées par l'employeur et date d'entrée en vigueur

Reasons for not taking corrective measures
Raisons pour lesquelles aucune mesure corrective n'a été prise

Supplementary preventative measures/Autres mesures de prévention

| Name of person investigating/Nom de la personne menant l'enquête | Signature/Signature | Date/Date |
|---|---|---|
| Title/Titre | Telephone Number/Numéro de téléphone | |

| Safety & Health Committee's or Representative's Comments/Observations du comité d'hygiène et de sécurité ou du représentant. | | |
|---|---|---|
| Committee Member or Representative's Name/Nom du membre du comité ou du représentant | Signature/Signature | Date/Date |
| Title/Titre | Telephone Number/Numéro de téléphone | |

Lab/Trav 369(COSH) (07/88)

COPIES 1 & 2 to R.S.O., COPY 3 to the **Safety and Health Committee** or Representative, COPY 4 to the Employer.
COPIES 1 et 2 à A.R.S., COPIE 3 au **Comité d'hygiène et de sécurité** ou au représentant, COPIE 4 à l'employeur.

# SCHEDULE II (section 15.10)

SCHEDULE II / ANNEXE II
SECTION 15.10 / ARTICLE 15.10

GUIDE TO COMPLETION ON OTHER SIDE / EXPLICATIONS AU VERSO

YEAR/ANNÉE **19** _____

| EMPLOYER IDENTIFICATION NUMBER / NUMÉRO D'IDENTIFICATION DE L'EMPLOYEUR | ADDRESS OF WORK PLACE / ADRESSE DU LIEU DE TRAVAIL | NUMBER OF DISABLING INJURIES / NOMBRE DE BLESSURES INVALIDANTES | NUMBER OF DEATHS / NOMBRE DE DÉCÈS | NUMBER OF MINOR INJURIES / NOMBRE DE BLESSURES LÉGÈRES | NUMBER OF OTHER HAZARDOUS OCCURRENCES / NOMBRE D'AUTRES SITUATIONS HASARDEUSES | TOTAL NUMBER OF EMPLOYEES / NOMBRE TOTAL D'EMPLOYÉS | NUMBER OF OFFICE EMPLOYEES / NOMBRE D'EMPLOYÉS DE BUREAU | TOTAL NUMBER OF HOURS WORKED / TOTAL DES HEURES TRAVAILLÉES |
|---|---|---|---|---|---|---|---|---|
|  |  |  |  |  |  |  |  |  |
|  |  |  |  |  |  |  |  |  |
|  |  |  |  |  |  |  |  |  |
|  |  |  |  |  |  |  |  |  |

If this address is incorrect, please make correct.

Si cette adresse est inexacte, veuillez la corriger.

SUBMITTING OFFICER'S NAME AND TITLE / NOM DE L'AUTEUR DU RAPPORT ET TITRE

SIGNATURE / SIGNATURE

DATE OF SUBMISSION / DATE DE PRÉSENTATION

TELEPHONE / TÉLÉPHONE

Lab./Trav.366 (COSH) (7/88)

123

## SCHEDULE II *(section 15.10) (page 2)*

### Guide to the completion of the Employer's Annual Hazardous Occurrence Report

**General Notes**

1. This report must be submitted not later than March 1 of each year for the 12 month period ending December 31 of the preceding year even if no hazardous occurrence has occurred.

2. A copy of this report must be kept by the employer for ten (10) years following its submission.

**Completing the Report**

1. Please type or print all your information.

2. If your employer name or Canadian head office address set out on this form is incorrect, please correct.

3. If your Report covers any subsidiaries, list them and their location on a separate piece of paper attached to the Report.

4. Explanation of the column headings

   **(a) Address of Work Place**

   Enter the address of each of your work places.

   If you have more than five work places in any one province and you employ less than 15 employees at each work place, you may group these work places together under a single address in this column. Each address must then be followed by the total number of work places in the grouping.

   Example: If you have 10 work places in a province with less than 15 persons in each work place.

   Employer Name
   123 Name of Street
   City, Province (10)
   Postal Code

   **(b) Number of Disabling Injuries**

   Enter the total number of disabling injuries that have occurred in each work place (or group of work places) during the year. If there were none, enter 0.

   "disabling injury" means an employment injury or an occupational disease that

   a) prevents an employee from reporting for work or from effectively performing all the duties connected with his regular work on any day subsequent to the day on which the disabling injury occurred, whether or not that subsequent day is a working day for that employee,

   b) results in the loss by an employee of a body member or part thereof or in the complete loss of the usefulness of a body member or part thereof, or

   c) results in the permanent impairment of a body function of an employee.

   **(c) Number of Deaths**

   Enter the total number of deaths resulting from hazardous occurrences that have occurred in each work place (or group of work places) during the year. If there were none, enter 0.

   **(d) Number of Minor Injuries**

   Enter the total number of minor injuries that have occurred in each work place (or group of work places) during the year. If there were none, enter 0.

   "minor injury" means an employment injury or an occupational disease for which medical treatment is provided and excludes a disabling injury.

   **(e) Number of Other Hazardous Occurrences**

   Enter the total number of other hazardous occurrences that have occurred in each work place (or group of work places) during the year. If there were none, enter 0.

   **(f) Total Number of Employees**

   Enter the average number of employees, including office and casual employees, you have at each work place (or group of work places). In the case of casual or part-time employees, compute the equivalent number of full time employees.

   e.g. 100 employees employed full-time equals 100 and 10 employees employed half time equals 5. Total for period 105.

   **(g) Number of Office Employees**

   Enter the total number of employees who are classified as office employees (clerks, stenographers, accountants, etc.) in each work place (or group of work places).

   **(h) Total Number of Hours Worked**

   Enter the approximate total number of hours worked, including any overtime, at each work place (or group of work places).

### Comment remplir le rapport annuel de l'employeur concernant les situations hasardeuses

**Dispositions générales**

1. Il faut présenter ce rapport, même si aucune situation hasardeuse ne s'est produite, au plus tard le 1ᵉʳ mars de chaque année, pour la période de 12 mois se terminant le 31 décembre précédent.

2. Un exemplaire du rapport doit être conservé par l'employeur pendant les dix (10) ans suivant la date de sa présentation.

**Comment remplir le rapport**

1. Veuillez inscrire tous les renseignements à la machine ou en caractères d'imprimerie.

2. Si le nom de l'employeur ou l'adresse de son siège social au Canada diffèrent de ce qui paraît sur ce formulaire, veuillez les corriger.

3. Si votre rapport vise des filiales, veuillez les énumérer, en ayant soin d'indiquer l'endroit où elles se trouvent, sur un feuillet séparé que vous joindrez au rapport.

4. Explication des rubriques :

   **(a) Adresse du lieu de travail**

   Inscrivez l'adresse de chacun des lieux de travail visés.

   Si vous comptez plus de cinq lieux de travail dans la même province et moins de 15 employés à chacun d'eux, vous pouvez grouper ces lieux de travail sous une seule adresse. Précisez ensuite le nombre de lieux de travail visés.

   Exemple : Si vous comptez 10 lieux de travail dans une province et moins de 15 employés à chaque lieu

   Nom de l'employeur
   123 nom de la rue
   Ville, Province (10)
   Code postal

   **(b) Nombre de blessures invalidantes**

   Inscrivez le nombre total de blessures invalidantes qui se sont produites à chaque lieu de travail (ou groupe de lieux de travail) au cours de l'année. Si aucune blessure invalidante ne s'est produite, inscrivez 0.

   «blessure invalidante» Blessure au travail ou maladie professionnelle qui, selon le cas :

   a) empêche l'employé de se présenter au travail ou de s'acquitter efficacement de toutes les fonctions liées à son travail habituel pour toute journée suivant celle où il a subi la blessure invalidante, qu'il s'agisse ou non d'une journée ouvrable pour lui;

   b) entraîne chez l'employé la perte d'un membre ou d'une partie d'un membre, ou la perte totale de l'usage d'un membre ou d'une partie d'un membre;

   c) entraîne chez l'employé une altération permanente d'une fonction de l'organisme.

   **(c) Nombre de décès**

   Inscrivez le nombre total de décès attribuables aux situations hasardeuses qui se sont produits à chaque lieu de travail (ou groupe de lieux de travail) au cours de l'année. Si aucun décès ne s'est produit, inscrivez 0.

   **(d) Nombre de blessures légères**

   Inscrivez le nombre total de blessures légères qu'ont subies les employés à chaque lieu de travail (ou groupe de lieux de travail) au cours de l'année. Si aucune blessure légère n'a été subie, inscrivez 0.

   «blessure légère» Toute blessure au travail ou maladie professionnelle, autre qu'une blessure invalidante, qui fait l'objet d'un traitement médical.

   **(e) Nombre d'autres situations hasardeuses**

   Inscrivez le nombre d'autres situations hasardeuses qui se sont produites à chaque lieu de travail (ou groupe de lieux de travail). Si aucune autre situation hasardeuse ne s'est produite, inscrivez 0.

   **(f) Nombre total d'employés**

   Inscrivez le nombre d'employés, y compris les employés de bureau et les employés occasionnels, que vous comptez à chaque lieu de travail (ou groupe de lieux de travail). Dans le cas des employés occasionnels ou à temps partiel, inscrivez le nombre estimatif équivalent d'employés à plein temps. Ainsi, 100 employés à plein temps plus 10 à mi-temps font un total de 105.

   **(g) Nombre d'employés de bureau**

   Inscrivez le nombre d'employés classifiés comme employés de bureau (commis, sténos, comptables, etc.) à chaque lieu de travail (ou groupe de lieux de travail).

   **(h) Total des heures travaillées**

   Inscrivez le total approximatif des heures travaillées par tous les employés, y compris les heures supplémentaires, à chaque lieu de travail (ou groupe de lieux de travail).»

**Annual Report**

15.10(1) Every employer shall, not later than March 1 in each year, submit to the Minister a written report setting out the number of accidents, occupational diseases and other hazardous occurrences of which the employer is aware affecting any employee in the course of employment during the 12 month period ending on December 31 of the preceding year.

(2) The report referred to in subsection (1) shall be in the form set out in Schedule II to this Part and shall contain the information required by that form.

**Retention of Reports and Records**

15.11 Every employer shall keep a copy of

(a)     each report submitted pursuant to section 15.9 or subsection 15.10(1) for a period of 10 years following the submission of the report to the safety officer or the Minister, and

(b)     the record or report referred to in subsection 15.6(1), 15.7(1) or 15.8(1) for a period of 10 years following the hazardous occurrence.

**PART XVI - FIRST AID**

**Interpretation**

16.1 In this Part,

"first aid attendant" means a holder of a basic first aid certificate or a standard first aid certificate; (*secouriste*)

"first aid station" means a place, other than a first aid room, at which first aid supplies or equipment are stored; (*poste de secours*)

"health unit" means a consultation and treatment facility that is in the charge of a person who is registered as a registered nurse under the laws of any province; (*service de santé*)

"isolated work place" means a work place that is more than two hours travel time from a hospital or a medical facility under normal travel conditions using the fastest available means of transportation; (*lieu de travail isolé*)

"medical facility" means a medical clinic or the office of a physician; (*installation médicale*)

"standard first aid certificate" means the certificate issued by an approved organization for successful completion of a two day first aid course. (*certificat de secourisme général*)

**General**

16.2(1)  Every employer shall establish written instructions that provide for the prompt rendering of first aid to an employee for an injury, an occupational disease or an illness.

(2)  A copy of the instructions referred to in subsection (1) shall be kept by the employer readily available for examination by employees.

**First Aid Attendants**

16.3(1)  Subject to subsection (3), at least one employee shall be trained and capable of providing artificial resuscitation, controlling a haemorrhage and rendering such other life-saving first aid as may be indicated by the nature of the work being done

(a)  at a work place at which at least four and not more than 14 employees are working at any time; and
(b)  at an isolated work place at which fewer than four employees are working at any time.

(2)  At a work place at which 15 or more employees are working at any time, at least one of the employees shall be a first aid attendant.

(3)  At a work place at which an employee is working on live high voltage electrical equipment,

(a)  a first aid attendant shall be readily available; or
(b)  at least one of the employees shall have the training necessary to provide resuscitation
    (i)  by mouth to mouth resuscitation, cardio-pulmonary resuscitation or other direct method, and
    (ii)  by the Holger-Nielsen Method or the Sylvester Method.

16.4(1)  A first aid attendant referred to in section 16.3 or paragraph 16.10(1)(a) shall

(a)  be assigned to a first aid station or first aid room;
(b)  be readily available and accessible to employees during working hours; and
(c)  render first aid to employees that are injured or ill at the work place.

(2)  The first aid attendant referred to in subsection (1)

(a)  shall work in close proximity to the first aid station or first aid room to which he is assigned; and

(b)     shall not be assigned duties that will interfere with the prompt and
        adequate rendering of first aid.

## First Aid Stations

16.5(1)  At least one first aid station shall be provided for every work place.

(2)  Every first aid station shall be

(a)     located at or near the work place;
(b)     available and accessible during all working hours;
(c)     inspected regularly and its contents maintained in a clean, dry and
        serviceable condition; and
(d)     clearly identified by a conspicuous sign.

(3)  Subsection (1) does not apply where a first aid room, health unit or medical
facility that meets the requirements of subsection (2) is provided by the
employer.

## Posting of Information

16.6(1)  Subject to subsection (2), the employer shall post and keep posted in a
conspicuous place accessible to every employee in each work place

(a)     information regarding first aid to be rendered for any injury, occupational
        disease or illness;
(b)     information regarding the location of first aid stations and first aid rooms;
(c)     at every first aid station and first aid room, a list of first aid attendants, the
        expiry date of their certificates and information regarding the places where
        they may be located; and
(d)     near the telephones, an up-to-date list of telephone numbers for use in
        emergencies.

(2)  At an isolated work place or in a motor vehicle, the information referred to
in subsection (1) shall be provided and retained with the first aid kit.

## First Aid Supplies and Equipment

16.7(1)  For each work place at which the number of employees working at any
time is the number set out in column I of an item of Schedule I to this Part, a
first aid kit that is of the type set out in column II of that item shall be provided.

(2)  For the purposes of subsection (1), a first aid kit of a type set out at the
head of column II, III, IV or V of Schedule II to this Part shall contain

(a) the first aid supplies and equipment set out in column I of Division 1 of that Schedule in the applicable number set out opposite those supplies and equipment in column II, III, IV or V; and

(b) where the first aid kit is for use in an isolated work place, the first aid supplies and equipment set out in column I of Divisions 1 and 2 of that Schedule in the applicable number set out opposite those supplies and equipment in column II, III, IV or V.

16.8(1)  Subject to subsection (2), where a hazard for skin or eye injury from a hazardous substance exists in the work place, shower facilities to wash the skin and eye wash facilities to irrigate the eyes shall be provided for immediate use by employees.

(2)  Where it is not practicable to comply with subsection (1), portable equipment that may be used in place of the facilities referred to in subsection (1) shall be provided.

**First Aid Rooms**

16.9(1)  A first aid room shall be provided where 200 or more employees are working at any time in a work place.

(2)  Subsection (1) does not apply

(a) where a health unit, medical facility or hospital at which medical treatment is provided without charge to employees is readily accessible; or

(b) where the number of employees working at any time does not exceed 400 and more than 70 per cent of those employees are normally employed in work that is relatively free from hazards to safety and health.

16.10(1)  Every first aid room provided in accordance with section 16.9 shall be

(a) under the supervision of a first aid attendant;

(b) located as close as practicable to the work place and within easy access to a toilet room;

(c) situated on a minimum floor area of 10 m² and constructed to allow for optimum ease of access to a person carrying a patient on a stretcher;

(d) maintained in an orderly and sanitary condition; and

(e) equipped with
   (i) a washbasin supplied with cold water and hot water that meets the standards set out in section 9.18 of Part IX,
   (ii) a storage cupboard and a counter,
   (iii) a separate cubicle or curtained-off area with a cot or bed equipped with a moisture-protected mattress and two pillows,
   (iv) a table and two or more chairs,

     (v)    a telephone and an up-to-date list of telephone numbers for use in emergencies,

     (vi)   the first aid supplies and equipment set out in column I of Schedule III to this Part in the applicable quantities set out in column II of that Schedule, and

     (vii)  for use at the scene of an accident, a Type A first aid kit and a flashlight that is appropriate for the environment of the work place.

(2) In every first aid room referred to in subsection (1),

(a) the temperature

     (i)    shall be maintained at not less than 21°C, measured 1 m above the floor, when the out of doors temperature is 21°C or less, and

     (ii)   to the extent that is reasonably practicable, where the out of doors temperature in the shade exceeds 24°C, shall not exceed the out of doors temperature; and

(b) there shall be at least one change of air per hour.

**Transportation**

16.11 Before assigning employees to a work place, the employer shall provide for that work place

(a) an ambulance service or other suitable means of transporting an injured employee to a health unit, medical facility, hospital or to the employee's residence;

(b) a first aid attendant to accompany an injured employee and to render first aid in transit if required; and

(c) a means of quickly summoning the ambulance service or other means of transportation.

**Records**

16.12(1) Where first aid is rendered in accordance with this Part, the employee who rendered the first aid shall

(a) enter in a first aid record the following information:

     (i)    the date and time of the reporting of the injury or illness,

     (ii)   the full name of the injured or ill employee,

     (iii)  the date, time and location of the occurrence of the injury or illness,

     (iv)  a brief description of the injury or illness,

     (v)   a brief description of the first aid rendered, if any, and

     (vi)  a brief description of arrangements made for the treatment or transportation of the injured or ill employee; and

(b) sign the first aid record beneath the information entered in accordance with paragraph (a).

(2) The information referred to in subsection (1) shall be entered

(a) where first aid was rendered to an employee at an isolated work place detached from the main party or on a snowmobile or other small vehicle, in the first aid record stored in the first aid kit at the site of the main party or work site; and

(b) in any other case, in the first aid record stored in the first aid kit.

(3) The employer shall keep a first aid record containing information entered in accordance with subsection (2) for two years after the date of that entry.

**SCHEDULE I** *(Subsection 16.7(1))*

**REQUIREMENTS FOR FIRST AID KITS**

| | Column I | Column II |
|---|---|---|
| **Item** | **Number of Employees** | **Type of First Aid Kit** |
| 1. | 2 to 5 (subject to item 5) | A |
| 2. | 6 to 19 | B |
| 3. | 20 or more | C |
| 4. | 1 detached from the main party in an isolated work place | D |
| 5. | 1 to 3 employees travelling by snowmobile or other small vehicle, other than a truck, van or automobile | D |

**SCHEDULE II** - *(Subsection 16.7(2))*

| | | | Type of First Aid Kit | | | |
|---|---|---|---|---|---|---|
| | | | A | B | C | D |
| | Column I | Column | II | III | IV | V |
| Item | Supplies and Equipment | | Quantities per Type of First Aid Kit | | | |
| | *Division 1* | | | | | |
| 1. | Antiseptic - wound solution, 60 ml or antiseptic swabs (10-pack) | | 1 | 2 | 3 | 1 |
| 2. | Applicator - disposable (10-pack) (not needed if antiseptic swabs used) | | 1 | 2 | 2 | - |
| 3. | Bag - disposable, waterproof, emesis | | 1 | 2 | 2 | - |
| 4. | Bandage-adhesive strips | | 12 | 100 | 100 | 6 |
| 5. | Bandage - gauze 2.5 cm x 4.5 m (not needed if ties attached to dressing) | | 2 | 6 | 8 | - |
| 6. | Bandage - triangular 100 cm folded and 2 pins | | 2 | 4 | 6 | 1 |
| 7. | Container - First Aid Kit | | 1 | 1 | 1 | 1 |
| 8. | Dressing - compress, sterile 7.5 cm x 12 cm approx. | | 2 | 4 | 8 | - |
| 9. | Dressing - gauze, sterile 7.5 cm x 7.5 cm approx. | | 4 | 8 | 12 | 2 |
| 10. | Forceps - splinter | | 1 | 1 | 1 | - |
| 11. | Manual - First Aid, English - current edition | | 1 | 1 | 1 | - |
| 12. | Manual - First Aid, French - current edition | | 1 | 1 | 1 | - |
| 13. | Pad with shield or tape for eye | | 1 | 1 | 1 | 1 |

| | | | Type of First Aid Kit | | | |
|---|---|---|---|---|---|---|
| | | | **A** | **B** | **C** | **D** |
| | Column I | Column | II | III | IV | V |
| Item | Supplies and Equipment | | Quantities per Type of First Aid Kit | | | |
| 14. | Record - First Aid (section 16.12) | | 1 | 1 | 1 | 1 |
| 15. | Scissors - 10 cm | | - | 1 | 1 | - |
| 16. | Tape - adhesive, surgical 1.2 cm x 4.6 m (not needed if ties attached to dressings) | | 1 | 1 | 2 | - |
| | *Division 2* **Additional Supplies and Equipment for Isolated Work Places** | | | | | |
| 17. | Antipruritic lotion 30 ml or swabs (10 pack) | | 1 | 1 | 1 | - |
| 18. | Bandage - elastic 7.5 cm x 5 m | | - | - | 1 | - |
| 19. | Blanket - bed size | | - | - | 1 | - |
| 20. | Blanket - emergency, pocket size | | - | - | - | 1 |
| 21. | Burn jelly or ointment, sterile, 5 ml | | 1 | 1 | 1 | - |
| 22. | Hand cleanser or cleansing towelettes, 1 pk. | | 1 | 1 | 1 | - |
| 23. | Splint set with padding - assorted sizes | | - | 1 | 1 | - |
| 24. | Stretcher | | - | - | 1 | - |

## SCHEDULE III *(Section 16.10)*

## FIRST AID ROOM SUPPLIES AND EQUIPMENT

|  | Column I | Column II |
|---|---|---|
| **Item** | **Supplies and Equipment** | **Quantity** |
| 1. | Depressor - tongue (25 pack) | 4 |
| 2. | Alcohol - isopropyl (500 ml) | 2 |
| 3. | Antiseptic - wound solution (250 ml) | 2 |
| 4. | Bandage - adhesive strips (100 pack) | 4 |
| 5. | Bandage with applicator - tubular, finger size | 1 |
| 6. | Bandage - gauze 10 cm x 4.5 m | 12 |
| 7. | Bandage - triangular, 100 cm folded and 2 pins | 12 |
| 8. | Brush - scrub, nail | 1 |
| 9. | Scissors - 10 cm | 1 |
| 10. | Stretcher - folding | 1 |
| 11. | Blanket - bed size | 2 |
| 12. | Basin - wash | 2 |
| 13. | Splint set with padding - assorted sizes | 1 |
| 14. | Bedding - disposable 2 sheets and 2 pillow cases | 12 |
| 15. | Gloves - disposable (100 pack) | 1 |
| 16. | Manual - First Aid, English, current edition | 1 |
| 17. | Manual - First Aid, French, current edition | 1 |
| 18. | Dressing - burn 10 cm x 10 cm, or burn jelly or ointment, 3 gm | 12 |
| 19. | Dressing, compress with ties, sterile, 7.5 cm x 7.5 cm | 24 |
| 20. | Dressing - field, sterile | 6 |

**FIRST AID ROOM SUPPLIES AND EQUIPMENT**

|  | Column I | Column II |
|---|---|---|
| **Item** | **Supplies and Equipment** | **Quantity** |
| 21. | Dressing - gauze squares, sterile, 5 cm x 5 cm (2 pack) | 100 |
| 22. | Forceps - splinter | 2 |
| 23. | Tray - instrument | 1 |
| 24. | Applicator, disposable (10 pack) | 8 |
| 25. | Waste receptacle - covered | 1 |
| 26. | Record - First Aid | 1 |
| 27. | Tape - adhesive, surgical 2.5 cm x 4.6 m | 1 |
| 28. | Bag - hot water or hot pack | 1 |
| 29. | Bag - ice or cold pack | 1 |
| 30. | Soap - liquid, with dispenser | 1 |
| 31. | Towels, package or roll of disposable, with dispenser | 1 |
| 32. | Bottle with solution - eye irrigation 200 ml | 4 |
| 33. | Pad with shield or tape for eye, cotton | 4 |
| 34. | Cups, box of disposable, with dispenser | 1 |

## PART XVII - SAFE OCCUPANCY OF THE WORK PLACE

### Interpretation

17.1 In this Part,

"emergency evacuation plan" means a written plan for use in an emergency, prepared in accordance with section 17.4.

### Application

17.2 This Part does not apply in respect of employees employed in the underground workings of mines.

**Fire Protection Equipment**

17.3(1)  Fire protection equipment shall be installed, inspected and maintained in every building in which there is a work place in accordance with the standards set out in Parts 6 and 7 of the *National Fire Code.*

(2)  For the purposes of interpreting the standards referred to in subsection (1), "acceptable" means "appropriate".

(3)  All fire protection equipment shall be maintained and repaired by a qualified person.

**Emergency Evacuation Plan**

17.4(1)  Where more than 50 employees are working in a building at any time, the employer or employers of those employees shall, after consultation with

(a)  the safety and health committee or safety and health representative of the employees, if either exists, and
(b)  the employers of any persons working in the building to whom the Act does not apply,

prepare an emergency evacuation plan.

(2)  An emergency evacuation plan referred to in subsection (1) shall contain

(a)  a plan of the building, showing
    (i)  the name, if any, and the address of the building,
    (ii)  the name and address of the owner of the building,
    (iii)  the names and locations of the tenants of the building,
    (iv)  the date of preparation of the plan,
    (v)  the scale of the plan,
    (vi)  the location of the building in relation to nearby streets and in relation to all buildings and other structures located within 30 m of the building,
    (vii)  the maximum number of persons normally occupying the building at any time,
    (viii)  a horizontal projection of the building, showing thereon its principal dimensions, and
    (ix)  the number of floors above and below ground level;
(b)  a plan of each floor of the building, showing
    (i)  the name, if any, and the address of the building,
    (ii)  the date of preparation of the plan,
    (iii)  the scale of the plan,
    (iv)  a horizontal projection of the floor, showing thereon its principal dimensions,
    (v)  the number of the floor to which the plan applies,
    (vi)  the maximum number of persons normally occupying the floor at any time,
    (vii)  the location of all fire escapes, fire exits, stairways, elevating devices, main corridors and other means of exit,
    (viii)  the location of all fire protection equipment, and

(ix)   the location of the main electric power switches for the lighting system, elevating devices, principal heating, ventilation and air-conditioning equipment and other electrical equipment;

(c)   a full description of the evacuation procedures to be followed in evacuating the building, including the time required to complete the evacuation; and

(d)   the names, room numbers and telephone numbers of the chief emergency warden and the deputy chief emergency warden of the building appointed by the employer or employers under section 17.7.

(3) An emergency evacuation plan referred to in subsection (1) shall be kept up-to-date and shall take into account any changes in the building or the nature of its occupancy.

(4) An employer referred to in subsection (1) shall keep a copy of the up-to-date emergency evacuation plan in the building to which it refers.

## Emergency Procedures

17.5(1) Every employer shall, after consultation with the safety and health committee or the safety and health representative of the employees, if either exists, and with the employers of any persons working in the building to whom the Act does not apply, prepare emergency procedures

(a)   to be implemented if any person commits or threatens to commit an act that is likely to be hazardous to the safety and health of the employer or any of his employees;

(b)   where there is a possibility of an accumulation, spill or leak of a hazardous substance in a work place controlled by him, to be implemented in the event of such an accumulation, spill or leak;

(c)   where more than 50 employees are working in a building at any time, to be implemented where evacuation is not an appropriate means of ensuring the safety and health of employees; and

(d)   to be implemented in the event of a failure of the lighting system.

(2) The emergency procedures referred to in subsection (1) shall contain

(a)   an emergency evacuation plan, where applicable;

(b)   a full description of the procedures to be followed;

(c)   the location of the emergency equipment provided by the employer; and

(d)   a plan of the building, showing

(i)   the name, if any, and the address of the building, and

(ii)   the name and address of the owner of the building.

## Instructions and Training

17.6(1) Every employee shall be instructed and trained in

(a)   the procedures to be followed by him in the event of an emergency; and

(b)   the location, use and operation of fire protection equipment and emergency equipment provided by the employer.

(2) Notices shall be posted at appropriate locations at a work place that set out the details of the evacuation procedures referred to in paragraph 17.4(2)(c) and the procedures referred to in paragraph 17.5(2)(b).

## Emergency Wardens

17.7(1) Where an employer or employers have prepared an emergency evacuation plan for a building, the employer or employers shall appoint

(a)    a chief emergency warden and a deputy chief emergency warden for that building; and

(b)    an emergency warden and a deputy emergency warden for each floor of the building that is occupied by employees of the employer or employers.

(2) The chief emergency warden and deputy chief emergency warden appointed for a building shall be employees who are normally employed in the building.

(3) The emergency warden and deputy emergency warden appointed for a floor in a building shall be employees who are normally employed on that floor.

17.8(1) Every emergency warden and every deputy emergency warden appointed under section 17.7 shall be instructed and trained in

(a)    his responsibilities under the emergency evacuation plan and the emergency procedures referred to in paragraph 17.5(1)(c); and

(b)    the use of fire protection equipment.

(2) A record of all instruction and training provided in accordance with subsection (1) shall be kept by the employer in the work place to which it applies for a period of two years from the date on which the instruction or training is provided.

## Inspections

17.9(1) In addition to the inspections carried out under section 17.3, a visual inspection of every building to which subsection 17.4(1) applies shall be carried out by a qualified person at least once every six months and shall include an inspection of all fire escapes, exits, stairways and fire protection equipment in the building in order to ensure that they are in serviceable condition and ready for use at all times.

(2) A record of each inspection carried out in accordance with subsection (1) shall be dated and signed by the person who made the inspection and kept by the employer in the building to which it applies for a period of two years from the date on which it is signed.

## Meetings of Emergency Wardens and Drills

17.10(1) At least once every year and after any change is made in the emergency evacuation plan or the emergency procedures referred to in paragraph 17.5(1)(c) for a building,

(a) all emergency wardens and deputy emergency wardens appointed under section 17.7 shall meet for the purpose of ensuring that they are familiar with the emergency evacuation plan and the emergency procedures and their responsibilities thereunder; and

(b) an evacuation or emergency drill shall be conducted for the employees in that building.

(2) The employer or employers shall keep a record of each meeting and drill referred to in subsection (1) in the building referred to in that subsection for a period of two years from the date of the meeting or the drill.

(3) The record referred to in subsection (2) shall contain

(a) in respect of each meeting,
    (i) the date of the meeting,
    (ii) the names and titles of those present, and
    (iii) a summary of the matters discussed; and
(b) in respect of each drill,
    (i) the date an time of the drill; and
    (ii) where applicable, the length of time taken to evacuate the building.

(4) The employer shall notify the local fire department for the building where an evacuation or emergency drill is to take place at least 24 hours in advance of the date and time of the drill.

**Fire Hazard Areas**

17.11(1) Subject to subsection (2), no person shall, in a fire hazard area,

(a) use any equipment, machinery or tool of a type that may provide a source of ignition; or
(b) smoke or use an open flame or other source of ignition.

(2) Where it is not reasonably practicable to avoid performing work involving the use of any equipment, machinery or tool that may provide a source of ignition in an area that has an atmosphere that contains or is likely to contain explosive concentrations of combustible dust or in an area where combustible dust has accumulated in a sufficient quantity to be a fire hazard, the following shall apply:

(a) the atmosphere and surfaces in the area where the work is to be performed and within that portion of the surrounding area that is accessible to sparks or pieces of hot metal produced by the work shall be substantially free of combustible dust;
(b) where any equipment, machinery or tool produces combustible dust that may reach the areas referred to in paragraph (1), the equipment, machinery or tool shall be made inoperative prior to and during the time the work is being performed;
(c) in so far as is practicable, the area where the work is to be performed shall be enclosed to prevent the escape of sparks or pieces of hot metal produced by the work;

(d) all openings in floors and walls through which sparks or pieces of hot metal produced by the work may pass shall be sealed or covered to prevent such passage;

(e) any combustible materials within the areas referred to in paragraph (a) shall be removed or, if this is not reasonably practicable, shall be covered with a non-combustible protective covering;

(f) floors and walls of combustible material within the areas referred to in paragraph (a) shall be protected from the fire hazard by
   (i) drenching the surfaces of the floors and walls with water, or
   (ii) covering the floors and walls with a non-combustible protective covering;

(g) the work shall be performed under the supervision of a qualified person, who shall remain in the work area while the work is performed and for 30 minutes thereafter; and

(h) there shall be readily available in the work area at least one hand-held portable fire extinguisher and
   (i) water hose at least 25 mm in diameter that is connected to a water supply line, or
   (ii) a supply of not less than 200 L of water and a bucket.

17.12 Signs shall be posted in conspicuous places at all entrances to a fire hazard area

(a) identifying the area as a fire hazard area; and
(b) prohibiting the use of an open flame or other source of ignition in the area.

## SAFETY AND HEALTH COMMITTEES AND REPRESENTATIVES REGULATIONS

### Short Title

1. These Regulations may be cited as the *Safety and Health Committees and Representatives Regulations*.

### PART I - SAFETY AND HEALTH COMMITTEES

#### Interpretation

2. In this Part,

"Act" means Part II of the Canada Labour Code. *(Loi)*

#### Selection of Members

3. The employer shall select the member or members of a safety and health committee to represent him from among persons who exercise managerial functions.

4. Where any employees at a work place are not represented by a trade union, those employees shall select, by majority vote, the member or members of the safety and health committee to represent them.

## Chairmen

5(1)  A safety and health committee shall have two chairmen selected from among the members of the committee, one being selected by the representatives of the employees and the other by the representatives of the employer.

(2)  The chairmen referred to in subsection (1) shall act alternately for such period of time as the safety and health committee specifies in its rules of procedure.

## Reselection of Members

6.  A person may be selected as a member of a safety and health committee for more than one term.

## Vacancy of Office

7.  Where a member of a safety and health committee resigns or ceases to be a member for any other reason, the vacancy shall be filled within 30 days after the next regular meeting of the committee.

## Quorum

8.  The quorum of a safety and health committee shall consist of the majority of the members of the committee, of which at least half are representatives of the employees and at least one is a representative of the employer.

## Minutes

9(1)  The minutes of each safety and health committee meeting shall be signed by the two chairmen referred to in subsection 5(1).

(2) The chairman selected by the representatives of the employer shall provide, as soon as possible after each safety and health committee meeting, a copy of the minutes referred to in subsection (1) to the employer and to each member of the safety and health committee.

(3)  The employer shall, as soon as possible after receiving a copy of the minutes referred to in subsection (2), post a copy of the minutes in the conspicuous place or places in which the employer has posted the information referred to in subsection 135(5) of the Act and keep the copy posted there for one month.

(4)  A copy of the minutes referred to in subsection (1) shall be kept by the employer at the work place to which it applies or at the head office of the employer for a period of two years from the day on which the safety and health committee meeting is held in such a manner that it is readily available for examination by a safety officer.

**Annual Report**

10. The chairman selected by the representatives of the employer shall

(a)    not later than March 1 in each year, submit a report of the safety and health committee's activities during the 12 month period ending on December 31 of the preceding year, signed by both chairmen referred to in subsection 5(1), in the form set out in the Schedule and containing the information required by that form, where the safety and health committee is established

      (i)    in respect of employees to whom the *On Board Trains Occupational Safety and Health Regulations* apply, to the regional safety officer of the Department of Transport (Railway Safety) Ottawa, Ontario, K1A ON5,

      (ii)    in respect of employees to whom the *Marine Occupational Safety and Health Regulations* apply, to the regional safety officer at the regional office of the Department of Transport (Marine) for the administrative region of that Department in which the employees are based,

      (iii)    in respect of employees to whom the *Aviation Occupational Safety and Health Regulations* apply, to the regional safety officer at the regional office of the Department of Transport (Aviation) for the administrative region of that Department in which the employees are based,

      (iv)    in respect of employees to whom the *Oil and Gas Occupational Safety and Health Regulations* apply, to the regional safety officer at the regional office of the Canada Oil and Gas Lands Administration, formed under the Department of Natural Resources and the Department of Indian Affairs and Northern Development, for the administrative region of that Administration in which the work place of those employees is situated, and

      (v)    in respect of employees to whom the *Canada Occupational Safety and Health Regulations* apply, to the regional safety officer at the regional office of the Department of Human Resources Development Canada for the administrative region of that Department in which the work place of those employees is situated; and

(b)    as soon as possible after submitting the report referred to in paragraph (a), post a copy of the report in the conspicuous place or places in which the employer has posted the information referred to in subsection 135(5) of the Act and keep the copy posted there for two months.

# SCHEDULE (section 10)

| Labour Canada / Travail Canada | SCHEDULE/ANNEXE (SECTION 10/ARTICLE 10) |
|---|---|

SAFETY AND HEALTH COMMITTEE REPORT
RAPPORT DU COMITÉ D'HYGIÈNE ET DE SÉCURITÉ

DEPARTMENT FILE NO. / N° DE DOSSIER DU MINSTÈRE

REGIONAL OFFICE/BUREAU RÉGIONAL

EMPLOYER IDENTIFICATION NUMBER/ NUMÉRO D'IDENTIFICATION DE L'EMPLOYEUR

EMPLOYER NAME AND MAILING ADDRESS / NOM ET ADRESSE POSTALE DE L'EMPLOYEUR

COMMITTEE EXEMPTION PURSUANT TO SUBSECTION 92(4) OF THE ACT / EXEMPTION DU COMITÉ EN VERTU DU PARAGRAPHE 92(4) DE LA LOI — YES OUI

POSTAL CODE/CODE POSTAL

NUMBER OF EMPLOYEES REPRESENTED BY COMMITTEE NOMBRE D'EMPLOYÉS REPRÉSENTÉS PAR LE COMITÉ

NUMBER OF TRADE UNION EMPLOYEE COMMITTEE MEMBERS NOMBRE DE MEMBRES REPRÉSENTANT LES EMPLOYÉS SYNDIQUÉS DANS LE COMITÉ

COMMITTEE NAME/WORK PLACE/ADDRESS IF DIFFERENT FROM ABOVE / NOM DU COMITÉ/LIEU DE TRAVAIL/ADRESSE SI DIFFÉRENTE DE CI-HAUT

NUMBER OF NON TRADE UNION EMPLOYEE COMMITTEE MEMBERS NOMBRE DE MEMBRES REPRÉSENTANT LES EMPLOYÉS NON SYNDIQUÉS DANS LE COMITÉ

NUMBER OF EMPLOYER COMMITTEE MEMBERS NOMBRE DE MEMBRES REPRÉSENTANT L'EMPLOYEUR DANS LE COMITÉ

POSTAL CODE/CODE POSTAL

TOTAL COMMITTEE MEMBERSHIP EFFECTIF TOTAL DU COMITÉ

CONTACT PERSON/PERSONNE RESSOURCE

TRADE UNION/NOM DU SYNDICAT

TELEPHONE NO / N° DE TÉLÉPHONE

| | MONTH/MOIS | JAN JANV | FEB FEV | MAR MARS | APR AVR | MAY MAI | JUNE JUIN | JULY JUILL | AUG AOÛT | SEPT | OCT | NOV | DEC DEC | TOTAL |
|---|---|---|---|---|---|---|---|---|---|---|---|---|---|---|
| MEETINGS RÉUNIONS | REGULAR ORDINAIRES | | | | | | | | | | | | | |
| | SPECIAL SPÉCIALES | | | | | | | | | | | | | |
| COMPLAINTS PLAINTES | RECEIVED REÇUES | | | | | | | | | | | | | |
| | RESOLVED RÉSOLUES | | | | | | | | | | | | | |
| | UNRESOLVED NON RÉSOLUES | | | | | | | | | | | | | |
| REFUSAL TO WORK REFUS DE TRAVAILLER | RECEIVED REÇUS | | | | | | | | | | | | | |
| | RESOLVED RÉSOLUS | | | | | | | | | | | | | |
| | UNRESOLVED NON RÉSOLUS | | | | | | | | | | | | | |
| INQUIRIES & INVESTIGATIONS ENQUÊTES ET INVESTIGATIONS | | | | | | | | | | | | | | |
| PROGRAMS, MEASURES AND PROCEDURES MONITORED. PROGRAMMES, MESURES ET PROCÉDURES SURVEILLÉS | | | | | | | | | | | | | | |
| SAFETY AND HEALTH HAZARDS IDENTIFIED RISQUES À LA SANTÉ ET À LA SÉCURITÉ DÉCELÉS | | | | | | | | | | | | | | |
| | RESOLVED RÉSOLUS | | | | | | | | | | | | | |
| | UNRESOLVED NON RÉSOLUS | | | | | | | | | | | | | |
| DISABLING INJURIES BLESSURES ENTRAÎNANT UNE INVALIDITÉ | | | | | | | | | | | | | | |
| MINOR INJURIES BLESSURES LÉGÈRES | | | | | | | | | | | | | | |
| TIME LOST DUE TO INJURIES PERTE DE TEMPS DUE AUX BLESSURES | | | | | | | | | | | | | | |

PLEASE HIGHLIGHT ANY SPECIAL PROGRAMS, INQUIRIES, UNRESOLVED ISSUES OR OTHER POINTS SIGNIFICANT TO THE COMMITTEE THAT OCCURRED DURING THE PREVIOUS 12 MONTHS ENDING DECEMBER 31, (ATTACH SHEET FOR ADDITIONAL INFORMATION).

DÉCRIRE TOUS PROGRAMMES, ENQUÊTES, QUESTIONS NON RÉSOLUES OU AUTRES FAITS PARTICULIERS SOULEVÉS AU COURS DES 12 MOIS SE TERMINANT LE 31 DÉCEMBRE ET POUVANT AVOIR UNE CERTAINE IMPORTANCE POUR LE COMITÉ (JOINDRE UNE FEUILLE ADDITIONNELLE AU BESOIN).

| EMPLOYEE CHAIRMAN / PRÉSIDENT REPRÉSENTANT LES EMPLOYÉS | PLEASE PRINT NAME/EN LETTRES MOULÉES | SIGNATURE/SIGNATURE | DATE/DATE |
|---|---|---|---|
| EMPLOYER CHAIRMAN/ PRÉSIDENT REPRÉSENTANT L'EMPLOYEUR | PLEASE PRINT NAME/EN LETTRES MOULÉES | SIGNATURE/SIGNATURE | DATE/DATE |

Lab/Trav 499 (Rev. 2/86)

Canadä

## PART II - SAFETY AND HEALTH REPRESENTATIVES

### Selection of Representatives

11.  Where none of the employees at a work place are represented by a trade union, those employees shall select, by majority vote, the safety and health representative for that work place.

### Term of Office

12.  The term of office of a safety and health representative shall be not more than two years.

### Reselection of Representatives

13.  A person may be selected as a safety and health representative for more than one term.

### Vacancy of Office

14.  Where a safety and health representative resigns or ceases to be a representative for any other reason, the vacancy shall be filled within 30 days after he resigns or ceases to be the representative.

## AVIATION OCCUPATIONAL SAFETY AND HEALTH REGULATIONS

These regulations apply to Public Service employees who are employed on aircraft that are in operation.  The regulations provide coverage similar to the *Canada Occupational Safety and Health Regulations*, modified to suit aircraft operations.

Administration of the *Canada Labour Code*, Part IV, and the aviation regulations has been delegated to Transport Canada, Aviation Group, and officials from that group have been appointed as safety officers and regional safety officers under Part IV. The *Aviation Occupational Safety and Health Regulations*, which may be obtained through Public Works and Government Services Canada (SOR/87-182), comprise the following:

| Part I | - | General |
|---|---|---|
| Part II | - | Levels of Sound |
| Part III | - | Electrical Safety |
| Part IV | - | Sanitation |
| Part V | - | Hazardous Substances |
| Part VI | - | Safety Materials, Equipment, Devices and Clothing |
| Part VII | - | Appliances and Machine Guards |
| Part VIII | - | Materials Handling |
| Part IX | - | Hazardous Occurrence Investigation, Recording and Reporting |

## MARINE OCCUPATIONAL SAFETY AND HEALTH REGULATIONS

The marine regulations apply where the duties of Public Service employees require them to work on ships registered in Canada, on Canadian Forces auxiliary vessels or in the loading and unloading of ships.  The regulations are similar to the

*Canada Occupational Safety and Health Regulations*, adapted to the marine environment.

Except for Canadian Forces auxiliary vessels, the Canadian Coast Guard has been delegated responsibility for administering Part IV of the *Canada Labour Code* and the marine regulations. For this purpose, officers of the Ship Safety Branch of the Coast Guard have been appointed as safety officers and regional safety officers. Administration of Part IV and of the marine regulations aboard Canadian Forces auxiliary vessels remains with Human Resources Development Canada.

The *Marine Occupational Safety and Health Regulations* may be purchased from Public Works and Government Services Canada (SOR/87-183) and consist of the following:

Part I      -      General
Part II     -      Temporary Structures
Part III    -      Elevating Devices
Part IV     -      Levels of Lighting
Part V      -      Levels of Sound
Part VI     -      Electrical Safety
Part VII    -      Sanitation
Part VIII   -      Hazardous Substances
Part IX     -      Confined Spaces
Part X      -      Safety Materials, Equipment, Devices and Clothing
Part XI     -      Tools and Machinery
Part XII    -      Materials Handling and Storage
Part XIII   -      First Aid

## WORKPLACE HAZARDOUS MATERIALS INFORMATION SYSTEM

The purpose of the Workplace Hazardous Materials Information System (WHMIS) is to ensure that the hazards of materials produced, imported or used in Canadian workplaces are identified by suppliers using standard criteria and that the hazard information is transmitted by suppliers to affected employers and employees. The number of illnesses and injuries resulting from the use of hazardous materials in the workplace is thus reduced through the provision of vital information to those who may be affected.

Hazardous materials, or controlled products as they are referred to in the legislation, are defined as pure substances or mixtures of substances that meet or exceed WHMIS hazard criteria. Suppliers use available information, data from existing data systems and pertinent technical literature to assess materials for use in the work place against the WHMIS criteria. Hazardous materials covered by WHMIS are grouped into distinct classes and sub-classes to identify their hazard potential and inform users of the specific dangers that may be expected.

First and foremost, WHMIS is a communication system composed of three interrelated and equally important elements:

- labels,
- Material Safety Data Sheets (MSDSs), and
- employee education programs.

The labelling system requires suppliers to provide cautionary labels on containers of hazardous materials as conditions of sale and importation, and employers must provide the supplier label, or an equivalent label containing the minimum information, in the workplace. The MSDSs are intended to supplement information on labels with a more detailed description of hazards, possible health effects and protective measures. This must include data required to provide effective engineering controls, safe work procedures, choice of personal protective equipment, procedures to be followed in the case of emergencies, and data for monitoring workplace conditions and the health of exposed employees.

A key element of WHMIS is the requirement for employers to establish educational and training programs for employees who may be exposed to hazardous materials in the workplace. These programs must ensure that employees have the information they need and that they are able to apply it for the safe use of such materials. The programs should include a description of all the mandatory aspects of WHMIS and of employer and employee responsibilities.

Within the federal jurisdiction, amendments have been made to Part II of the *Canada Labour Code* and to the occupational safety and health regulations to incorporate changes made under WHMIS. In the Public Service of Canada, amendments were made to the *Occupational Safety and Health Policy* and the *Dangerous Substances Directive* in response to these changes.

## CHAPTER 1 - POLICIES

### Introduction

Public Service policies* on occupational safety and health (OSH) issues are established by the Treasury Board pursuant to the *Financial Administration Act*, the *Canada Labour Code*, Part II, the *Non-smokers' Health Act* and their regulations. The policies apply to that part of the Public Service for which the Treasury Board is the employer under Part I, Schedule I of the *Public Service Staff Relations Act*.

Through its policies, the Treasury Board provides direction to departments on their mandatory OSH responsibilities. As the administrative arm of the Board, the Treasury Board Secretariat monitors and evaluates departmental performance.

* The Clothing Directive is the subject of consultation by the National Joint Council and is deemed to be part of collective agreements.

### CHAPTER 1-1 - OCCUPATIONAL SAFETY AND HEALTH

**This policy has been designated by the Treasury Board as a key policy for the management of human resources.**

### Policy objective

- To promote a safe and healthy workplace for Public Service employees.
- To reduce the incidence of occupational injuries and illnesses.

### Policy statement

It is the policy of the government to provide employees with a safe and healthful working environment and with occupational health services as defined in this policy.

### Application

This policy applies to departments and agencies listed in Schedule I, Part I of the *Public Service Staff Relations Act*.

### Policy requirements

1. Departments must:

- implement the requirements of the *Canada Labour Code*, Part II, and its regulations and comply with authorized Human Resources Development Canada directions;
- establish and maintain effective occupational safety and health (OSH) programs consistent with Treasury Board policies, standards and procedures;

- comply with Health Canada directives on the occupational safety and health of employees;
- assign departmental OSH personnel according to the size, complexity and operating risks of the department;
- post at a place or places accessible to all employees the documents listed in appendix B;
- provide OSH training and information to employees; and
- provide employee assistance services.

2. Health Canada, under delegation from Treasury Board, must:

- administer the Public Service Health Program as described in the attached appendix A; and
- through on-site investigations, monitor departmental compliance with legal standards (Treasury Board OSH standards and regulations pursuant to Part II of the *Canada Labour Code*), promote their full implementation and, in accordance with Treasury Board procedures, direct corrective action if necessary.

**Notes**

All professional Public Service occupational health personnel associated with this program will be under the direct control of Health Canada.

Public Service Health Program staff cannot be required to disclose medical information on employees and must observe strict professional ethics on confidentiality.

3. Employees must:

- learn and follow the OSH provisions of the workplace;
- use the OSH equipment and devices provided;
- take reasonable precautions to protect themselves and other employees;
- comply with authorized OSH instructions; and
- report all accidents and hazards.

**Monitoring**

The Treasury Board Secretariat will monitor departmental performance by:

- analyzing information contained in Human Resources Development Canada and Health Canada statistical reports, Annual Management Reports and Multi-Year Human Resource Plans;
- reviewing internal and external audits and evaluations on the application of the policy;
- reviewing overall departmental application of the policy or particular elements of it, on a periodic basis.

Performance assessment will be based on:

- the frequency and severity of occupational injuries and illnesses;
- the existence and use of systems and processes to review and analyze the frequency and severity of occupational injuries and illnesses;
- the allocation of sufficient resources;
- the quality and extent of training and information activities in occupational safety and health.

## References

*Canada Labour Code*, Part II and its regulations:

- *Canada Occupational Safety and Health Regulations;*
- *Marine Occupational Safety and Health Regulations;*
- *Aviation Safety and Health Regulations;*
- *Safety and Health Committee and Representatives Regulations.*

## Enquiries

Enquiries about this policy should be directed to the responsible officers in departmental headquarters who, in turn, may seek interpretations from:

Safety, Health and Employee Services Group
Staff Relations Division
Human Resources Policy Branch
Treasury Board Secretariat

## CHAPTER 1-1 - APPENDIX A - DEFINITIONS AND ROLES

*Employer* (employeur)

Under the *Canada Labour Code* (CLC), Part II, this term means a person who employs one or more employees and includes an employers' organization and any person who acts on behalf of an employer. In the Public Service context the term includes an agency acting on behalf of the Treasury Board, a department or any person who acts in a supervisory or managerial capacity on behalf of a department.

*Treasury Board Occupational Safety and Health (OSH) standards* (Normes de sécurité et de santé au travail du Conseil du Trésor)

The expression is self-explanatory. However, it should be noted that most of these standards are:

- based on the CLC Part II regulations;
- consulted upon in the National Joint Council and, therefore, have the force of collective agreements.

Since Treasury Board OSH standards and regulations pursuant to Part II of the *Canada Labour Code* are legal requirements, in the case of differences the most stringent requirement is the operative one.

**The Public Service Health Program** (Le Programme de santé des fonctionnaires fédéraux)

This program is administered by Health Canada under delegation from the Treasury Board. The Public Service Health Program is a corporate resource and should be used by departments accordingly. Its services are essentially preventative and they should not interfere with or replace those available through private physicians and community health agencies.

The program deals with:

– health assessments of employees and emergency medical services;
– environmental health investigations and surveys;
– occupational health nursing services in departments;
– advice to the Treasury Board on occupational safety and health matters, including advice on the development and monitoring of occupational and environmental standards, procedures and other directives for the prevention of occupational illness and injury; and on the provision of occupational health services within the Public Service;
– advice to departments and managers on occupational safety and health training; first aid training, facilities, services and supplies; the selection and use of personal protective equipment and clothing; the adaptation and selection of work; and the rehabilitation and retraining of employees disabled by work injuries or illnesses;
– health advice and education for employees, as well as on-the-job medical care with the cooperation and consent of the employee's private physician;
– employee assistance services, as specified in the Public Service Employee Assistance Program Directive;
– research and special studies.

**Human Resources Development Canada as regulator** (Développement des ressources humaines Canada, organisme de réglementation)

The Public Service became subject to CLC Part II (Part IV as it was then known) on March 31, 1986, by an amendment to the *Financial Administration Act*. Prior to that Human Resources Development Canada provided certain OSH services to departments as a corporate resource.

Human Resources Development Canada is now (as it always has been for other employers under the federal jurisdiction) the regulator for the Public Service on OSH matters. Its responsibility is to monitor and enforce CLC Part II. With this in mind, it is instructive to read in CLC Part II the duties assigned to employers and the penalties for contraventions.

To carry out its role, Human Resources Development Canada has designated safety officers and regional safety officers. Their duties include inspections, accident and refusal-to-work investigations and the stipulation of corrective measures.

It should be noted that Human Resources Development Canada has also appointed officials from:

- Transport Canada's Aviation Group, to act as safety officers and regional safety officers in the administration of CLC Part II and the *Aviation Occupational Safety and Health Regulations* (for public servants employed in aircraft in operation);
- The Coast Guard's Ship Safety Branch, to act as safety officers and regional safety officers in the administration of CLC Part II and the *Marine Occupational Safety and Health Regulations* (for public servants working on ships registered in Canada, or in the loading and unloading of ships).

## CHAPTER 1-1 - APPENDIX B - POSTING OF DOCUMENTS AND INFORMATION

Following is a list of the items that must be posted according to policy requirements:

- a copy of the *Canada Labour Code*, Part II;
- any printed notices or other material prescribed by Human Resources Development Canada or the Treasury Board Secretariat;
- a copy of a general policy statement worded this way:

> "A high priority in the Public Service of Canada is providing working conditions conducive to the safety and health of employees.

> "This department is committed to promoting occupational safety and health and provides programs in both occupational safety and health and employee assistance. It also provides training and information in these areas to all employees.

> "Please see your supervisor if you need more information."

## CHAPTER 1-2 - EMPLOYEE ASSISTANCE PROGRAM

### Policy objective

To foster and maintain the well-being and productivity of employees by providing confidential assistance or short-term counselling to those who are experiencing personal or work-related problems.

### Policy statement

The government recognizes that it is possible in the work environment to identify employees with problems, including those related to substance abuse, and to motivate them to seek assistance or treatment at an early stage. Although there are numerous organizations in the community to help people with various problems, assistance is often required to determine the most appropriate resource. To this end, the government makes available to employees a confidential and voluntary Employee Assistance Program (EAP) without prejudice to job security or career progression.

## Application

This policy applies to the departments and agencies listed under Part I of Schedule I of the *Public Service Staff Relations Act*.

## Policy requirements

Departments must ensure that:

–   an EAP coordinator is appointed;
–   employees are provided with EAP services that conform to this policy, including definitions in Appendix A;
–   the confidentiality and privacy of EAP information is maintained in accordance with Appendix B;
–   EAP practitioners abide by the Code of Ethics in Appendix C;
–   employees are kept informed about the program and how to access its services;
–   managers and supervisors are provided with education on matters related to the program and informed of their responsibility to refer employees with work performance problems related to personal difficulties to their departmental EAP;
–   where employees are likely to be involved in critical incidents because of the nature of their work, a procedure is established to provide an initial debriefing session as soon as possible after such an incident and EAP follow-up if required;
–   leave credits for treatment and rehabilitation are used in accordance with existing collective agreements and Public Service policies;
–   reasonable access to departmental EAP services is provided to dependants of employees who live, have lived, or will live outside Canada under the provisions of the Foreign Service directives, or who are subject to the Isolated Posts Directive, or in exceptional circumstances such as a work related hostage-taking incident; and
–   employee representatives are consulted and are provided with the opportunity to develop jointly and to participate fully in the application of the departmental EAP, including its referral, educational and promotional aspects.

Where EAP services are provided by a non-governmental agency or another department, departments must ensure that the services conform to this policy. Moreover, departments must continue to assess the quality and accountability of the program.

Contracts must specify that personal information collected for EAP purposes is deemed to be under the control of the department and is consequently subject to the *Privacy Act*.

**Responsibilities**

On request, Health Canada, through the Public Service Health Program, shall:

– provide professional and technical advice about the EAP to departments;
– provide or arrange for medical diagnosis or initial counselling on alcoholism or other health problems and refer clients to community agencies for subsequent treatment; and
– provide consultative services to departments and the Public Service Commission with respect to the health-medical aspects of training programs.

Health Canada shall ensure the continuous professional development of its occupational health personnel, including nurses and physicians, on substance abuse counselling and rehabilitation techniques.

**Monitoring**

To assess the effectiveness of this policy, the Treasury Board Secretariat requires that departments submit a report on program structure and a statistical summary of its EAP activities. The Secretariat requests the first report for the two year period ending March 31, 1993 and subsequently every two years.

**References**

*Treasury Board Manual*

Occupational Safety and Health Volume,
- OSH Policy (Chapter 1-1)

Security Volume

Materiel, Services and Risk Management Volume,
- Indemnification of Servants of the Crown (Part III, Chapter 2)

Human Resources Volume,
- Conflict of Interest and Post Employment (Chapter 3-1);
- Harassment in the workplace (Chapter 3-2)

*Provision of Legal Services to Public Servants*, T.B. Circular 1983-52

*General Records Disposal Schedules of the Government of Canada*, PAC 86/001

Policy Guide: *Access to Information Act* and *Privacy Act*, Treasury Board Secretariat, 1992

*Access to Information Act*

*Canadian Human Rights Act*

*Official Languages Act*

*Privacy Act*

*Public Service Employment Act*

**Enquiries**

Enquiries about this policy should be directed to the responsible officers in departmental headquarters who, in turn, may seek interpretations from:

**On EAP, health and safety policy issues**

> Safety, Health and Employee Services Group
> Staff Relations Division
> Human Resources Policy Branch
> Treasury Board Secretariat

**On staff relations issues**

> Employer Representation Group
> Staff Relations Division
> Human Resources Policy Branch
> Treasury Board Secretariat

**On health issues and EAP program advice**

> Occupational and Environmental Health Directorate
> Medical Services Branch
> Health Canada

## CHAPTER 1-2 - APPENDIX A - DEFINITIONS

*Client:* an employee or a dependant, as defined in the policy, who accepts the assistance of EAP (client).

*Critical incident:* a traumatic event that produces a strong emotional reaction that could affect the ability to cope (incident critique).

*Employee Assistance Program (EAP):* a program to identify, provide short-term counselling and referral service to employees with personal or work-related problems to resources within the Public Service or the community, when appropriate, and provide follow-up. These services do not replace those provided by the Public Service Health Program (programme d'aide aux employés [PAE]).

*EAP coordinator:* an individual designated by a department to manage or coordinate the EAP (coordonnateur du PAE).

*EAP practitioner:* an individual qualified by training or certification in the techniques of assessment of problems, particularly in respect of substance abuse, and of intervention. This includes, but is not limited to, social workers, occupational health nurses and physicians, and volunteer peer referral agents trained in EAP (intervenant du PAE).

*Referral:* an oral or written recommendation to use departmental EAP services or other comparable services to assist in resolving personal or work-related problems that may affect performance (orientation).

*Short-term counselling:* discussions leading to identification of clients' problems and referral to appropriate resources, normally up to three sessions (conseils à court terme).

## CHAPTER 1-2 - APPENDIX B - CONFIDENTIALITY AND PRIVACY

### 1. General

1.1 In conformity with this policy, departments shall have policies and procedures that safeguard client information in records or gathered in counselling sessions.

1.2 Individual client case files are required only where documentation is necessary for practitioners to carry out their EAP functions. When information is recorded, it should be kept to a minimum and it may include dates, the nature of problems, minimal progress notes, recommended referrals and non-medical reports related to a client's work capability or limitations. Such individual case files shall be designated as PROTECTED — EAP.

1.3 At the outset of the initial interview, the practitioner shall advise the client, orally or in written format, of the confidentiality policy and its limitations and the following information if a file is created:

– the type of information that the file may contain;
– the length of time the file will be kept before it is destroyed;
– the client's right of access to the file under the *Privacy Act* to review and correct information or have notations attached; and
– the possibility of periodic reviews by the Office of the Privacy Commissioner in accordance with the procedure in paragraphs 4.1, 4.2 and 4.3 below.

1.4 The practitioner cannot promise confidentiality in the following situations:

– the circumstances set out in subsection 8(2) of the *Privacy Act* which include court subpoenas;
– suspected cases of child abuse; or
– a threat of suicide or illegal activity.

When such information is received, the practitioner is advised to consult the departmental EAP Coordinator and legal counsel immediately.

1.5 Employee Assistance Program case files are a personal information bank that must conform with *Privacy Act* requirements on the collection, use, disclosure, retention and disposal of personal information and must be described in the annual publication *Info Source (Sources of Federal Employee Information).*

1.6 Personal information collected by an agency providing EAP services under contract to a department is deemed to be personal information under the control of that department and is therefore subject to the *Privacy Act.*

1.7 When assessing or auditing their EAP, departments must protect the confidentiality of personal information as described in paragraph 4.2 below.

## 2. Physical security

2.1 The government Security policy and standards specify that personal information receive enhanced protection. Particularly sensitive personal information in Employee Assistance Program case files requires additional protection. The files must be protected by particular storage and transmittal standards in accordance with the appropriate sections of the Security policy and standards of the *Treasury Board Manual* : Appendix A, Security organization and administration standards; Appendix B, Physical security standards; and Appendix C, Information technology security standards.

2.2 All client case files and sensitive program information require locked storage space. Secure cabinets should be approved by the departmental security office. Only authorized personnel have access to EAP case files.

2.3 Agencies providing EAP services under contract to departments shall provide a level of security for EAP information equivalent to that described in paragraphs 2.1 and 2.2 above.

## 3. Release of information

3.1 No information may be released about clients' files except with their written consent or where the law requires or permits release of that information (see paragraph 4). Written consent shall include the following:

- the name of the person or agency to whom the information is to be released;
- the specific information to be released;
- the date of consent;
- the client's signature; and
- the expiry date of the consent.

NOTE: Practitioners should refer to the Employee Privacy Code and their departmental procedure for guidance on employee access to the data banks listed in *Info Source.*

## 4. Compliance review under Privacy Act

4.1 Under the *Privacy Act*, the Office of the Privacy Commissioner may conduct periodic compliance reviews of EAP case files to provide independent verification that privacy rights are being protected and to ensure that the information in these files is collected, used, retained and disposed of in accordance with the *Privacy Act.*

4.2 During these EAP case file reviews by officers from the Office of the Privacy Commissioner, the EAP practitioner who has custody of the files will be present. Personal identifiers on the file such as client's name, title and address, must be masked to protect the identity of the employee. It is preferable to create a numbered filing system that is cross-referenced to a master file kept separately.

4.3 Only when a violation of the *Privacy Act* is suspected are EAP staff required to disclose the identity of the client to the Privacy Commissioner or his staff in order that they may communicate this fact to the client.

## 5. Disposal of records

5.1 PROTECTED-EAP case files must be retained and disposed of in accordance with schedules approved by the National Archivist and issued in *General Records Disposal Schedules of the Government of Canada*, PAC 86/001. Individual case files must be destroyed two years after the date of the client's most recent contact with EAP. PROTECTED — EAP case files may be destroyed earlier, with the client's consent, or at the client's request. In situations where the client or the EAP practitioner involved is leaving the department, or in the event that a contract is cancelled or not-renewed, the EAP case files may only be transferred to the new service provider with the client's consent. In other situations, EAP case files are to be destroyed.

5.2 Employee Assistance Program case files should be disposed of in an appropriate manner, which may involve burning under controlled conditions, pulping, pulverizing, shredding or another destruction method using equipment appropriate to the level of sensitivity of the information involved. The departmental information management staff provides guidance on the measures to apply.

## CHAPTER 1-2 - APPENDIX C - CODE OF ETHICS

### Introduction

This Code sets out the ethical attitudes expected of Employee Assistance Program (EAP) practitioners in the Public Service regardless of their formal preparation, place of work or population focus.

### Respect for the dignity and rights of persons

Fundamental to the principle of respect for the dignity of persons is the belief that all persons have a right to appreciation of their innate worth as human beings.

Every EAP client must be respected regardless of race, national or ethnic origin, colour, religion, age, sex, marital status, family status, disability, conviction for which a pardon has been granted, political affiliation, social or economic status, or physical or mental capabilities or characteristics or other similar grounds.

Clients of EAP have a right to consent to matters that affect their treatment and that are within their control. Moreover, EAP is directed at decreasing the dependency of clients who seek advice and sponsoring positive personal control.

A practitioner must:

1. respect all clients regardless of their personal characteristics.

2. respect the client's right of self-determination by encouraging active involvement in decisions related to treatment and referral;

3. respect the client's right to privacy and confidentiality by collecting only personal information that is relevant and by sharing it only to the extent required for referral, with the informed consent of the client.

4. respect the client's right of informed consent by providing all the information necessary so that a reasonable person in similar circumstances would be able to make a reasoned and informed choice;

5. obtain, except in those instances where the law requires or permits the release of information*, signed consent forms from the client in all situations where the release of personal information is requested.

6. honour commitments made to the client;

7. not exploit relationships with clients to enhance the practitioner's own self-worth or position in the department, agency or community;

8. encourage respect for the dignity of others and avoid practices that are inconsistent with the legal, civil or human rights of others;

9. respect the social norms and moral attitudes of the community in which the practitioner works; and

10. respect the right of the client to discontinue participation in the program at any time.

* The Canadian social system provides for the protection of legal, civil and human rights. However, the needs and rights of others in the same social context mitigate the recognition and expression of these rights.

**Conflict of interest**

Practitioners must conform to the government's Conflict of interest and post-employment policy.

A practitioner must:

1. remain within the scope of the program;

2. resist exploiting of the professional relationship with the client to further any social, political, economic, personal or business interest;

3. refer a client to another resource when the practitioner, for a personal reason, cannot provide service to the client;

4. inform concerned parties of possible or actual conflicts of interest;

5. initiate steps that precipitate a reasonable solution without causing undue harm to the client or the organization; and

6. remain neutral in conflicts between the client, the union, or management.

**Competency**

A responsible counsellor recognizes the need to make continuous efforts to upgrade and refine skills. Practitioners should acknowledge their limitations and provide services that are consistent with their skills.

A practitioner must:

1. offer services that are within his or her established competence and the program's defined parameters. When the problems of clients are beyond these limits, practitioners must refer clients to an appropriate resource;

2. seek consultation with fellow practitioners or other appropriate resources in managing cases when the practitioner encounters clients or situations that are beyond the program's parameters or the practitioner's expertise;

3. ensure that appropriate referrals are made to recognized resources;

4. continually evaluate his or her own background, experiences and values so as to assess their influence on interactions with others and attend, in accordance with departmental policies, educational programs directed at improving performance; and

5. accurately represent his or her own professional qualifications, competence and purposes of the program.

**CHAPTER 1-2 - APPENDIX D - GUIDELINES ON EMPLOYEE ASSISTANCE PROGRAM SERVICES**

**Counselling service**

1. A response to a request for service should occur as soon as possible, preferably within one working day.

2. The first counselling session should occur within two weeks unless unforeseen circumstances arise.

3. Follow-up should be an integral part of the EAP service, usually by meeting with the client. Sometimes a letter or telephone call is approrpriate.

4. The program normally offers up to three sessions for each new case.

5. Where employees are likely to be involved in critical incidents because of the nature of their work, the department must develop a procedure for initial counselling and follow-up. The same can apply to cases that involve threats of violence or suicide.

## Advisory service

A response to a request for advisory services from managers, supervisors or unions should occur as soon as possible and, at the latest, within a week.

## Education service

1. EAP training sessions for managers and supervisors should cover, at a minimum:

- the scope and limits of EAP, the confidentiality aspects, the *Privacy Act* provisions and the procedure for accessing the program;
- the administrative role with respect to EAP and referrals;
- the concepts and methods that allow for early detection of problems that interfere with job performance;
- constructive methods to deal with employees experiencing performance problems due to personal or behavioural problems; and
- support approaches to assist the employee.

2. Every new employee should receive a brochure, information sheet or other appropriate format containing information that describes the scope and means of accessing the EAP services.

3. The departmental program should keep all employees informed of the service, its location and how to access it.

## Prevention activities

Consistent with the program design and the organizational mandate, preventive programs should be held to educate employees about personal problems related to life style and work environment, and the possible responses. This may involve cooperation with the Public Service Health Services and the departmental safety and health committee or the safety and health representative.

## Notes

As indicated in the Occupational safety and health policy, the Treasury Board has delegated the administration of the Public Service Health Program to Health Canada.

The Committees and representatives directive encourages safety and health committees to establish and promote programs to educate employees represented by the committee.

## CHAPTER 1-3 - CLOTHING DIRECTIVE

## GENERAL

### Collective agreement

This directive is deemed to be part of collective agreements between the parties to the National Joint Council (NJC), and employees are to be afforded ready access to this directive.

### Grievance procedure

In cases of alleged misinterpretation or misapplication arising out of this directive, the grievance procedure, for all represented employees within the meaning of the *Public Service Staff Relations Act*, will be in accordance with Section 7.0 of the *National Joint Council By-Laws*. For unrepresented employees the departmental grievance procedure applies.

### Effective date

This directive was effective on January 1, 1992.

### Purpose and scope

It is the policy of the government to provide appropriate items of clothing to employees where the nature of the work is such that special protection is required for reasons of occupational safety, health or cleanliness or where special identification at the local, national or international level will aid in the effective performance of duties and in meeting program objectives.

Departments and agencies shall review their existing clothing policies to ensure that they comply with this directive.

It should be noted that this directive combines and replaces the former Protective Clothing policy and the Uniform Clothing policy.

This directive is intended to assist departments in ensuring that their practices are economical, equitable and reasonably consistent with those throughout the Public Service and are comparable with those for similar occupations outside the Public Service.

### Application

This directive applies to all departments and agencies listed in Schedules A and B of the *Financial Administration Act* and to branches designated as departments for the purposes of the Act, with the exception of Royal Commissions.

### Authorities

This policy was approved by the President of the Treasury Board under Section 7 of the *Financial Administration Act*, following consultation within the National Joint Council.

This directive supersedes all previous Treasury Board authorities on the provision of protective and uniform clothing but does not affect those authorities dealing with allowances or the provisions contained in collective agreements.

The President of the Treasury Board has delegated authority to approve exceptions to the directive. Requests for such exceptions should be made in the form of a letter to the Deputy Secretary of the Human Resources Policy Branch.

Such requests should be signed by departmental officials who have authority to sign submissions and should contain the same information as submissions.

Deputy heads have the authority to issue necessary items of clothing.

## Responsibilities

Public Works and Government Services Canada (PWGSC) provides clothing advisory services to departments and agencies, through the Clothing Advisory Section of the Consumer Products Branch.

The services provided by the Clothing Advisory Section are listed in Appendix A.

It is the responsibility of each department:

- to identify the situations where the provision of clothing is necessary or desirable;
- to determine that the type of clothing provided is adequate and suitable;
- to develop and to maintain up-to-date clothing standards; and
- to ensure that the appropriate consultation, as specified in this directive, takes place.

Departments are required to incorporate controls to ensure that practices are consistent with the policy directives. The internal controls shall include the maintenance of a record containing the following information:

- the number of employees provided with clothing;
- the composition of standard clothing issues;
- the value of clothing issued (in total and by unit);
- the average cost per employee provided with clothing;
- the value of clothing allowances (in total and individually);
- copies of relevant departmental bulletins or directives.

Organizational identifiers on clothing, for example, shoulder flashes, may be subject to the requirements of the Federal Identity Program (see Chapter 470, *Administrative Policy Manual*).

## Union management consultation

Departments shall consult with employee representatives at the local, regional or national level, as appropriate, regarding the application of this directive, and prior to any planned changes in existing practices.

Departments and agencies should be aware of the consultation provisions of the relevant collective agreements when applying this directive.

When clothing serves for both identification and personal protection, departments shall ensure that workplace safety and health committees or safety and health representatives, if any, assist in the determination of personal protective equipment and clothing requirements. (See Personal Protective Equipment Safety directive.)

## Consultation with the Clothing Advisory Section

Departments shall consult with the Clothing Advisory Section:

- before introducing new items of clothing or replacing existing issues;
- to ensure that the quality and quantity of clothing to be provided to employees performing similar functions in similar working environments are reasonably consistent from department to department;
- to ensure fabrics selected for protection meet good industrial safety practices, and fabrics selected for uniforms meet the PWGSC criteria;
- when new uniforms are being introduced (consultation shall begin not later than two years prior to introduction);
- when clothing purchases are expected to exceed $10,000.

A department that finds the PWGSC recommendations unacceptable shall submit the dispute to the President of the Treasury Board, as provided for in the authorities section.

## Inquiries

All inquiries regarding this directive should be routed through departmental headquarters.

For interpretation of specific policy statements contained in this directive, designated members of the departmental headquarters should contact the:

Safety, Health and Employee Services Directorate,
Staff Relations Division,
Human Resources Policy Branch,
Treasury Board Secretariat.

## Credit revenue

Unless authority to credit revenue to the vote has been obtained by either vote-netting authority or a revolving fund authority, departments and agencies must credit the proceeds of sales to non-tax revenue.

## References

Directives referred to in this directive, such as the Travel or the Personal Protective Equipment directive may be found in either the *Treasury Board Manual*, Personnel Management volumes, or the *National Joint Council Agreements*, Volumes 1 and 2.

## Part I - Protective Clothing

### Acquisition

### 1.1 Selection

1.1.1 In the selection of clothing the prime consideration shall be the protection of employees; however considerations of serviceability and ease of care should be a factor. Care labels, as designated by the Department of Industry Canada, should be attached to each new item of apparel. The selection of protective clothing, including protective footwear, shall be suitable for the gender of the user.

### Provision

### 1.2 General

1.2.1 Protective clothing is normally:

– provided free of charge to employees,
– replaced free of charge under prescribed conditions, and
– worn over the employee's personal clothing.

1.2.2 Protective clothing is maintained and laundered by the employer. In exceptional cases, however, where this clothing is provided on an individual basis and the employer permits the employee to wear it away from the workplace at the employee's request, the wearer is responsible for maintenance and laundering.

1.2.3 All protective clothing may be withdrawn when conditions in the work area improve, e.g. when risks from hazards or contamination are eliminated or where wet or dirty conditions no longer prevail.

1.2.4 Protective and special clothing for occupational safety, health and cleanliness is provided free of charge to employees and replaced free of charge under prescribed conditions.

1.2.5 Protective clothing shall be provided to employees when there is a requirement for protection of employees to serve the following purposes:

– occupational safety,
– occupational health,
– occupational cleanliness.

1.2.6 Bulletins shall be issued to employees when the wearing of protective clothing is required. Such bulletins normally will identify and enumerate clothing commodities, state the employee's responsibility for the clothing received, and specify the manner of accounting for clothing when the employee is no longer eligible to receive or retain it (e.g. on promotion, demotion, separation or due to a change in the physical working conditions).

1.2.7 Normally, protective clothing which is provided to employees shall be worn only on duty and will not be worn away from the workplace. When employees are provided with specific items of protective clothing to be worn on duty, substitute items shall not be worn.

## 1.3 **Quantities**

1.3.1 The quantity of each commodity to be provided initially to each employee shall be based on conditions of wear and tear and the expected wear-life of each commodity.

1.3.2 When employees desire additional items of protective clothing, over and above the amount of authorized issue, the department or agency may make reasonable amounts of such clothing available for employees to purchase.

## 1.4 **Replacement**

1.4.1 Replacement items of protective clothing shall be exchanged free of charge for existing items when these are no longer serviceable.

1.4.2 Employees will be responsible for the replacement of clothing that is lost while in their custody, unless the loss was beyond their control.

## 1.5 **Pool clothing**

1.5.1 Protective clothing may be pooled when the frequency of use by employees does not justify individual provision. Pool clothing is worn over the employee's own clothing; as a general rule, clothing which is worn next to the skin is provided on an individual basis. For example, parkas which are pooled should have detachable hoods, which shall be issued on an individual basis – alternatively, individual head coverings shall be issued for wear inside the hood.

1.5.2 Quantities of pool clothing shall be adequate to provide a range of sizes and also to permit rotational cleaning.

1.5.3 Cleaning and upkeep shall be scheduled on a regular basis.

## 1.6 **Clothing allowance**

1.6.1 As a general principle, the Treasury Board prefers the direct provision of clothing to the payment of clothing allowances, except for safety footwear. At the same time, Treasury Board does not wish to preclude payment of such allowances in cases where the practice is established or the economy of introducing a new allowance can be clearly demonstrated.

1.6.2 The allowance shall cover the full cost of such protective clothing.

1.6.3 No new allowances or changes in existing allowances shall be introduced without the prior authorization of the Treasury Board.

1.6.4 Without specific Treasury Board approval, no allowances shall be paid for:

– repair, cleaning, pressing and laundering;
– personal clothing.

## 1.7 Safety equipment

1.7.1 Safety equipment is provided, as required, in addition to protective clothing.

1.7.2 Prescription safety lenses are not provided except in situations where eye protection is required, and

(a) the nature of the work is such that the protective prescription lenses are installed in specialized protective frames such as in goggles and other eye protection which is not normally worn off the job, or

(b) it is impractical to wear protection over glasses because of distortion.

1.7.3 Departments may provide basic safety glasses.

## 1.8 Clothing for identification and security

1.8.1 Clothing provided for protection may also serve the purposes of management for identification of employees; thus duplication of issues for protection and identification should be avoided.

1.8.2 In areas where security is an issue, management shall ensure that all outer protective clothing bears permanent, clearly visible and easily identifiable markings.

## 1.9 Clothing for occupational safety

1.9.1 Protective clothing is required for the safety of the employee in order that job-related duties may be performed with the minimum risk of being physically injured.

1.9.2 The following are examples of protective clothing, which shall be provided as required:

– head: cap, hearing protectors, eye protectors, face masks, respirators;
– torso: impervious or insulating apron or clothing;
– arms and hands: impervious or insulating gloves, mitts or gauntlets;
– legs and feet: impervious leggings and purpose-designed footwear;
– body: pressure suits for flying or diving, diving suits, flotation jackets, anti-static jackets, outer clothing conforming to the Electrical safety directive, paragraph 19.

1.9.3 Insulation clothing shall be provided for duty in hazardous weather conditions where the type of personal outer clothing normally worn while working outdoors is inadequate to protect the employee from physical and health harm in the particular working environment.

1.9.4 Insulation clothing shall also be provided for duty in hazardous weather conditions when there is risk of damaging or soiling to the employee's personal insulation clothing.

1.9.5 Insulation clothing designed to prevent hypothermia shall be provided to individuals when their duties involve significant risks of immersion in cold water.

## 1.10 Clothing for occupational health

1.10.1 Protective clothing is provided to:

- protect the employee from environmental hazards and extreme weather conditions;
- protect the health of the employee on duty;
- preserve the cleanliness of food or other materials and products which the employee is handling or dispensing to the public or other persons;
- prevent the spread of contamination or diseases;
- protect the employee from the risk of disease that may be contracted in the performance of job-related duties.

1.10.2 The following are examples of items of protective clothing, which shall be provided as required:

- head: cap, eye protection;
- torso: apron or smock or coveralls, shirts and pants, soil-resistant parka, snowmobile suit;
- arms and hands: impervious gloves or gauntlets;
- internal organs: respirators, masks;
- legs and feet: purpose-designed footwear.

## 1.11 Clothing for occupational cleanliness

1.11.1 Cleanliness clothing shall be provided when:

- the nature of the duties performed by the employee is such as to present a risk of significant or permanent damage to the employee's personal clothing, or
- the working environment results in a significant degree of soiling of the employee's personal clothing, e.g. coveralls provided to mechanics, lab coats for laboratory use.

1.11.2 The following are examples of items which shall be provided as required:

- head: cap;
- torso: apron or smock or coveralls, shirts and pants, soil-resistant parka;
- body: snowmobile suits;
- legs and feet: purpose-designed footwear.

**Footwear**

### 1.12 Requirement

1.12.1 Specific types of footwear may be required for protection of employees for the following purposes:

- occupational safety,
- occupational health,
- occupational cleanliness.

### 1.13 Footwear provided on a loan basis

1.13.1 Except where otherwise specified, footwear is provided to the employee on a loan basis, free of charge, and is replaced free of charge under prescribed conditions.

1.13.2 When feasible, footwear should be worn over the employee's personal footwear.

### 1.14 Purpose-designed footwear

1.14.1 Purpose-designed footwear is provided by the employer.

1.14.2 Purpose-designed footwear is defined as specialized footwear other than that prescribed in 1.15.1, meeting the following criteria:

- designed and constructed to meet requirements based on the unique occupation of the wearer or unusual nature of the environment;
- possesses specialized protective features additional to, or other than, those found in regular protective footwear;
- significantly higher in cost than regular protective footwear; and
- generally unsuitable for wear away from the workplace for which it was designed to be worn.

1.14.3 All questions regarding the definition and assignment of footwear types to the purpose-designed category shall be referred to the Treasury Board Human Resources Policy Branch for resolution.

1.14.4 Environmental factors which would be expected to call for purpose-designed footwear, and the design features one would expect to find in that footwear, are:

- dangerous liquids: footwear either constructed of impermeable materials or specially treated to protect the wearer's feet from contact with dangerous or corrosive liquids or other dangerous substances, or where feet may be immersed in any liquid;
- explosive-electrical hazards: footwear made with non-sparking and/or non-conducting materials (except metal box toe) for use by workers subject to explosion or electrical hazards;
- physical hazards: footwear designed to protect against a harmful degree of physical stress resulting from requirements of an unusual nature as

may be encountered in such activities as mountain climbing, logging, skiing, pole climbing, riding horses, operating chainsaws, etc.;

– temperature extremes: thermo-insulated footwear for extreme cold, where the regular insulated winter work footwear is inadequate.

## 1.15 Protective footwear

1.15.1 Protective footwear is as defined by the Canadian Standards Association Standard CSA Z195-M1981, *Protective Footwear*.

## 1.16 Allowance for protective footwear

1.16.1 An allowance will be paid to employees upon presentation of proof of purchase of protective footwear meeting the CSA standard. For the purpose of this paragraph "employee" means a person employed in the Public Service other than a person employed on a casual or temporary basis, unless he or she is employed for a period of six months or more.

1.16.2 Employees occupying a position where protective footwear is mandatory, but who are required to wear protective footwear on an infrequent, periodic or intermittent basis in the performance of their regular duties, are eligible for the allowance.

1.16.3 It is expected that the frequency of replacement will be governed by the nature of the work, therefore replacement may occur at shorter than yearly intervals.

1.16.4 This allowance will not be paid for protective footwear purchased from the employer when this footwear is sold "at cost".

1.16.5 The allowance is based on the average retail cost difference between regular work footwear and protective footwear meeting the CSA standard.

1.16.6 The amount of this allowance, effective January 1, 1993, is $34.92. The allowance will be paid each time the employee submits proof of purchase.

## 1.17 Occasional requirement for protective footwear

1.17.1 The employer will provide protective footwear and purpose-designed footwear, on a loan basis for the time required, to an employee when:

– the position has not been identified as requiring protective footwear; i.e. the employee is not eligible for the allowance; and
– the employee is obliged, infrequently, to perform work which has been identified as requiring the wearing of protective footwear, but which is not associated with the employee's regular duties where the work is normally performed.

1.17.2 Pool protective footwear shall be disinfected after loan to an individual.

**Part II - Uniform Clothing**

## Acquisition

### 2.1 Selection

2.1.1 Selection of clothing shall be based as much as possible on comfort, serviceability and ease of care. Natural fabrics, natural fabric blends and fabrics not requiring dry-cleaning are the preferred choice.

2.1.2 When departments and agencies are reviewing their uniform policy, and the current or planned uniform requires dry-cleaning, employee representatives, at the local, regional or national level, as appropriate, and the Clothing Advisory Group shall assist in the selection of the uniforms.

2.1.3 Uniforms which require dry-cleaning shall only be selected when easy-care uniforms are clearly unsuitable and the Clothing Advisor agrees.

2.1.4 Care labels, as designated by the Department of Industry Canada, should be attached to each new item of clothing.

2.1.5 Sun protection shall be made available to employees who wear uniforms outdoors in summer. This means the provision of summer-weight long pants and summer-weight long-sleeved shirts for sun protection in addition to skirts, shorts and short-sleeved shirts. Employees should have the option of choosing, from the clothing provided, the combination they prefer.

2.1.6 Normally, it will be advantageous to have clothing for identification manufactured from all-season fabric, requiring minimum care. Commercially available items in standard sizes are more economical than custom-tailored special designs.

## Provision

### 2.2 General

2.2.1 Clothing shall be issued to employees when there is a requirement for identification of employees to serve the following purposes:

–   identification of the employee's occupation in emergency control,
–   public identification of the employee's authority to enforce laws or direct the public,
–   identification of the employee's service function,
–   identification of an employee's authority to access and work in a secure area. (Identification clothing may supplement the primary form of identification.)

2.2.2 Items of wearing apparel provided free of charge to employees and replaced free of charge under prescribed conditions are supplied for the following purposes:

- clothing for occupational identification, consisting of items that are of the same pattern or material or colour and worn as required locally under local management;
- distinctive uniform clothing to fulfill requirements for the national or international image of federal employees, or items that are of the same pattern, material and colour and worn as required throughout a sector such as a unit, branch or division of a department or agency in accordance with strictly enforced orders.

2.2.3 Regular shoes of a specific type or colour, which serve only to provide coordination with clothing, are not considered essential to identify the employee. Departments shall not provide regular shoes free of cost, nor shall they demand that employees wear specific types or colours of shoes. Departments may, however, specify that the footwear be of a type generally considered as acceptable and to coordinate with the uniforms provided.

Departments may, however, utilize the provisions of 2.5.2 to make such footwear available to employees for purchase at cost.

2.2.4 Bulletins shall be issued to employees when the wearing of uniform clothing is required. Such bulletins normally will identify and enumerate clothing commodities, state the employee's responsibility for clothing received and specify the manner of accounting for clothing when the employee is no longer eligible to receive or retain it (e.g. on promotion, demotion, separation or due to a change in working conditions). Employees will be responsible for the replacement of clothing that is lost while in their custody, unless the loss was beyond their control.

2.2.5 Normally, clothing which is issued to employees shall be worn only on duty and will not be worn away from the workplace. When employees are provided with specific items of clothing for wear on duty, substitute items shall not be worn. Clothing which is issued to employees may be worn in public to travel to and from work when the safe storage of personal clothing is not possible.

2.2.6 When, as a condition of employment, an employee receives any item of clothing as an individual issue, that employee will be expected to wear and maintain it in a clean, pressed and repaired condition, in accordance with departmental directives and in accordance with care labels permanently attached to each garment.

## 2.3 Quantities

2.3.1 The quantity of each commodity to be provided initially to each employee shall be based on conditions of wear and tear and the expected wear-life of each commodity.

## 2.4 Replacement

2.4.1 Replacement items of clothing shall be issued free of charge when existing items are no longer serviceable.

## 2.5 **Personal clothing**

2.5.1 Employees will normally be expected to provide, wear and maintain personal clothing as appropriate and necessary for their duties. Personal clothing does not include items which are designated as essential for identification within the context of this policy.

2.5.2 In special circumstances departments may make arrangements for employees to purchase reasonable amounts of personal clothing for use while on duty.

2.5.3 Generally, items of personal clothing may be made available for employees to purchase when:

(a)   the department is providing clothing and employees are responsible for wearing items of personal clothing that foster neatness and uniform appearance and complement clothing which is provided;

(b)   the employees request items of personal clothing that are not essential for identification, but the department considers that it would be beneficial, in order to improve the general appearance and comfort of employees while on duty;

(c)   employees desire additional items of clothing, over and above the amount of authorized issue.

Such a service will be provided only when there are positive assurances that employees will purchase and use any items of personal clothing that are made available under this arrangement.

2.5.4 Departments may purchase from Public Works and Government Services Canada a number of items at cost for resale to employees. These may include, but shall not be restricted to:

–   jacket, blazer and windbreaker,
–   trousers (work pants) and skirt,
–   shirt or blouse, turtleneck shirt,
–   tie,
–   socks,
–   gloves or mitts,
–   topcoat or other similar type of raincoat,
–   parka (non-distinctive),
–   belt,
–   scarf,
–   footwear,
–   maternity clothing.

## 2.6 **Clothing allowance**

2.6.1 As a general principle, the Treasury Board prefers the direct issue of clothing to the payment of clothing allowances. At the same time, Treasury Board does not wish to preclude payment of such allowances in cases where the practice is established or the economy of introducing a new allowance can be clearly demonstrated.

2.6.2  No new allowances or changes in existing allowances shall be introduced without the prior authorization of the Treasury Board.

2.6.3  No allowances shall be paid for:

– repair, cleaning, pressing and laundering;
– personal clothing.

## Identification

### 2.7  Requirement

2.7.1  Uniforms or other identifying items may be provided to employees. In general, the type of identification provided should correspond with identification requirements. Deputy heads are responsible for determining the requirements for identification items except when the design of a uniform is changed. In this case prior Treasury Board approval must be obtained.

2.7.2  The introduction of new uniforms, or changes to present departmental uniform policy, shall be subject to Treasury Board authorization.

2.7.3  The requirement of management for identification of the employee shall be determined by the degree to which the identification will aid in the effective performance of duties. Employees may be identified by the use of readily available identity cards or by a card at their work station in an office or other setting where special clothing would not be required.

2.7.4  Where the use of employees' full names represents a security problem, departments should use alternate forms of identification.

2.7.5  The amount of identification depends upon the continual contact of the employee with either the local, national or international population and the requirement for promotion of Canada-wide departmental services and promotion of a Canadian image. There are three distinguishing conditions under which identification of the employee may be required:

(a)  when identification of the employee is required by management, either permanently or in an emergency, to control emergency equipment and direct persons during an emergency. Such employees must be readily identifiable by the local public;
(b)  when identification of the employee is required by management to provide a sign of vested authority in directing, inspecting or enforcing specific laws and regulations;
(c)  when identification of the employee is required by management to provide an appropriate identification of the employee's function.

2.7.6  Items such as shirts, which are normally considered as personal clothing, may be provided as clothing when essential for a distinctive and consistent image as part of the identifying clothing.

2.7.7 Outer identifying clothing is provided only when the employee is required to wear it while on duty outdoors for a significant portion of the working period.

2.7.8 Clothing provided for identification may also serve to protect employees. When possible, avoid duplication of issue for identification and protection.

2.7.9 In some situations only one "identifier" will be required; in others, a combination of two or more may be necessary.

2.7.10 Identification clothing consistent with job requirements should be provided to probationers and casual or part-time employees. Items for identification may differ from those provided to full-time employees with the same job requirements (e.g. armband instead of headgear and tunic). The scale of issuance may also vary.

## 2.8 Local image

2.8.1 Clothing is provided when required for continual identification of employees, while on duty at the local level, when in continual direct contact with the local public whom they are serving.

2.8.2 Clothing for local image includes the following identifiers to wear with personal clothing:

– identification card, badge (i.e. for attachment to personal clothing),
– armband,
– headgear,
– smock or coveralls with identification markings,
– identification vest.

## 2.9 National or international image

2.9.1 Clothing is provided when required for identification of an employee while on duty as an official representative of the federal government and when formal identification of vested authority is required to aid the employee in the effective performance of duties. The appearance of the employee must be readily distinguishable from other employees working in the area and must also enhance the image of Canada.

2.9.2 Clothing for national or international image consists of uniform clothing of a distinctive design and includes:

– headgear,
– tunic,
– pants and skirt,
– outer identifying clothing including one of: parka, pea jacket, ski jacket, cape, overcoat, rainwear,
– badges or rank insignia that could vary with department and unit.

## 2.10 **Identification of members of the Canadian police and Armed Forces**

2.10.1  The departments concerned are authorized by legislation to provide personnel with uniform clothing for identification nationally and internationally. Special clothing for protection may also be provided to aid in the effective performance of duties.

2.10.2  The department normally shall indicate that, for the period of engagement (usually specified on enlistment), personnel will wear clothing provided for identification, operations and protection as required and make suitable arrangements for replenishment, either on exchange or through allowances. This excludes personal underclothing, sleepwear and clothing used primarily for social functions.

2.10.3  Clothing shall be provided when required for identification while on duty by a member of the police or Armed Forces of the federal government, and when formal identification of vested authority is required to aid the effective performance of duties.

2.10.4  The components of uniforms required will continue to be defined by departments, taking into account the guidelines which have been promulgated on the provision of clothing.

## CHAPTER 1-3 - APPENDIX A - CLOTHING ADVISORY SECTION

The Clothing Advisory Section of the Consumer Products Branch, Public Works and Government Services Canada, will:

(a)  provide information on commercially available commodities and advise on materials, apparel, their components and their availability;

(b)  produce information bulletins to be distributed to materiel managers dealing with all aspects of apparel purchasing, expected future costs and the latest available technology in the apparel fields;

(c)  advise on or produce purchase descriptions and specifications, including quality assurance requirements in both official languages;

(d)  evaluate the design of present and proposed apparel;

(e)  produce and arrange for new design apparel;

(f)  maintain contact with the Treasury Board Secretariat with respect to the Federal Identity Program. (Departments and agencies may consult FIP officials directly when this approach is desirable);

(g)  arrange for the production of samples;

(h)  assist in cost-benefit analyses against actual field performance of clothing commodities using commodity performance reports;

(i)  arrange for the testing of materials and apparel;

(j)  arrange for outside inspection services to be carried out at a plant or a consignee point;

(k)  promote the use of common terminology;

(l)  assist departments and agencies to follow Treasury Board guidelines on the provision of clothing to federal employees, with:

    –  guidance in the procurement of clothing according to the guidelines set out by the Treasury Board, the Federal Identity Program and according to national objectives, e.g. domestic purchases, regional considerations, scale of issue, economics, design, functionalism, protection, etc.;

–   assistance in fabric selection consistent with the demand for standardized fabrics;

–   forecast of fabric required to meet anticipated scale of issue, cost-benefit of maintaining inventories of fabric and garments, average allowance for normal maintenance;

–   cost estimates related to current prices falling within budgetary limitations as set out in the departmental objectives;

–   assistance with requisitions that clearly state to contracting officers the precise requirements, including purchase descriptions provided by the Clothing Advisory Section;

–   a critical path from first advice to garment delivery, showing involvement of all parties;

–   arrangement for consolidation and distribution of all clothing items;

(m)   act as the design authority when requested;

(n)   produce and arrange for computerized nested paper patterns and cutting markers.

## CHAPTER 1-4 - WORKPLACE FITNESS PROGRAMS

### Policy objective

To allow departments to establish workplace fitness programs.

### Policy statement

Where there is significant employee interest and a reasonable cost option has been determined to the satisfaction of the deputy head, departments may sponsor workplace fitness programs.

### Application

This policy applies to voluntary fitness programs in departments and agencies listed under Schedule I, Part I, of the *Public Service Staff Relations Act*.

### Policy requirements

Departments electing to sponsor workplace fitness programs must:

–   accept responsibility for the payment of all costs of the fitness program, including fit-up, operating and maintenance costs for the facilities associated with the program. In addition, **departments that do not pay rent** for their accommodation must pay to the custodian department providing their accommodation, the market value of the space dedicated to the fitness program. One time costs, and ongoing costs related to the space shall be paid directly to the custodian department in the first year. For the following years, an amount equivalent to the annual market value of the space utilized, including any associated operating costs, must be transferred to the custodian department through the MYOP process;

- ensure that user fees cover, or recover, incremental operating costs (e.g. special maintenance and security, exercise equipment, insurance, instructor training, staff salaries, etc.) where these are significant;
- ensure that work to set up facilities is authorized by custodian departments and any lease arrangement with third parties for space is made with custodian departments;
- ensure that these programs meet reasonable safety guidelines as advised by Fitness Canada (e.g. health history screening, consent forms indicating that participants accept the normal risks inherent in and incidental to fitness activities, competent instructors, the availability of first aid); and,
- ensure that commercial contractors, if any, have adequate liability insurance.

## Monitoring

Treasury Board Secretariat will periodically review the extent to which departments are respecting the requirements of this policy.

## References

Risk Management policy
Policy on Indemnification of Servants of the Crown
Volunteers policy
Claims policy

All of the above policies are in the Material, services and risk management volume of the *Treasury Board Manual*.

Program management guidelines and recommendations are available from Fitness Canada.

## Enquiries

Enquiries about this policy should be directed to the responsible officers in departmental headquarters who, in turn, may seek interpretation from:

### On general policy issues

Safety, Health and Employee Services Group
Staff Relations Division
Human Resources Policy Branch
Treasury Board Secretariat

### On real property issues

Bureau of Real Property Management
Administrative Policy Branch
Treasury Board Secretariat

**On financial issues**

> Program Branch
> Treasury Board Secretariat

**On risk management issues**

> Materiel and Risk Management Group
> Administrative Policy Branch
> Treasury Board Secretariat

**For information and advice on workplace fitness programs contact**

> Fitness Canada
> 10th Floor, Journal Tower South
> 365 Laurier Avenue West
> Ottawa, Ontario
> K1A 0X6

## CHAPTER 1-5 - SMOKING IN THE WORKPLACE

### Policy objective

To promote a safe and healthful working environment free, to the extent possible, of tobacco smoke.

### Policy statement

In the light of evidence on the health hazards of tobacco smoke and the employer's responsibility under the *Non-smokers' Health Act*, smoking is prohibited in Public Service workplaces.

### Application

This policy applies to all departments and other portions of the Public Service listed in Part I of Schedule I of the *Public Service Staff Relations Act*.

### Policy requirements

1. Departments must ensure that:

- no person smokes in the workplace (as defined in appendix A);
- all employees and the public are informed about the smoking prohibition in all parts of the workplace;
- appropriate signage, which meets legislative requirements (available from Public Works and Government Services Canada), is visibly displayed;
- reasonable measures are taken to minimize the effects of tobacco smoke coming from locations controlled by other employers or individuals who are not subject to this policy or the *Non-smokers' Health Act*;
- non-smoking employees are protected from exposure to tobacco smoke in residential training centres and other situations where the employer provides living accommodation or recreational facilities;

– consultation takes place with employees, through safety and health committees or safety and health representatives, if any, before designating smoking rooms in the limited situations described in the Guidelines.

2. Employees must:

– refrain from smoking in any part of the workplace, including hospitality rooms, except in smoking rooms designated by departments in accordance with this policy and the *Non-smokers' Health Act*; and
– request visitors and clients who are smoking to refrain from doing so in the workplace.

## Responsibilities

Departments and employees, in addition to the Treasury Board as employer, are subject to legal sanctions for failure to comply with the *Non-smokers' Health Act*.

Human Resources Development Canada and, in the case of transportation services, Transport Canada are responsible for enforcing the legislation.

## Monitoring

The Treasury Board Secretariat will assess departmental performance based on the results of investigations of complaints related to departmental application of this policy.

## References

*Financial Administration Act*, Section 7

*Non-smokers' Health Act*

*An Act to amend the Non-smokers' Health Act*

*Non-smokers' Health Regulations*

## Enquiries

Enquiries about this policy should be directed to the responsible officers in departmental headquarters who, in turn, may seek interpretations from:

### On safety and health issues

Safety, Health and Employee Services Group
Human Resources Policy Branch
Treasury Board Secretariat

**On staff relations issues**

Collective Bargaining Services
Staff Relations Division
Human Resources Policy Branch
Treasury Board Secretariat

**On smoking cessation programs**

Public Service Health Units in Departments
Occupational and Environmental Health Directorate
Medical Services Branch
Health Canada

Tobacco Programs Unit
Health Services and Promotion Branch
Health Canada

## CHAPTER 1-5 - APPENDIX A - DEFINITIONS

***Smoking***: to smoke, hold or otherwise have control over an ignited product manufactured from tobacco and intended for use by smoking (fumer).

***Smoking room*** (fumoir): a room or area that is designated as a smoking room by the department (excluding areas specified in the definition of workplace) in consultation with the safety and health committee or representative and is:

(a) enclosed by walls, a ceiling and floor:
(b) clearly identified as a smoking room;
(c) equipped with ashtrays or non-combustible covered receptacles for waste disposal;
(d) independently ventilated unless it is reasonably impractical to do so. In buildings constructed after December 31, 1989, designated smoking rooms must be separately ventilated with the air exhausted to the outside without being re-circulated within any workspace.

(b) and (d) do not apply to ships and motor vehicles.

***Workplace***: any indoor or enclosed space, under the employer's control, in which employees perform the duties of their employment. This includes any adjacent corridor, lobby, stairwell, elevator, cafeteria, washroom or other common area frequented by such employees during the course of their employment (lieu de travail).

Commercially-leased space for cafeterias is not considered to be under the employer's control. However, where the employer enters into a management contract with a person to operate a cafeteria on the employer's behalf, the cafeteria is deemed to be under the employer's control.

## CHAPTER 1-5 - APPENDIX B - GUIDELINES

### Smoking rooms

Departments may designate the following as smoking rooms:

- motor vehicles or separate Buildings that are normally occupied by one person and that do not share the ventilation system with any other workspace, e.g. lighthouses;
- a proportion of living accommodations or recreational facilities that are provided by the department for employees;
- hospitality rooms where official functions for non-employees are being held.

Departments may require employees to work in a designated smoking room if this is part of their duties;

### Smoking cessation courses

Departments may provide for interested employees smoking cessation programs that are acceptable to Health Canada.

## CHAPTER 1-6 - HUMAN IMMUNODEFICIENCY VIRUS (HIV) AND ACQUIRED IMMUNODEFICIENCY SYNDROME (AIDS)

### Policy objective

To provide guidance and facilitate understanding of the human immunodeficiency virus (HIV) and the acquired immunodeficiency syndrome (AIDS) in the workplace.

### Policy statement

In most work environments in the Public Service, employees with HIV infection or AIDS do not pose a health risk to others. As with other serious illnesses and disabilities, these employees are encouraged to remain productive as long as they are able. They must not be subject to discriminatory practices.

Employees of the Public Service are not required to undergo mandatory tests for HIV infection.

### Policy requirements

Departments must ensure that

- the rights and benefits of employees with HIV infection or AIDS are respected;
- the occupational safety and health of employees with a potential risk of exposure to the HIV is protected;
- all employees are informed of existing information, education, counselling and evaluation services in the Public Service with respect to HIV infection and AIDS.

## Application

The policy applies to all departments and other portions of the Public Service listed in Part I of Schedule I of the *Public Service Staff Relations Act*.

## Monitoring

Treasury Board Secretariat will monitor departmental performance by:

- reviewing the overall departmental application of the HIV and AIDS policy, the Occupational safety and health policy and the Employee assistance program policy;
- reviewing audit and evaluation reports, both internal and external, on the application of the policy.

## References

### Policies, directives and standards

Access to information and privacy, policies and guidelines, *Treasury Board Manual*, August 1992

Harassment in the workplace (TBM, Human Resources volume, chapter 3-2)

Disability Insurance Plan (TBM, Insurance and Related Benefits volume, chapter 3-3)

Long-term Disability Insurance (TBM, Insurance and Related Benefits volume, chapter 3-4)

Telework, TBM, Human Resources volume, chapter 2-4

Security volume, TBM

Occupational safety and health (TBM, Occupational Safety and Health volume, chapter 1-1)

Employee assistance program (TBM, OSH volume, chapter 1-2)

Clothing directive (TBM, OSH volume, chapter 1-3)

Dangerous substances directive (TBM, OSH volume, chapter 2-2)

Occupational health evaluations standard (TBM, OSH volume, chapter 2-13)

Personal protective equipment directive (TBM, OSH volume, chapter 2-14)

Procedures for accident investigation and reporting (TBM, OSH volume, chapter 4-1)

**Legislation**

*Canada Labour Code, Part II*

*Canadian Human Rights Act*

*Constitution Act 1982 (Canadian Charter of Rights and Freedoms)*

*Government Employees Compensation Act*

*Privacy Act*

*Public Service Employment Act*

*Public Service Staff Relations Act*

**Other**

AIDS and the *Privacy Act*, the Privacy Commissioner of Canada, 1989

The Human Rights Commission Policy on AIDS, May 1988

Bloodborne Pathogens in Health-Care Settings: Risk for Transmission, *Canada Communicable Disease Report*, Volume 18-24, Laboratory Centre for Disease Control, Ottawa, December 1992

Update: Universal Precautions for Prevention of Transmission of Human Immunodeficiency Virus, Hepatitis B Virus, and Other Blood-Borne Pathogens in Health Care Settings, *Canada Diseases Weekly Report* Volume 14-27, Federal Centre for Aids and Laboratory Centre for Disease Control, Ottawa, July 1988

Recommendations for Prevention of HIV Transmission in Health-Care Settings, *Canada Diseases Weekly Report* Volume 13S3, Federal Centre for AIDS and Laboratory Centre for Disease Control, Ottawa, November 1987

Public Service Health Bulletin No. 18 — Information on HIV Infection and AIDS for Public Service Employees, Health and Welfare Canada 1988 (under revision)

AIDS in the Workplace — Risky Business, film by Health and Welfare Canada 1988

AIDS — A Summary of Occupational Health Concerns, Canadian Centre for Occupational Health and Safety, Hamilton, 2nd edition

Counselling Guidelines for Human Immunodeficiency Virus Serologic Testing, Canadian Medical Association, March 1993

Act Now: Managing HIV and AIDS in the Canadian Workplace, Canadian AIDS Society, 1991

**Enquiries**

Enquiries about the policy should be directed to the responsible officers in departmental headquarters, who in turn, may seek interpretation from the following:

**For specific interpretations or direct questions on:**

- **Safety and health issues:**

  Safety, Health and Employee Services Group
  Human Resources Policy Branch
  Treasury Board Secretariat

- **Staff relations issues:**

  Collective Bargaining Services
  Staff Relations Division
  Treasury Board Secretariat

**For information and advice on HIV infection, and AIDS consult:**

- Occupational and Environmental Health Services Directorate
  Medical Services Branch
  Health Canada

- National AIDS Secretariat
  Health Canada

- National AIDS Clearinghouse
  Canadian Public Health Association

## APPENDIX A - DEFINITIONS

*Acquired Immunodeficiency Syndrome (AIDS)* — the terminal stage of a viral infection that affects the body's immune system, its natural resistance to disease. As a result, persons with AIDS are susceptible to life-threatening illnesses including rare forms of pneumonia, skin cancer and brain deterioration (syndrome d'immuno-déficience acquise (SIDA).

*Casual contact* — activities that bring someone in contact with another person or a common object (such as shaking hands, hugging, being close to someone who is coughing, sneezing or crying; touching common objects such as money, paper, door-knobs, telephones, or toilet seats; swimming in a public pool) (contact occasionnel).

*Employees with a potential risk* — employees whose work brings them in contact with human blood or other body fluids. This includes health-care workers such as nurses, doctors, dentists and laboratory workers; emergency and rescue personnel such as ambulance attendants, firefighters and first-aid attendants; and law enforcement personnel such as police officers and institutional guards. Consult references in the policy for other occupational groups (employés exposés au virus).

***Human Immunodeficiency Virus (HIV)*** — the virus that leads to AIDS. Many HIV-infected persons have few or no symptoms of illness (virus d'immuno-déficience humaine (VIH)).

## APPENDIX B - GUIDELINES

### Introduction

HIV is a fragile virus that cannot be spread by casual contact or by usual workplace activities. HIV infection is a chronic progressive disease transmitted by: sexual activity with an HIV-infected person; receiving HIV-contaminated blood or other body fluids into the body, for example by sharing needles and syringes contaminated with blood from an HIV-infected person; and an HIV-infected mother in the case of the unborn or newborn. No effective treatment for HIV infection or AIDS is available. However, new treatments for AIDS-related illnesses, are allowing persons with AIDS to live longer.

### Precautions for employees with a potential risk of exposure

As indicated in the policy, departments must ensure the safety and health protection of employees who have a potential risk of exposure to a causative agent.

(a) Where there exists a potential risk of exposure to HIV as well as to other blood-borne infectious agents, departments must establish and enforce infection control procedures recommended by Health Canada and provide appropriate protective clothing and equipment in accordance with the Clothing policy, the Personal protective equipment directive and the *Canada Labour Code*, Part II.

(b) Workplace exposure to blood or other body fluids that are potentially contaminated with the HIV shall be reported in accordance with TB Procedures for accident investigation and reporting (TBM, chapter 4-1).

(c) In addition, ongoing education and training in infection control must be provided for these potential risk employees in accordance with the Occupational safety and health policy for the Public Service and the *Canada Labour Code*, Part II.

### Employee rights and benefits

As indicated in the policy, departments must ensure that the rights and benefits of employees with HIV infection or AIDS are respected as well as those of co-workers and clients. They must ensure that:

(a) Any form of discrimination of an employee who has or is presumed to have HIV infection or AIDS is prohibited.

(b) The harassment of an employee with HIV infection or AIDS is handled in accordance with the procedure outlined in the Policy on harassment in the workplace.

(c) Records containing HIV or AIDS related personal information are designated as **PROTECTED** and handled in accordance with the provisions of the *Privacy Act* and the confidentiality requirements outlined in the Occupational safety and health policy, the Employee assistance program policy, and the Security policy of the Government of Canada. Reference may also be made to the Treasury Board Policy Guide with respect to access, collection, disclosure and accuracy of personal information.

However, employees, who wish to apply for disability benefits under the Disability Insurance Plan of the Public Service and the Long-term disability insurance portion of the Public Service Management Insurance Plan, will be required to provide complete medical information to the appropriate insurance company.

Similarly, employees, who wish to apply for compensation under the *Government Employees Compensation Act*, will be required to provide complete medical information to the appropriate provincial Workers' Compensation authority.

(d)   Work-related benefits such as sick leave provisions, medical and disability benefits, return-to-work privileges and leave for family-related responsibilities continue to be provided in accordance with existing Public Service policies and/or collective agreements.

(e)   As is the case with other health conditions, reasonable employment accommodation is made for employees with HIV or AIDS. Where possible and in accordance with the *Public Service Employment Act*, a suitable alternative position is offered when the employee can no longer meet the requirements of the position. The Occupational and Environmental Health Directorate, Health Canada, may assist in determining suitable employment.

(f)   Employees with HIV or AIDS are made aware of the CONFIDENTIAL counselling, information and referral services that are available, on a voluntary basis, through the occupational health services provided by Health Canada and through departmental employee assistance programs.

**Education and information**

As indicated in the policy, it is a departmental responsibility to inform employees of existing workplace services related to HIV and AIDS. Some available services are:

(a)   For those employees posted or travelling abroad on government business, departments should provide up-to-date reliable information on prevention measures either through departmental resources or the Occupational and Environmental Health Directorate (OEHD) of Health Canada.

(b)   Managers and other employees concerned about dealing with employees with HIV or AIDS should consult with the occupational health services for their department (form part of OEHD), their departmental employee assistance program practitioner or their safety and health committee.

(c)   As a means of alleviating workplace concerns and preventing further spread of HIV generally, departments may arrange for employee education sessions through the Occupational and Environmental Health Directorate of Health Canada.

## Testing

The Public Service does not require mandatory HIV testing as a condition of employment.

The Occupational and Environmental Health Directorate of Health Canada may provide voluntary HIV screening tests for employees assuming responsibilities on behalf of the government in foreign countries. Proof of HIV sero-negativity is required for entry into some countries.

The tests are also available to employees occupying potential risk positions or returning from countries where the virus is prevalent, if a risk of exposure has been identified.

Testing is carried out according to the principles of informed consent, pre-test and post-test counselling, and confidentiality of results.

## CHAPTER 2 - DIRECTIVES AND STANDARDS

### Introduction

Occupational safety and health (OSH) directives and standards are designed to protect employees from the specific hazards of their employment. They are based on good employer OSH practices and legal requirements, i.e., the *Canada Labour Code*, Part II and the pursuant Canada Occupational Safety and Health Regulations.

Directives have been the subject of consultation by the National Joint Council (NJC) and form a part of Public Service collective agreements. Collectively, they are subject to ongoing consultation and change; individually they are reviewed by the NJC every 6 years.

## CHAPTER 2-1 - BOILERS AND PRESSURE VESSELS DIRECTIVE

### General

### Collective agreement

This standard is deemed to be part of collective agreements between the parties to the National Joint Council and employees are to be afforded ready access to this standard.

### Grievance procedure

In cases of alleged misinterpretation or misapplication arising out of this standard, the grievance procedure, for all represented employees within the meaning of the *Public Service Staff Relations Act*, will be in accordance with Section 7.0 of the *National Joint Council By-Laws*. For unrepresented employees the departmental grievance procedure applies.

## Application

This standard incorporates the minimum requirements of the *Canada Labour Code*, Part II and applicable regulations issued pursuant to that legislation, and applies to all departments and other portions of the Public Service, as defined in Part I of Schedule I of the *Public Service Staff Relations Act*.

## Scope

Notwithstanding the scope of other federal government codes or standards concerning boilers and pressure vessels this standard is primarily concerned with Occupational safety. This standard shall have application in all government owned buildings occupied by Public Service employees. Where Public Service employees occupy buildings not owned by the federal government it shall be applied to the maximum extent that is reasonably practical. Privately owned facilities occupied by the Public Service are expected to comply with the applicable provincial or territorial requirements.

## Definitions

In this standard:

*Authorized inspection agency* means the provincial or territorial or other inspection agency which:

(a) employs qualified inspectors;
(b) meets the requirements of an "Authorized Inspection Agency" as defined in Section 3.2 of the ASME Boiler and Pressure Vessel Code;
(c) does not contract with any department or agency of the Public Service for the operation, repair or maintenance of boilers, pressure vessels or piping systems (organisme d'inspection autorisé);

*boiler* means a fired pressure vessel including flanged nozzles, screwed or welded connections in which gas or vapour may be generated or a gas, vapour or liquid may be put under pressure by heating (chaudière);

*code* means the CSA Standard B51-M1981 Code for the Construction and Inspection of Boilers and Pressure Vessels. The English version of which is dated March 1981 and the French version of which is dated September 1981 as amended to May 1984 (code);

*design* means the plans, patterns, drawings and specifications used for the fabrication of a boiler, pressure vessel or piping system (plan);

*designated inspection agency* means the provincial, territorial or other inspection agency engaged by the Minister to inspect boilers, pressure vessels or piping systems for specified geographic areas (organisme d'inspection désigné);

*fitting* means a regulating, controlling or measuring device subject to internal pressure and attached to a boiler, pressure vessel or piping system and includes a gauge cock, water column, feedwater level controller, pipe fittings and safety, stop-check, blowdown, continuous blowdown, soot blower, feed water, water treatment, drain vent and isolating valves (accessoire);

*major repairs* means repairs that may affect the strength of a boiler, pressure vessel or piping system (réparations importantes);

*maximum allowable working pressure* means the maximum allowable working pressure set out in the record of inspection (pression de service maximale autorisée);

*maximum temperature* means the maximum temperature set out in the record of inspection (température maximale);

*minister* means the Minister of Public Works and Government Services (ministre);

*operating authority* means the Public Service department or agency responsible for the operation and/or maintenance of a boiler, pressure vessel or piping system (autorité exploitante);

*piping system* means an assembly of pipes, pipe fittings, valves, safety devices, pumps, compressors and other fixed equipment which contains a gas, vapour or liquid and is connected to a boiler or pressure vessel (tuyauterie);

*pressure* means pressure in kilopascals measured above the prevailing atmospheric pressure (pression);

*pressure vessel* means a vessel, other than a boiler, that is used for containing, storing, distributing, processing or otherwise handling any gas, vapour or liquid under pressure and includes any piping system attached to the vessel (récipient soumis à une pression interne);

*provincial or territorial inspection agency* means the agency responsible for inspection, certification and registration of boilers, pressure vessels and piping systems under provincial or territorial jurisdiction in the geographical area in which a boiler, pressure vessel or piping system of the Public Service is located (organisme d'inspection provincial ou territorial);

*qualified inspector* means a person recognized under the laws of the province or territory in which the boiler, pressure vessel or piping system is located as qualified to inspect boilers, pressure vessels and piping systems (inspecteur compétent);

*qualified person* means, in respect of a specified duty, a person who because of his/her knowledge, training and experience is qualified to perform that duty safely and properly (personne compétente);

*regional director* means an officer designated by the Minister to administer the safety inspection program in the area in which a Public Service occupancy or establishment is located (directeur régional);

*record of inspection* means a record prepared by a qualified inspector in accordance with section 48 of this standard (dossier d'inspection);

*seal* means to take any measure necessary to prevent the unauthorized operation or use of a boiler, pressure vessel or piping system (fermer);

*welding* means welding in connection with the fabrication, alteration or repair of a boiler, pressure vessel or piping system (soudure).

## Requirements

### 1.1 Specific application and exclusion

1.1.1 Subject to this section every boiler, pressure vessel and piping system in the Public Service shall comply with the requirements relating to design, construction, installation, operation, maintenance, repair and inspection set out in this standard and in clauses 3.8, 3.9, 4.8 to 5.1, 5.3 to 6.3, 7.1 and 8.1 of the Code to the extent essential to the safety and health of employees.

1.1.2 This standard does not apply to:

(a) a heating boiler that has a heating surface of 3 square metres (30 square feet) or less;
(b) a pressure vessel that has a capacity of 40 L (1-1/2 cubic feet) or less;
(c) a pressure vessel that is installed for use at a pressure of 100 kPa (15 pounds per square inch) or less;
(d) a pressure vessel that has an internal diameter of 150 mm (6 inches) or less;
(e) a pressure vessel that has an internal diameter of 600 mm (24 inches) or less and is used for the storage of hot water;
(f) a pressure vessel that has an internal diameter of 600 mm (24 inches) or less and is connected to a water-pumping system containing gas that is compressed to serve as a cushion; and
(g) a refrigeration plant operating under pressure that has a capacity of 18 kW or less of refrigeration.

### 1.2 Design, construction and installation

1.2.1 Design, construction and installation of every boiler, pressure vessel and piping system used in the Public Service shall conform to the requirements of 1.1 of this standard.

1.2.2 Solid fuel fire tube boilers operating at a pressure over 103 kPa shall be provided with a fusible plug which meets the standards set out in Appendix A-19 to A-20.8 of Section 1 of the American Society of Mechanical Engineers Boiler and Pressure Vessel Code date July 1, 1983.

1.2.3 Every boiler and pressure vessel shall have at least one safety valve or other equivalent fitting to relieve pressure at or below the maximum allowable working pressure.

1.2.4 Where two or more boilers or pressure vessels are connected to each other and are used at a common operating pressure they shall each be fitted with one or more safety valves or other equivalent fittings to relieve pressure at or below the maximum allowable working pressure of the boiler or pressure vessel that has the lowest maximum allowable working pressure.

1.2.5 Every steam boiler that is not under continuous attendance by a qualified person shall be equipped with a low-water fuel cut-off device which serves no other purpose.

1.2.6 Where an automatically fired hot-water boiler is installed in a forced circulation system and is not under continuous attendance by a qualified person the boiler shall be equipped with a low-water fuel cut-off device.

1.2.7 Where two or more hot water boilers of the coil or fin-tube type are installed in one system, a low-water fuel cut-off device is not required on each boiler if:

(a)    the low-water fuel cut-off device is installed on the main water outlet header; and,
(b)    a flow switch which will cut off the fuel supply to the burner is installed in the outlet piping of each boiler.

1.2.8 Low-water fuel cut-off devices and flow switches shall be installed in such a manner that:

(a)    they cannot be rendered inoperative; and,
(b)    they can be tested under operating conditions.

## 1.3 Marking and identification

1.3.1 The operating authority shall ensure that the provisions of the Code are complied with in respect of the marking and identification of any boiler, pressure vessel or piping system used in the Public Service.

1.3.2 Where an existing boiler, pressure vessel or piping system lacks marking and identification data as required by the Code, the operating authority shall arrange for its inspection in accordance with 1.4 and shall ensure that the required marking and identification is provided.

## 1.4 Inspection and certification of new installations and major repairs

1.4.1 The operating authority shall ensure that the provincial or territorial inspection agency has access to all plans and specifications relating to a new installation or major repair of a boiler, pressure vessel or piping system.

1.4.2 Subject to this section no boiler, pressure vessel or piping system shall be operated or used following installation or major repair until the boiler, pressure vessel or piping system has been inspected and certified by the provincial or territorial inspection agency.

1.4.3 Where the provincial or territorial inspection agency is not prepared to provide the inspection and certification services referred to in this section the operating authority shall ensure that the new installation, or major repair, is inspected by an authorized inspection agency and that documentation acceptable to Human Resources Development Canada is obtained certifying that the newly installed or repaired boiler, pressure vessel or piping system complies with the requirements of this standard and with the Code to the extent essential for the safety and health of employees.

## 1.5 **Operation**

1.5.1 No person shall operate or use, or permit to be operated or used, a boiler, pressure vessel or piping system:

(a) unless it has been inspected by a qualified inspector in accordance with the requirements set out in 1.8 and a valid record of inspection has been issued in respect of that boiler, pressure vessel or piping system;

(b) unless every operator thereof is qualified in accordance with this section; and

(c) at a pressure higher than its maximum allowable pressure.

1.5.2 The operating authority shall ensure that:

(a) every boiler, pressure vessel or plant has at least one safety valve or other approved equivalent fitting to relieve pressure at or below its maximum allowable pressure;

(b) where two or more boilers or pressure vessels are connected to each other in a plant for use at a common operating pressure, they are each fitted with one or more safety valves or other approved equivalent fittings to relieve pressure at or below the maximum allowable pressure of the weakest boiler or pressure vessel in the plant as shown on the certificate of inspection for that boiler or pressure vessel;

(c) no person alters, interferes with or renders inoperative any fitting attached to a boiler, pressure vessel or plant, except for the purpose of adjusting or testing the fitting, and on instructions from the inspection agency.

1.5.3 The standards for control and supervision of the operation of boilers, pressure vessels and piping systems located in a province or territory are those standards established under the applicable provincial or territorial statute or ordinance.

1.5.4 Subject to the provisions of this section, the qualifications and requirements of an operator of a boiler, pressure vessel or piping system are those qualifications and requirements established under the applicable provincial or territorial statute or ordinance.

1.5.5 Any person employed as an operator who holds a valid Certificate of Qualification issued by any province or territory or a federal agency authorized to do so is considered qualified to operate a boiler, pressure vessel or piping system in any province or territory for which an equivalent certificate is required.

## 1.6 Repair, alteration and maintenance

1.6.1 Every boiler, pressure vessel and piping system shall be maintained and repaired by qualified persons.

1.6.2 All repairs and welding of boilers, pressure vessels and piping systems shall be carried out in accordance with the standards referred to in clauses 5.1, 6.1 and 7.1 of the Code.

1.6.3 Where, in the course of maintenance and repairs an employee is required to enter a boiler or pressure vessel the operating authority shall ensure that the requirements of the *Hazardous Confined Spaces Safety Standard* are adhered to.

## 1.7 Inspections general

1.7.1 A qualified inspector shall inspect every boiler, pressure vessel and piping system:

(a)   after installation;
(b)   after any welding, alteration or repair is carried out on it; and,
(c)   in accordance with this section, 1.8 and 1.9.

1.7.2 The Minister shall designate a provincial, territorial or other authorized inspection agency to carry out safety inspections of boilers, pressure vessels and piping systems for specified geographical areas in accordance with this section, 1.8 and 1.9 of this standard.

1.7.3 The designated inspection agency shall assign qualified inspectors to perform safety inspection of boilers, pressure vessels and piping systems in its geographical area.

1.7.4 Qualified inspectors employed by the designated inspection agency shall be furnished with accreditation by the Minister identifying them as safety inspectors authorized to carry out the inspections referred to in this section, 1.8 and 1.9 and, on producing his or her credentials shall, at any reasonable time, be permitted access to public Service facilities in order to inspect any boiler, pressure vessel or piping system.

1.7.5 Operating authorities shall provide to the Regional Director, a list of all boilers, pressure vessels and piping systems in their charge which are subject to the requirements of this standard and shall provide prompt notification of any additions or deletions to this list.

1.7.6 The operating authority shall ensure that, during any inspection of a boiler, pressure vessel or piping system, there is a person in attendance who is capable of taking all the necessary precautions to ensure the safety of the person making the inspection.

1.7.7  Subject to this section, 1.8 and 1.9, every boiler, pressure vessel and piping system in use shall be inspected by a qualified inspector as frequently as is necessary to ensure that it is safe for its intended use.

1.7.8  To the extent essential to the safety and health of employees every boiler in use in a workplace shall be inspected:

(a)  externally not less than once each year; and,
(b)  internally not less than once every two years.

1.7.9  Every pressure vessel in use in a workplace, other than a buried pressure vessel, shall be inspected:

(a)  externally not less than once each year; and
(b)  subject to this section, internally not less than once every two years.

1.7.10  Where a pressure vessel is used to store anhydrous ammonia, the internal inspection may be replaced by an internal inspection conducted once every five years if at the same time a hydrostatic test at a pressure equal to one and one half times the maximum allowable working pressure is conducted.

1.7.11  The factor of safety for a high pressure lap-seam riveted boiler shall be increased by at least 0.1 each year after 20 years of use and, if the boiler is relocated at any time, it shall not be operated at a pressure greater than 102 kPa.

## 1.8  Inspection and testing – Halon systems

1.8.1  Subject to this section Halon 1301 and Halon 1211 containers shall not be recharged without a test of container strength and a complete visual inspection being carried out if more than five years have elapsed since the date of the last test and inspection.

1.8.2  Subject to this section Halon 1301 and Halon 1211 containers that have been continuously in service without discharging may be retained in service for a maximum of 20 years from the date of the last test and inspection at which time they will be emptied, retested, subjected to a complete visual inspection and re-marked before being put back in service.

1.8.3  Where Halon 1301 or Halon 1211 containers have been subjected to unusual corrosion, shock or vibration, a visual inspection and a test of container strength shall be carried out.

1.8.4  A Halon 1301 or Halon 1211 container shall be tested by non-destructive test methods such as hydrostatic testing and shall be thoroughly dried before being filled.

## 1.9  Buried pressure vessels

1.9.1  Where a pressure vessel is buried the installation shall conform to the standards set out in clauses A1.1(a) to (g), (i) to (k) and (n) of Appendix A to the boiler code.

1.9.2 Before backfilling is done over a pressure vessel, notice of the proposed backfilling shall be given to the Regional Safety Officer of Human Resources Development Canada.

1.9.3 Where test plates are used as indication of corrosion of a buried pressure vessel the test plates and, subject to this section, the pressure vessel shall be completely uncovered and inspected by a qualified inspector at least once every three years.

1.9.4 Where the test plates on an inspection referred to in subsection 1.8.4 show no appreciable corrosion, the pressure vessel may be completely uncovered and inspected at intervals exceeding three years if the operating authority notifies the Regional Safety Officer of Human Resources Development Canada of the condition of the test plates and of the proposed inspection schedule for the pressure vessel.

1.9.5 Every buried pressure vessel shall be completely uncovered and inspected at least every 15 years.

## 1.10 **Records**

1.10.1 Where a boiler, pressure vessel or piping system has been inspected in accordance with sections 1.7 to 1.9 and the boiler, pressure vessel or piping system has been found to comply with the requirements of this standard, and of the Code to the extent essential for the safety and health of employees he/she shall forthwith issue a record of inspection.

1.10.2 Every record of inspection shall be signed by the qualified inspector who carried out the inspection and shall include:

(a)    the date of the inspection;
(b)    the identification and location of the boiler, pressure vessel or piping system that was inspected;
(c)    the maximum allowable working pressure and the maximum temperature at which the boiler or pressure vessel may be operated;
(d)    a certification that the boiler, pressure vessel or piping system meets the requirements of this standard and of the Code to the extent essential for the safety and health of employees;
(e)    a declaration as to whether in the opinion of the person carrying out the inspection, the boiler, pressure vessel or piping system is safe for its intended use; and,
(f)    any other observations that the person considers relevant to the safety and health of employees.

1.10.3 The operating authority shall keep every record referred to in this section for a period of ten years after the inspection is made, at the workplace in which the boiler pressure vessel or piping system is located.

1.10.4 Where a qualified inspector finds, on inspection of the boiler, pressure vessel or piping system, a condition that makes the operation of that boiler, pressure vessel or piping system unsafe the inspector shall notify the operating authority of the unsafe condition and direct that the use of the boiler, pressure vessel or piping system is prohibited. The inspector shall also direct that the boiler, pressure vessel or piping system is to be sealed in the manner prescribed and shall cancel the existing record of inspection and shall so advise the Regional Director.

1.10.5 Where the use of a boiler, pressure vessel or piping system has been prohibited it shall not be returned to service until repairs, if practical, have been completed and the boiler, pressure vessel or piping system has been inspected and certified by the provincial or territorial inspection agency in accordance with 1.4.

1.10.6 Where the use of a boiler, pressure vessel or piping system has been prohibited, and, in the opinion of the inspection agency, the boiler, pressure vessel or plant is not capable of being repaired or the operating authority does not wish to have it repaired, the operating authority shall specify a method of disposal that will effectively prevent further use of the boiler, pressure vessel or piping system in the Public Service.

1.10.7 The operating authority shall immediately notify the authorized inspection agency upon the discovery of any condition in a boiler, pressure vessel or piping system which may make the operation of the boiler, pressure vessel or piping system unsafe.

### 1.11 Accident reporting and investigation

1.11.1 The operating authority shall ensure that any accident or hazardous occurrence involving a boiler, pressure vessel or piping system is investigated and reported to Human Resources Development Canada.

1.11.2 Any damage to a boiler, pressure vessel or piping system which results in fire or the rupture of the boiler, pressure vessel or piping system shall be reported to a Safety Officer of Human Resources Development Canada within 24 hours.

1.11.3 No person shall disturb, destroy or alter any wreckage of a ruptured boiler, pressure vessel or piping system unless permission to do so is given by a Safety Officer of Human Resources Development Canada.

1.11.4 Notwithstanding this section, the wreckage of a ruptured boiler, pressure vessel or piping system may be moved to the extent necessary to allow the safe removal of an injured person.

## CHAPTER 2-2 - DANGEROUS SUBSTANCES DIRECTIVE

### Application

1. This standard applies to all Public Service departments and agencies, as defined in Part I of Schedule I of the *Public Service Staff Relations Act*.

2. This standard does not apply in respect of the transportation of dangerous substances over a public highway.

3. Regulations, including those concerning the use of radioactive substances, which are issued pursuant to the *Atomic Energy Control Act*, shall where applicable, take precedence over the provisions of this standard.

4. Requirements to safeguard the fire and explosion hazards of "dangerous substances" are the responsibility of the Fire Commissioner of Canada and are covered in Standards and Requirements issued by his Office under the authority of the *Government Property Fire Prevention Regulations*.

## Definitions

5. In this standard:

*dangerous substance* means any substance, that because of a property it possesses, is dangerous to the safety or health of any person who is exposed to it (substances dangereuses);

*environmental health officer* means a person so designated by Health Canada (agent d'hygiène du milieu);

*piping system* means an assembly of pipe, pipe fitting and valves, together with any pumps, compressors and other fixed equipment to which it is connected, that is used for transferring a dangerous liquid or gaseous substance from one location to another (système de tuyauterie);

*qualified person* means a person who because of his knowledge, training and experience is qualified to perform safely and properly a specified job (personne qualifiée);

*radiation emitting device* means any device that is capable of producing and emitting energy in the form of

(a)     electromagnetic waves having frequencies greater than ten megacycles per second (ten megahertz); and
(b)     ultrasonic waves having frequencies greater than ten kilocycles per second (ten kilohertz) (dispositif émettant des radiations);

*restricted area* means an area where explosives, flammable liquids, or flammable gases are stored, handled or processed or where the atmosphere contains or is likely to contain explosives concentrations of combustible dust or other combustible suspended material (secteur de danger);

*safety officer* means a person designated by the Minister of Human Resources Development pursuant to Part II of the *Canada Labour Code* (agent de sécurité).

## Substitution of non-dangerous substances

6. A dangerous substance or radiation emitting device shall not be used if it is reasonably practicable to use a substance or device that is not dangerous.

7. Where it is necessary to use a dangerous substance or a radiation emitting device and more than one kind of such substance or device is available, to the extent that it is reasonably practicable, the one that is least dangerous is to be used.

**Isolation and confinement**

8. Where operations involve the use of a dangerous substance or a radiation emitting device in any area, the use of that substance or device and any hazard resulting from that use are to be confined within that area, to the extent that is reasonably practicable.

9. Where operations require the storing of dangerous substances in any area, they are to be stored, to the extent that is reasonably practicable, in a manner that will prevent the transmission of the effect of an explosion, fire or other accident in that area to any adjacent area.

10. A dangerous substance shall not be stored near another substance if the potential danger of the dangerous substance is likely to be increased thereby.

11. To the extent that is reasonably practicable, the quantity of a dangerous substance in any area where it is being used, processed or manufactured should not exceed

(1)    the quality that is consistent with good industrial safety practice; or
(2)    the amount required for that area for one work day, whichever is the lesser.

**Control of airborne contaminants**

12. Any dangerous substance that may be carried by the air is to be confined as closely as is reasonably practicable to its source.

13. Subject to paragraph 14, each department shall ensure that the concentration of any dangerous substance that may be carried by the air in any area where an employee is working

(1)    does not exceed the threshold limit value recommended by the American Conference of Governmental Industrial Hygienists in its pamphlet "Threshold Limit Values for Airborne Contaminants 1976", as amended from time to time; or
(2)    conforms with any standard that follows good industrial safety practice, and is recommended by Human Resources Development Canada or Health Canada.

14. Except in respect of any dangerous substance that is assigned a Ceiling "C" value by the American Conference of Governmental Industrial Hygienists, it is permissible for the concentration of a dangerous substance that may be carried by the air in the area where an employee is working to exceed the threshold limit value described in paragraph 13 for a period of time calculated according to a formula that

(1)    is prescribed by the American Conference of Governmental Industrial Hygienists; or

(2)    is recommended by Human Resources Development Canada or Health Canada.

15. Where the atmosphere of any area in which an employee is working is subject to contamination by a dangerous substance, the atmosphere is to be sampled and tested by a qualified person as frequently

(1)    as may be necessary to ensure that the level of contamination does not at any time exceed the safe limits prescribed by paragraphs 13 and 14; or

(2)    as may be recommended by Human Resources Development Canada or Health Canada.

16. The sampling and testing referred to in paragraph 15 shall comply with

(1)    a method recommended by the American Conference of Governmental Industrial Hygienists, the American Society for Testing and Materials, the Fire Commissioner of Canada; or

(2)    any other sampling and testing method that follows good industrial safety practices, and is recommended by Human Resources Development Canada, Health Canada or the Fire Commissioner of Canada.

17. A record of each test made pursuant to paragraph 15 shall be retained for at least three years.

18. Every record referred to in paragraph 17 shall

(1)    be signed by the person who carried out the test;

(2)    be available at all reasonable times for examination by a safety officer or an environmental health officer; and

(3)    include the following data:
    (a)    the date, time and location of the test,
    (b)    the number of persons normally occupying the area tested,
    (c)    the dangerous substance for which the test was made,
    (d)    the type of testing equipment used,
    (e)    the result obtained, and
    (f)    the name and occupation of the person who made the test.

19. Where it is not reasonably practicable to prevent harmful exposure to a dangerous substance or radiation emitting device, personal protective equipment that will reduce such exposure to a safe level shall be worn and used.

**Warning and training of employees**

20. Each employee whose safety or health may be endangered by exposure to a dangerous substance or radiation emitting device is to be informed of the danger.

21. An employee shall not use or handle, or be permitted to use or handle, a dangerous substance or radiation emitting device where such use or handling would expose the employee to danger unless the employee has been instructed and trained

(1)    in the proper method to follow in order to minimize or control the danger; and
(2)    in the emergency procedures to follow in the event of an accident involving that substance or device.

22. The method referred to in paragraph 21 shall

(1)    be set out in writing;
(2)    follow good industrial safety practice; and
(3)    be readily available for examination by any employee to whom it applies, and a safety officer or an environmental health officer.

23. A record of any training provided to employees relating to paragraph 21 should be retained for at least three years and be available for examination at all reasonable times.

### Signs

24. Where a dangerous substance or radiation emitting device is handled, stored or used in any area in any manner that is dangerous to the safety or health of an employee who might be in that area, signs are to be posted to warn persons entering the area of that danger.

### Containers

25. Departments shall ensure that

(1)    every portable container for a dangerous substance that is used on its premises complies with a portable container specification prescribed for that dangerous substance in the Canadian Transport Commission *Regulations for the Transportation of Dangerous Commodities by Rail*, or with a portable container specification recommended by Human Resources Development Canada or Health Canada;
(2)    every stationary storage container for a dangerous substance that is used on its premises complies with a stationary storage container specification prescribed for that dangerous substance pursuant to a law of the province or territory in which the container is located, or with a stationary storage container specification recommended by Human Resources Development Canada or Health Canada;
(3)    every container for a radiation emitting device that is used on its premises complies with a container specification prescribed for that radiation emitting device by the Radiation Protection Bureau of Health Canada.

26. Every container of a dangerous substance that is used is, with respect to its contents, to be labelled, marked or tagged in accordance with

(1)  the Canadian Transport Commission *Regulations for the Transportation of Dangerous Commodities by Rail*;
(2)  the Manufacturing Chemists Association *Guide to Precautionary Labelling of Hazardous Chemicals*;
(3)  the requirements of the *Hazardous Products (Hazardous Substances) Regulations of Canada*, or any other labelling standard that identifies the dangerous substance in the container by its common name, and lists the principal danger or dangers of that substance.

## Ventilation

27.  Where there are a number of substances in the air in different areas of a workplace, a combination of which might cause a hazard, the air is to be exhausted from those areas in such a manner that the various substances are not combined.

28.  Exhaust and inlet ducts for ventilation systems are to be located and arranged so as to ensure that air contaminated with dangerous substances does not enter areas occupied by employees.

## Housekeeping

29.  Departments shall ensure that:

(1)  premises and equipment are, to the extent that is reasonably practicable, designated, constructed and maintained in a manner that will
     (a)  prevent the dust and waste from dangerous substances from accumulating in dangerous quantities, and
     (b)  facilitate the easy removal of the dust and waste referred to in paragraph 29(1)(a);
(2)  all dust, waste material and any spill of a dangerous substance is
     (a)  removed from its premises in such a manner and as frequently as will ensure a safe and healthful environment for employees, and
     (b)  disposed of in a manner that does not endanger the health and safety of any employee.

## Emergency equipment

30.  To the extent that is reasonably practicable, the following shall be provided:

(1)  emergency shower and eye washing equipment, where there is a danger of skin or eye injury from corrosive substances;
(2)  a fire blanket and a suitable portable fire extinguisher, where there is a danger of fire due to the presence of flammable liquids or gases, and meeting the requirements of the Fire Commissioner of Canada;
(3)  rescue equipment, where there is a danger that a toxic substance may be released into, or an oxygen deficient atmosphere created in, an area that would render any employee incapable of escaping without assistance;
(4)  a warning and detection system where the seriousness of any danger so requires; and

(5)     such other emergency equipment as may be necessary to ensure a standard of protection that is consistent with good industrial safety practice.

31. All equipment described in paragraph 30 shall be of a type and quantity that:

(1)     is recommended by the Canadian Standards Association, the American National Standards Institute or the Fire Commissioner of Canada; or
(2)     conforms with any other standard that is recommended by Human Resources Development Canada, Health Canada, or the Fire Commissioner of Canada.

## Combustible dusts

32. Combustible dust collectors are to be designed, installed, operated and maintained in accordance with the requirements of the Fire Commissioner of Canada.

33. The exterior surface temperature of pipes or ducts exposed to combustible dusts and insulation used on those pipes or ducts shall comply with the requirements of the Fire Commissioner of Canada.

## Restricted area

34. Measures and precautions concerning smoking, or any procedure or equipment the use of which in a restricted area may cause ignition or explosion of a dangerous substance, shall be in compliance with the requirements of the Fire Commissioner of Canada.

## Compressed air

35. An employee shall not use or be permitted to use compressed air for cleaning or any other purpose

(1)     where that use will result in a concentration of a dangerous substance in the atmosphere that is in excess of the prescribed safe limits referred to in paragraphs 13 and 14, or
(2)     in such a manner that the air is directly forcible against the body of the employee or any other person.

36. To the extent that is reasonably practicable, compressed air shall be used only with ventilated hoods or booths, or in areas where employees are protected fully from any dangerous substance or flying particles.

## General design of workplaces

37. To the extent that is reasonably practicable, the design and construction of every place in which a dangerous substance is manufactured, handled, stored, processed or used, shall be such that:

(1)   in an emergency, employees may be quickly evacuated;
(2)   where an accident is likely to result in a spill or leak of a dangerous substance or in a fire, the effect of such a spill, leak or fire on the safety and health of any employee is minimized; and
(3)   where a dangerous substance may explode, pressure resulting from any such explosion will be relieved in a manner that will prevent the explosive pressure from exceeding one pound per square inch (seven kilopascals).

38. Paragraph 37 does not apply to the handling, storing or using of a dangerous substance in a vehicle.

## Piping systems

39. Every piping system is to be:

(1)   adequate for its intended purpose, having regard to the corrosiveness, pressure, temperature and other properties of the dangerous substances that is being conveyed; and
(2)   fitting with valves and other control and safety devices sufficient to ensure the safe operation, repair and maintenance of the system.

40. Every valve and other control or safety device that is essential to the safe operation, repair or maintenance of a piping system is to be marked, tagged or otherwise identified by a system that follows good industrial safety practice and will assist in the safe use of the valve or other control or safety device.

41. Every person who operates, maintains or repairs a piping system or any part thereof is to be aware of the location of every valve and other control or safety device connected with that system and is to be trained in its proper and safe use.

## Radiation emitting devices

42. Every radiation emitting device to which any employee is exposed is to be:

(1)   registered with the Radiation Protection Bureau, Health Canada; and
(2)   designed, constructed, installed, maintained and used in accordance with a standard that is acceptable to the Radiation Protection Bureau.

## Electrical safety

43. Where dangerous substances are present in hazardous qualities in a location, all electrical facilities used in that location shall comply with:

(1)   the Canadian Standards Association *Canadian Electrical Code* Standard C22.1-1975, as amended from time to time; or

(2)    any other safety standard that conforms with good industrial safety
       practice and is recommended by Human Resources Development
       Canada.

44. Where there is a danger of ignition or explosion of a dangerous substance due to static electricity, such hazard shall be controlled in accordance with the requirements of the Fire Commissioner of Canada.

### Explosives

45. An employee shall not use, or be permitted to use dynamite, blasting caps or other explosives used in blasting unless he has in his possession a blaster's certificate that is issued:

(1)    under the authority of a provincial, territorial or municipal authority; or
(2)    by a qualified person recommended by the Regional Director of Human
       Resources Development Canada.

46. Where explosives are being used in an area that has been designated by a person in charge as a danger area, no unauthorized person, except a safety officer, shall enter that area.

47. Warning signs or guards shall be placed at the main entrances to any area referred to in paragraph 46 to warn persons of the danger in that area.

### Medical examinations

48. Medical examinations for employees exposed to dangerous substances shall be administered as required in accordance with the Occupational health evaluation standard, chapter 2-13.

49. Where recommended by Health Canada, appropriate records are to be maintained in respect to an employee's exposure to dangerous substances which may have an accumulative effect on the health of the employee.

### CHAPTER 2-3 - ELECTRICAL DIRECTIVE

#### Grievance procedure

In cases of alleged misinterpretation of misapplication arising out of this standard, the grievance procedure, for all represented employees, within the meaning of the *Public Service Staff Relations Act*, will be in accordance with Section 7.0 of the *National Joint Council By-Laws*. For unrepresented employees, the departmental grievance procedure applies.

This standard is deemed to be part of collective agreements between the parties to the National Joint Council and employees are to be afforded ready access to this standard.

#### Application

1. This standard incorporates the minimum requirements of the *Canada Labour Code*, Part II and applicable regulations issued pursuant to that legislation, and

applies to all departments and other portions of the Public Service, as defined in Part I of Schedule I of the *Public Service Staff Relations Act*.

## Exceptions

2. This standard does not apply to hearing aids, watches or other electrically powered devices that have an amperage and voltage that are not dangerous to employees.

## Definitions

3. In this standard:

***ampacity*** means current-carrying capacity expressed in amperes (ampacité);

***Canadian Electrical Code*** means

(a)  CSA Standard C22.1-1986, *Safety Standard for Electrical Installations*, dated January, 1986; and
(b)  CSA Standard C22.3 No. 1-M1979, *Overhead Systems and Underground Systems*, dated April, 1979 *(Code canadien de l'électricité)*;

***control device*** means a device that will safely disconnect electrical equipment from its source of energy (dispositif de commande);

***electrical equipment*** means equipment for the generation, distribution or use of electricity (outillage électrique);

***electrical shock*** is the effect produced on the body, and in particular the nerves, by an electrical current passing through it. The magnitude of the shock depends on current flow, usually measured in milliamperes (mA) rather than by voltage (measured in volts). It is possible to have extremely high voltages with little current and no injury occur in the voltage discharge. Appendix A shows the health effects of various currents.

A shock from DC has less muscular contraction effect than one from AC, the tendency being to cause the victim to violently withdraw from contact if possible, so that the period of contact is usually short (choc électrique).

***guarantee of isolation*** means, in respect of electrical equipment, a guarantee in writing by the person in charge that it is isolated (attestation de coupure de la source);

***guarantor*** means a person who gives a guarantee of isolation (garant);

***guarded*** means that an electrical equipment is covered, shielded, fenced, enclosed or inaccessible by location or otherwise protected in a manner that will prevent or reduce danger to any person who might touch or go near that equipment (protégé);

**high voltage** means a voltage of seven hundred and fifty-one volts or more between any two conductors or between any conductor and ground (haute tension);

**locked out** means, in respect of any electrical equipment, that the equipment has been rendered inoperative and cannot be operated or energized without the consent of the person who rendered it inoperative (verrouillé);

**person in charge** means a qualified person who supervises employees performing work in order to ensure the safe and proper conduct of an operation or of the work of employees (responsable);

**qualified person** means a person who, because of knowledge, training and experience, is licensed or otherwise qualified to perform safely and properly a specified job (personne qualifiée);

**qualified electrician** means a person who, because of knowledge, training and experience, is licensed and otherwise qualified to perform safely and properly a specified job (électricien compétent);

**safety ground** or safety grounding means a system of conductors, electrodes and clamps, connections or devices that electrically connect an isolated electrical facility to ground for the purpose of protecting employees working on the facility from dangerous electrical shock (prise de terre de sécurité);

**safety officer** means a person designated as a safety officer pursuant to the *Canada Labour Code*, Part II, and includes a regional safety officer (agent de sécurité);

**voltage** means the greatest root-mean-square voltage between any two conductors of an electrical circuit, or between any conductor of a circuit and ground and, in respect to a direct current electrical circuit, means the greatest voltage between any two conductors of the circuit or between any conductor of the circuit and ground (tension).

Departments shall ensure that requirements specified in the *Canada Occupational Safety and Health Regulations*, Part VIII, Electrical Safety Regulation, issued pursuant to the *Canada Labour Code,* Part II, are applied at every workplace occupied by employees.

### Design, construction, installation, operation, use, repair, maintenance and alteration

4. The design, construction and installation of all electrical equipment shall meet the standards set out in the *Canadian Electrical Code* to the extent that is essential for the safety and health of employees.

The operation and maintenance of all electrical equipment shall meet the standards set out in the *Canadian Electrical Code*.

5.  Where practicable, plans and specifications in respect of new electrical facilities and/or major alterations to existing facilities, including plans relating to the installation or relocation of equipment and the location and siting of work areas, shall be submitted to the appropriate municipal or provincial agency for review and comment prior to the commencement of such work.

**General precautions**

6.  No employee shall be permitted to install, modify, adjust, test, operate, repair or do any other similar work on electrical equipment, and no employee shall do any such work, unless:

(a)   the employee is a qualified person and the qualified person
   (i)    shall use such insulated protective equipment and tools that are necessary to prevent injury; and
   (ii)   shall be instructed and trained in the use of the insulated protective equipment and tools; or
(b)   the employee has been instructed and trained in
   (i)    the safe use of the tools and equipment required to do the work, and the safety precautions necessary to avoid injury to himself or herself and other employees;
   (ii)   uses such insulated protective equipment and tools necessary to prevent injury; and
   (iii)  does such work under the direct supervision of a qualified person.

7.  Where electrical equipment is not live, but is capable of becoming live, no employee shall work on that equipment unless it is completely isolated by a locking device and a safety ground is properly connected to that equipment and it is tagged as locked out.

8.  Where electrical equipment is live or is not properly isolated, no employee shall work on the equipment unless:

(a)   the employer has provided detailed instructions in procedures that are safe for work on live conductors and live equipment; and
(b)   if so required by paragraph 16, a safety watcher is present.

9.  Where employees are working on or near electrical equipment that is live, or is capable of becoming live, the person in charge shall ensure that the electrical equipment is guarded and warning signs attached, or that other measures acceptable to a safety officer are taken to protect persons from injury.

10.  Subject to paragraph 11, where it is not practicable for electrical equipment referred to in paragraph 9 to be guarded, the department shall take measures to protect the employee from injury by insulating the equipment from the employee or the employee from ground.

11.  Where live electrical equipment is not guarded or insulated in accordance with paragraphs 8 or 9 or where the employee referred to in paragraph 9 is not insulated from ground, no employee shall work so near to any live part of the electrical equipment, that is within a voltage range listed in column I of an item

of Table 1, that the distance between the body of the employee or any thing with which the employee is in contact and the live part of the equipment is less than:

(1)   the distance set out in column II of that item, where the employee is not a qualified person, or
(2)   the distance set out in column III of that item, where the employee is a qualified person.

12.   No employee shall work near a live part of any electrical equipment referred to in paragraph 11 where there is a hazard that an unintentional movement by the employee would bring any part of his or her body or any thing with which he or she is in contact closer to that live part than the distance referred to in that paragraph.

13.   A legible sign with the words "Danger — High Voltage" and "Danger — Haute Tension" in letters that are not less than 50 mm in height on a contrasting background shall be posted in a conspicuous place at every approach to live high voltage electrical equipment.

## Consent to work on high voltage electrical equipment

14.   No employee shall be permitted to work on any high voltage electrical equipment without the written consent of the person in charge of that equipment, except where the operation of the equipment is necessary to prevent loss of life, serious injury or extensive damage to property or equipment.

15.   No employee, other than a qualified person, shall enter alone or be permitted to enter any part of an electrical vault or station in which live high voltage electrical equipment is installed without the consent of the person in charge of that equipment.

## Safety watcher

16.(1)   Where an employee is working on or near live electrical equipment and, because of the nature of the work or the condition or location of the workplace, it is necessary for the safety of the employee that the work be observed by a person not engaged in the work, the employer shall appoint a safety watcher:

(a)   to warn all employees in the workplace of the hazard, and
(b)   to ensure that all safety precautions and procedures are complied with.

16.(2)   A safety watcher shall be:

(a)   a qualified person informed of the duties of a safety watcher and of the hazards involved in the work being carried out;
(b)   trained and instructed in procedures to follow in the event of an emergency;
(c)   authorized to stop immediately any work that is considered dangerous; and

(d)   free of any other duties that might interfere with the duties as a safety watcher.

## Coordination of work

17. Where an employee is working on or in connection with electrical equipment, that employee and every other person who is so working, including every safety watcher, shall be fully informed by the person in charge with respect to the safe coordination of their work.

## Protective clothing and equipment

18. No employee shall work on electrical equipment unless that employee uses such protective and insulated clothing and equipment as is necessary.

Determination of the protective clothing and equipment to be used shall be in accordance with the Personal protective equipment directive, chapter 2-14 and the Clothing directive as outlined in chapter 1-3.

19. Unless otherwise specified in writing by a safety officer, no employee shall work on or near live high voltage electrical equipment unless the employee is wearing outer clothing with full-length sleeves fastened at the wrists and that is fabricated from tightly woven natural wool, non-flammable material or some other material that is equally resistant to ignition.

## Testing of insulated clothing, equipment and tools

20. Every article of insulated protective clothing, insulated equipment and insulated device or tool referred to in this standard shall be so designed, constructed and maintained as to be safe, adequate and reliable under all conditions of intended use. Unless each article has been certified by a recognized testing agency before initial use, it shall be tested by a qualified person. Thereafter, it shall be tested annually by an approved method or more frequently as is necessary to ensure it retains its integrity.

21. Each time an article or piece of protective clothing, insulated equipment or an insulated device or tool passes a test, it shall be clearly marked to show the date of the test.

22. Where any protective clothing, equipment, device or tool fails a test, it shall be immediately removed from the service for which it was designed and tested, and so marked, tagged or disabled as to prevent its use until it has been repaired and passed the test.

23. Tests of rubber insulating gloves and mitts shall follow a procedure that complies with Canadian Standards Association Standard Z259.4-M1979, *Rubber Insulating Gloves and Mitts.*

24. Protective clothing, equipment, devices and tools shall be inspected by the user prior to use to ensure that each such item is safe for its intended use.

## Poles and elevated structures

25. No employee shall climb or be permitted to climb any pole or elevated structure used to support an electrical equipment unless qualified to do so, and until the employee has examined and tested that pole or structure and determined, from such examination, that the pole or structure is safe for climbing.

26. Where it appears that a pole or elevated structure requires temporary supports to be safe for climbing, such supports shall be installed, and pikepoles alone shall not be used.

27. No employee shall work or be permitted to work on a pole or elevated structure unless the employee is qualified, properly equipped, and instructed and trained in the rescue of persons who may be injured in such work and a safety watcher, determined to be required in accordance with paragraph 16, is present.

28. Every pole or elevated structure that is embedded in the ground and is used to support electrical equipment shall meet the standards set out in:

(1) CSA Standard CAN3-015-M83, *Wood Utility Poles and Reinforcing Studs*, dated January, 1983; or
(2) CSA Standard A14-M1979, *Concrete Poles*, dated September, 1979.

29. No employee shall climb or be permitted to climb or work from a pole or structure referred to in paragraph 28 that is located so near another structure or object, or has affixed to it any thing that is not part of the electrical equipment, which interferes with the safe climbing of the pole or structure or the safe conduct of work therefrom.

## Isolation of electrical facilities

30. Where work is to be performed on electrical equipment, and the equipment requires isolation to permit work or live tests to be performed thereon, or its isolation is changed or terminated, the special requirements contained in paragraphs 41 to 48 shall apply.

## Capacitors

31. Where a capacitor that has an ampacity and voltage that is dangerous to employees is disconnected from its source of electrical energy, no person shall short-circuit or apply a safety ground to the capacitor within five minutes of the time it was disconnected, unless the capacitor is already equipped with an adequate short-circuiting and grounding device.

32. Measures shall be taken to ensure that no person shall contact the terminals of a capacitor referred to in paragraph 31 unless the terminals are short-circuited and safety-grounded and a safety watcher, determined to be required in accordance with paragraph 16, is present.

33. The short circuit and safety ground on the capacitor referred to in paragraph 32 shall remain in position until any work on the capacitor that involves contact by an employee is completed, and all persons are clear of the work area.

## Battery rooms

34. Departments shall ensure that every room or area in which storage batteries that discharge flammable gases are electrically charged is adequately ventilated to prevent the accumulation of flammable gases, is as free as possible from all sources or causes of ignition, and is operated and maintained in accordance with good industrial safety practice.

35. Each battery charging room or area shall be marked at the entrance thereto with a sign containing the words "Danger — No Smoking or Open Flame" and "Défense de fumer et d'utiliser une flamme nue" or other similar words in letters not less than 50 mm in height on a contrasting background. An approved warning symbol conveying the same meaning as the words specified for the aforementioned sign may be used in lieu.

## Switches and control devices

36. The access to every electrical switch, control device or meter shall always be free from obstruction, and control devices shall be so designed and located as to permit quick and safe operations at all times.

37. High voltage electrical switches or other control devices shall not be installed, operated or used for any purpose other than that for which that equipment was specifically designed and approved.

38. Where, for safety reasons, it is necessary that any electrical switch or other device controlling a supply of electrical energy is to be operated only by certain authorized persons, the switch or other device shall be fitted with a locking device or controlled and tagged in such a manner that no unauthorized person can operate it.

## Conductive equipment

39. Metal rules, measuring tapes, metallic fish wire, wire-reinforced fabric tape, wire-bound hydraulic hoses, portable metal or metal-reinforced ladders or any similar electrically conductive equipment shall not be used so near to live electrical equipment that such conductive equipment may become live.

## Lightning protection

40. Lightning protection devices shall comply with Canadian Standards Association Standard B72-M87, *Installation Code for Lightning Protection Systems*.

## Isolation of electrical facilities

41. Before an employee isolates electrical equipment or changes or terminates the isolation of electrical equipment, the department shall issue written

instructions with respect to the procedures to be followed for the safe performance of that work.

42. The instructions referred to in paragraph 41 shall be signed by the person in charge and shall specify:

(1) the date and hour when the instructions are issued;
(2) the date and hour of the commencement and of the termination of the period during which the instructions are to be followed;
(3) the name of the employee to whom the instructions are issued; and
(4) where the instructions are in respect of the operation of a control device that affects the isolation of the electrical equipment:
  (a) the device to which the instructions apply, and
  (b) where applicable, the correct sequence of procedures.

43. A copy of the instructions referred to in paragraph 41 shall be shown and explained to the employee.

44. The instructions referred to in paragraph 41 shall be kept readily available for examination by employees for the period referred to in paragraph 42(2) and thereafter shall be kept by the department for a period of one year at the location nearest to the workplace in which the electrical equipment is located.

45. Subject to paragraph 48, no work on or live test of isolated electrical equipment shall be performed unless:

(1) isolation of the equipment has been confirmed by test; and
(2) the person in charge has determined on the basis of visual observation that every control device and every locking device necessary to establish and maintain the isolation of the equipment:
  (a) is set in the safety position with the disconnecting contacts of control devices safely separated or, in the case of a draw-out type electrical switch gear, is withdrawn to its full extent from the contacts of the electrical switch gear;
  (b) is locked out;
  (c) bears a distinctive tag or sign designed to notify persons that the operation of the control device and the movement of the blocking device is prohibited during the performance of the work or live test; and
  (d) is, where physically possible, locked or blocked in the safe position in such a manner that the position cannot be changed without the consent of the person in charge of the work or test;
(3) where it is appropriate and to the extent that it is reasonably practicable, isolation of the facility is confirmed by a test; and
(4) to the maximum possible extent, no person can inadvertently make the facility live while the work or test is in progress.

46. Where more than one employee is performing any work on or live test of isolated electrical equipment, a separate tag or sign for each such employee shall be attached to each control device and locking device referred to in paragraph 45.

47. The tag or sign referred to in paragraphs 45(2)(c) or 46 shall:

(1) contain the words "DO NOT OPERATE — DÉFENSE D'ACTIONNER" or display a symbol conveying the same meaning;
(2) show the date and hour that the control device and the locking device referred to in paragraph 45(2) were set in the safe position or were withdrawn to their full extent from the contacts;
(3) show the name of the employee performing the work or live test;
(4) where used in connection with a live test, be distinctively marked as a testing tag or sign;
(5) be removed only by the employee performing the work or live test; and
(6) be used for no purpose other than the purpose referred to in paragraph 45(2)(c).

48. Where, because of the nature of the work in which the electrical equipment is being used, it is not practicable to comply with paragraph 45, no work on or live test of electrical equipment shall be performed unless a guarantee of isolation referred to in paragraphs 49 and 55 is given to the person in charge.

**Guarantees of isolation for electrical equipment**

49. No employee shall give or receive a guarantee of isolation unless he or she is authorized by his or her department to give or receive a guarantee of isolation.

50. No more than one employee shall give a guarantee of isolation for a piece of electrical equipment for the same period of time.

51. Before an employee performs work on or a live test of isolated electrical equipment, the person in charge shall receive from the guarantor:

(1) a written guarantee of isolation, or
(2) where it is not practicable for him or her to receive a written guarantee of isolation, an oral guarantee of isolation.

52. A written guarantee of isolation referred to in paragraph 51(1) shall be signed by the guarantor and by the person in charge and shall contain the following information:

(1) the date and hour when the guarantee of isolation is given to the person in charge;
(2) the date and hour when the electrical equipment will become isolated;
(3) the date and hour when the isolation will be terminated if known;
(4) the procedures by which isolation will be assured;
(5) the name of the guarantor and the person in charge; and
(6) a statement as to whether live tests are to be performed.

53. Where an oral guarantee of isolation referred to in paragraph 51(2) is given, a written record thereof shall forthwith:

(1) be made by the guarantor; and

(2)    be made and signed by the person in charge.

54.  A written record referred to in paragraph 53 shall contain the information referred to in paragraph 52.

55.  Every written guarantee of isolation and every written record referred to in paragraph 53 shall be:

(1)    kept by the person in charge readily available for examination by the employee performing the work or live test until the work or live test is completed;
(2)    given to the department when the work or live test is completed; and
(3)    kept by the department for a period of one year after completion of the work or live test at a location nearest to the workplace in which the electrical equipment is located.

56.  Where a written guarantee of isolation or a written record of an oral guarantee of isolation is given to a person in charge and the person in charge is replaced at the workplace by another person in charge before the guarantee has terminated, the other person in charge shall sign the written guarantee of isolation or written record of the oral guarantee of isolation.

57.  Where the employees working on isolated electrical equipment are divided into two or more crews, each of which is supervised by a person in charge of work on the facility, each such person in charge shall obtain a guarantee of isolation before the crew is permitted to begin work.

58.  Before an employee gives a guarantee of isolation for electrical equipment that obtains all or any portion of its electrical energy from a source that is not under his or her direct control, the employee shall obtain a guarantee of isolation in respect of the source from the person who is in direct control thereof and is authorized to give the guarantee in respect thereof.

59.  Where electrical energy is supplied to electrical equipment from two or more sources which are under the control of other departments or employers, they may cooperatively agree that a guarantee of isolation for that electrical equipment may be given in respect of each source of energy which shall be designated in writing by the other parties or on behalf of one of the parties as the party responsible for giving the guarantee.

60.  The party having been designated pursuant to paragraph 59 as responsible for giving the guarantee may:

(1)    act as the guarantor; or
(2)    designate in writing one or more of its employees to act as the guarantor.

61.  Every agreement referred to in paragraph 59 shall state:

(1)    the identity of the equipment to which the agreement applies;
(2)    the period during which the agreement will remain in effect;
(3)    the date of the agreement; and

(4)   the name of the guarantor or guarantors, as the case may be, and shall be signed by the parties thereto.

62.  A copy of every agreement referred to in paragraph 59 in respect of any guarantee of isolation shall be readily available to the persons affected by the guarantee while the agreement remains in effect and thereafter be retained by the guarantor for at least one year and be readily available for examination by any such person or by a safety officer.

**Live test**

63.  No employee shall give a guarantee of isolation for the performance of a test on isolated electrical equipment where an auxiliary power source makes the equipment live unless:

(1)   any other guarantee of isolation given in respect of the electrical equipment for any part of the period for which the guarantee of isolation is given is terminated;
(2)   every person to whom the other guarantee of isolation referred to in sub-paragraph (1) was given has been informed of its termination; and
(3)   any live test to be performed on the electrical equipment will not be hazardous to the safety or health of the person performing the live test.

64.  For the purposes of this paragraph, where a guarantee of isolation for the performance of a live test of isolated electrical equipment is given to a person in charge of the test, that person shall, while the test is being performed, be deemed to be the person in charge of the tests and of any other work that is being performed on the equipment while the guarantee is in effect.

65.  Every person performing a live test shall warn all persons who, during or as a result of the test, are likely to be exposed to a hazard.

**Termination of guarantee of isolation**

66.  Every person in charge shall, when work on or a live test of isolated electrical equipment is completed:

(1)   inform the guarantor thereof; and
(2)   make and sign a record in writing containing the date and hour when he or she so informed the guarantor and the name of the guarantor.

67.  On receipt of the information referred to in paragraph 66, the guarantor shall make and sign a record in writing containing:

(1)   the date and hour when the work or live test was completed; and
(2)   the name of the person in charge.

68.  Each record made pursuant to paragraph 67 shall show:

(1)   the day and hour, according to the 24-hour clock, when the guarantee of isolation terminated;

(2) the name of the guarantor or any person who has assumed the guarantor's responsibilities;

(3) the person to whom the guarantee of isolation was given; and

(4) the date and hour in accordance with the 24-hour clock that the guarantor was notified that the guarantee was no longer required.

69. The records referred to in paragraph 66 and 67 shall be kept by the department for a period of one year after the date of signature thereof at a location nearest to the workplace in which the electrical equipment is located.

**Safety grounding**

70. No employee shall attach a safety ground to electrical equipment unless he or she has tested the electrical equipment and has established that it is isolated.

71. Paragraph 70 does not apply in respect of electrical equipment that is grounded by means of a grounding switch that is an integral part of the equipment.

72. Subject to paragraph 73, no work shall be performed on any electrical equipment in an area in which is located:

(1) a grounding bus,

(2) a station grounding network,

(3) a neutral conductor,

(4) temporary phase grounding, or

(5) a metal structure

unless the equipment referred to in subparagraphs (1) to (5) is connected to a common grounding network.

73. Where, after the connections referred to in paragraph 72 are made, a safety ground is required to ensure the safety of an employee working on the electrical equipment referred to in that paragraph, the safety ground shall be connected to the common grounding network.

74. No safety ground shall be attached or disconnected from isolated electrical equipment except in accordance with the following requirements:

(1) the safety ground shall, to the extent that is practicable, be attached to the pole, structure, apparatus or other thing to which the electrical equipment is attached;

(2) all isolated conductors, neutral conductors and all non-insulated surfaces of the electrical equipment shall be short-circuited, electrically bonded together and attached by a safety ground to a point of safety grounding in a manner that establishes equal voltage on all surfaces that can be touched by persons who work on the electrical equipment;

(3) the safety ground shall be attached by means of mechanical clamps that are tightened securely and are in direct contact with bare metal;

(4) the safety ground shall be so secured that none of its parts can make contact accidentally with any live electrical equipment;

(5) the safety ground shall be attached and disconnected using insulated protection equipment and tools;

(6) the safety ground shall, before it is attached to isolated electrical equipment, be attached to a point of safety grounding; and

(7) the safety ground shall, before being disconnected from the point of safety grounding, be removed from the isolated electrical equipment in such a manner that the employee avoids contact with all live conductors.

75. For the purposes of paragraph 74(2), a "point of safety grounding" means:

(1) a grounding bus, a station grounding network, a neutral conductor, a metal structure, or an aerial ground (static wire); or

(2) one or more metal rods that are not less than 16 mm in diameter and are driven not less than 1 m into undisturbed compact earth at a minimum distance of 4.5 m from the base of the pole, structure, apparatus or other thing to which the electrical equipment is attached or from the area where the persons on the ground work and in a direction away from the main work site.

76. Every conducting part of a safety ground on isolated electrical equipment shall have sufficient current carrying capacity to conduct the maximum current that is likely to be carried on any part of the equipment for such time as is necessary to permit operation of any device that is installed on the electrical equipment so that, in the event of a short circuit or other electrical current overload, the electrical equipment is automatically disconnected from its source of electrical energy.

77. Where there is a dispute regarding the term "qualified person" for purposes of an occupational safety and health standard, the following procedure shall be implemented:

(a) The employee shall raise the matter directly with the person in charge.

(b) The person in charge shall review the employee's qualifications and decide upon the employee's status as a qualified person.

(c) If the employee is dissatisfied with the decision, the matter shall be referred to the safety and health committee established for the employee's workplace.

(d) The safety and health committee shall review the matter and make appropriate recommendations to the person in charge.

(e) If the safety and health committee does not consider itself competent to deal with the case, it shall recommend an acceptable third party to the person in charge.

(f) The person in charge shall, pursuant to (d) or (e), take the recommendations into consideration, render a final management decision and undertake the appropriate action.

If the employee does not agree with the final decision which has been rendered, a grievance may be initiated pursuant to the NJC redress procedure.

# CHAPTER 2-3 - APPENDIX A - EFFECTS OF ELECTRICAL CONTACT

Major burns of increasing extent and severity.

Increasing risk of burns at exit and entry points.

75 to 200 milliamperes: Risk of death from Ventricular Fibrillation if current pathway goes through heart.

Risk of severe breathing difficulties.

Severe shock.

Risk of breathing difficulties due to muscular contractions.

Cannot release hand grip on conductor due to muscular contractions.

Painful sensation (electric shock).

Increasingly unpleasant sensation or shock.

Mild sensation.

Threshold of sensation.

Varies from person to person and points of skin contact.

# TABLE 1 - EFFECTS OF ELECTRICAL CONTACT

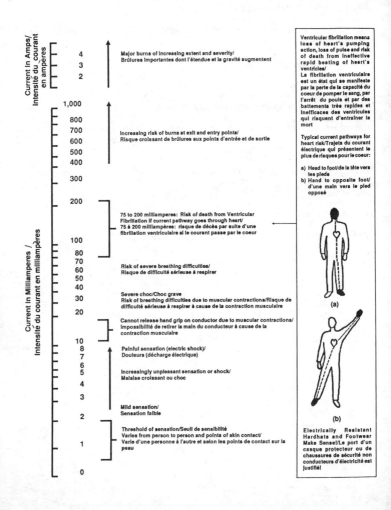

Current in Amps/
Intensité du courant
en ampères

4
3
2

Major burns of increasing extent and severity/
Brûlures importantes dont l'étendue et la gravité augmentent

Current in Milliamperes /
Intensité du courant en milliampères

1,000
800
700
600
500
400

Increasing risk of burns at exit and entry points/
Risque croissant de brûlures aux points d'entrée et de sortie

300

200

75 to 200 milliamperes: Risk of death from Ventricular
Fibrillation if current pathway goes through heart/
75 à 200 milliampères: risque de décès par suite d'une
fibrillation ventriculaire si le courant passe par le coeur

100
80
70
60
50
40

Risk of severe breathing difficulties/
Risque de difficulté sérieuse à respirer

30

Severe choc/Choc grave
Risk of breathing difficulties due to muscular contractions/Risque de
difficulté sérieuse à respirer à cause de la contraction musculaire

20

Cannot release hand grip on conductor due to muscular contractions/
Impossibilité de retirer la main du conducteur à cause de la
contraction musculaire

10
8
7

Painful sensation (electric shock)/
Douleurs (décharge électrique)

6
5

Increasingly unpleasant sensation or shock/
Malaise croissant ou choc

4
3

Mild sensation/
Sensation faible

2

Threshold of sensation/Seuil de sensibilité
Varies from person to person and points of skin contact/
Varie d'une personne à l'autre et selon les points de contact sur la
peau

1

0

Ventricular fibrillation means
loss of heart's pumping
action, loss of pulse and risk
of death from ineffective
rapid beating of heart's
ventricles/
La fibrillation ventriculaire
est un état qui se manifeste
par la perte de la capacité du
coeur de pomper le sang, par
l'arrêt du pouls et par des
battements très rapides et
inefficaces des ventricules
qui risquent d'entraîner la
mort

Typical current pathways for
heart risk/Trajets du courant
électrique qui présentent le
plus de risques pour le coeur:

a) Head to foot/de la tête vers
les pieds
b) Hand to opposite foot/
d'une main vers le pied
opposé

(a)

(b)

Electrically Resistant
Hardhats and Footwear
Make Sense!/Le port d'un
casque protecteur ou de
chaussures de sécurité non
conducteurs d'électricité est
justifié!

218

## CHAPTER 2-4 - ELEVATING DEVICES DIRECTIVE

### General

### Collective agreement

This standard is deemed to be part of collective agreements between the parties to the National Joint Council and employees are to be afforded ready access to this standard.

### Grievance procedure

In cases of alleged misinterpretation or misapplication arising out of this standard, the grievance procedure, for all represented employees within the meaning of the *Public Service Staff Relations Act*, will be in accordance with Section 7.0 of the *National Joint Council By-Laws*. For unrepresented employees, the departmental grievance procedure applies.

### Application

This standard incorporates the minimum requirements of the *Canada Labour Code,* Part II and applicable regulations issued pursuant to that legislation, and applies to all departments and other portions of the Public Service, as defined in Part I of Schedule I of the *Public Service Staff Relations Act*.

### Scope

Notwithstanding the scope of other federal government codes or standards concerning elevating devices this standard is primarily concerned with occupational safety. This standard shall have application in all government owned buildings occupied by Public Service employees. Where Public Service employees occupy buildings not owned by the federal government it shall be applied to the maximum extent that is reasonably practical. Privately owned facilities occupied by the Public Service are expected to comply with the applicable provincial or territorial requirements.

### Definitions

In this standard:

***authorized inspection agency*** means the provincial or territorial or other inspection agency which:

(a)   employs qualified inspectors;
(b)   meets the requirements of an "Authorized Inspection Agency" as defined in Section 3.2 of the ASME *Boiler and Pressure Vessel Code*;
(c)   does not contract with any department or agency of the Public Service for the operation, repair or maintenance of boilers, pressure vessels or piping systems (organisme d'inspection autorisé);

*code* means:

(a) for elevators, dumbwaiters, escalators and moving walks: CSA Standard CAN3-B44-M85 *Safety Code for Elevators* dated November 1985, other than clause 9.1.4;

(b) for manlifts, CSA Standard B311-M1979 *Safety Code for Manlifts*, the English version of which is dated October 1979 and the French version of which is dated July 1984, and Supplement No. 1-1984 to B311-M1979, the English version of which is dated April 1981 and the French version of which is dated December 1981 (code);

*design* means the plans, patterns, drawings and specifications of an elevating device (plan);

*elevating device* means a fixed mechanical device for moving passengers or freight and includes an elevator, dumbwaiter, manlift, escalator, inclined lift, moving sidewalk or other similar device (appareil de levage);

*major alterations* means those alterations set out in Section 10 of the Code (modifications importantes);

*maximum carrying capacity* means, with respect to an elevating device, the load that the elevating device is designed and installed to lift safely (capacité maximale de transport);

*minister* means the Minister of Public Works and Government Services (ministre);

*operating authority* means the Public Service department or agency responsible for the operation and/or maintenance of an elevating device (autorité exploitante);

*qualified person* means a person who, because of knowledge, training and experience, is qualified to perform safely and properly assigned duties with respect to the inspection and testing of safety devices or the inspection, repair and maintenance of elevating devices (personne compétente);

*record of inspection* means a record prepared by a safety inspector in accordance with section 1.3.6 of this standard (dossier d'inspection);

*regional director* means an officer designated by the Minister to administer the safety inspection program in the area in which a Public Service occupancy or establishment is located (directeur régional);

*safety device* means any device intended to aid in preventing the unsafe operation or use of an elevating device or manlift (dispositif de sécurité);

*safety inspector* means an employee of an inspection agency accredited by the Minister to perform safety inspections of elevating devices in Public Service facilities (inspecteur de sécurité);

*seal* means to take any measures necessary by a qualified person to prevent the unauthorized operation or use of an elevating device (fermer).

**Requirements**

### 1.1 Installation and alteration

1.1.1  No person shall undertake the installation or major alteration of an elevating device unless the design has been approved by the provincial or territorial jurisdictional authorities.

1.1.2  No elevating device or safety device shall be constructed, installed, or altered, unless the design, construction, installation, or alteration, as the case may be, meets the requirements of this standard and the Code.

1.1.3  Where a safety inspector is satisfied that an elevating device complies with the Code and has found on inspection that it is safe to operate to the extent that is essential for the safety of employees he or she shall forthwith issue a record of inspection and shall so advise the Regional Director.

### 1.2 Operation

1.2.1  No person shall operate or use an elevating device or permit an elevating device to be operated, used or placed in service:

(a)  unless a valid record of inspection has been issued in respect of that elevating device; or
(b)  with a load in excess of that which it was designed and installed to move safely.

1.2.2  Subject to subsection (c):

(a)  no elevating device shall be used or placed in service while any safety device attached thereto is inoperative;
(b)  no safety device attached to an elevating device shall be altered, interfered with or rendered inoperative;
(c)  subsections (a) and (b) do not apply to an elevating device or safety device which is being inspected, tested, repaired or maintained by a qualified person.

1.2.3  No person shall operate or use an elevating device or a manlift, or permit an elevating device or manlift to be operated or used, while a safety device connected thereto is inoperative, except for authorized testing purposes.

### 1.3 Inspection and testing

1.3.1  The Regional Director is responsible for ensuring that safety inspections are performed by qualified inspection agencies in accordance with the requirements set out in the Code and this standard.

1.3.2 Every elevating device and safety device attached thereto shall be inspected and tested by a safety inspector to determine that the requirements of section 6 are met:

(a) before the elevating device and safety device attached thereto is placed in service;
(b) after a major alteration to the elevating device or the safety device; and
(c) at least once every twelve months or more frequently if necessary to protect the safety and health of employees.

1.3.3 Safety inspectors shall be provided with accreditation by the Minister identifying them as persons qualified and authorized to perform safety inspections of elevating devices in accordance with this standard.

1.3.4 Operating authorities shall provide to the Regional Director a list of all elevating devices in their charge which are subject to the requirements of this standard and shall provide prompt notification of any additions or deletions to this list.

1.3.5 The operating authority shall, when requested by a safety inspector conducting an inspection or test pursuant to this standard, provide that person with an assistant who is capable of taking all precautions necessary to ensure that inspector's safety during the inspection or test and to otherwise assist in the safe conduct of the inspection or test.

1.3.6 A record of each inspection and test shall be provided by the safety inspector to the operating authority and shall:

(a) be signed by the inspector who carried out the inspection or test;
(b) include the date of inspection or test and the identification of the elevating device or safety device that was inspected;
(c) set out the observations of the person carrying out the inspection or test of the elevating device or safety device on their safety.

1.3.7 Every record of inspection shall be retained by the operating agency in the workplace in which the elevating device is located, for a period of two years after the date on which it is signed and shall be available at all reasonable times for inspection.

1.3.8 Where a safety inspector finds, on inspection, that an elevating device is not safe to operate the inspector shall:

(a) immediately seal the elevating device and so inform the operating agency that the use of the elevating device is prohibited;
(b) take possession of or cancel the certificate of inspection, if any; and
(c) advise the regional director.

1.3.9 The operating authority shall, upon discovery of any defect or condition in the elevating device that may render it unsafe to operate, immediately take the device out of service until repairs have been completed, inspected and a new record of inspection issued.

1.3.10 The operating authority shall ensure that maintenance and repair of elevating devices or safety devices attached thereto is performed by a qualified person in accordance with standards which comply with good industrial safety practice.

1.3.11 Maintenance or repairs on an elevating device shall only be performed by a qualified person.

## 1.4 Accident investigations and reporting

1.4.1 The operating authority shall ensure that, within 72 hours of any accident or other occurrence involving the use of an elevating device which endangers the safety or health of any person, a record is made of the accident or occurrence, which record shall be kept for at least ten years and be available for examination by a person entitled to do so.

1.4.2 Every accident or hazardous occurrence involving an elevating device shall be investigated in accordance with the requirements of *Canada Occupational Safety and Health Regulations*, part XV.

1.4.3 Subject to subsection 1.4.4, no person shall disturb, destroy or alter any wreckage of an elevating device or manlift without the permission of a Safety Officer of Human Resources Development Canada.

1.4.4 The wreckage of an elevating device or a manlift may be moved to the extent necessary to permit the safe removal of an injured person.

## CHAPTER 2-5 - FIRST-AID DIRECTIVE

### Grievance procedure

In cases of alleged misinterpretation or misapplication arising out of this standard, the grievance procedure, for all represented employees, within the meaning of the *Public Service Staff Relations Act*, will be in accordance with Section 7.0 of the *National Joint Council By-Laws*. For unrepresented employees, the departmental grievance procedure applies.

This standard is deemed to be part of collective agreements between the parties to the National Joint Council and employees are to be afforded ready access to this standard.

### Application

1. This standard incorporates the minimum requirements of the *Canada Labour Code*, Part II, and applicable regulations issued pursuant to that legislation and applies to all departments and other portions of the Public Service, as defined in Part I of Schedule I of the *Public Service Staff Relations Act*.

**Definitions**

2. In this standard:

*field party* means a field survey or field operations party, or a party operating in an area which is generally more than two hours travel time by usually available transportation from the nearest medical facility. However, in any unusual circumstances, a department may apply this term to parties operating at locations less than two hours travel time from such a facility (équipe de travail sur le terrain);

*first-aid* means emergency primary treatment or care that conforms with the recommended practice of the St. John Ambulance Association and that is provided by a department or agency in respect of an injury or illness of an employee arising out of or in the course of employment (premiers soins);

*first-aid attendant* means a person who is qualified in accordance with paragraph 9 and who provides first-aid services on a voluntary basis in conjunction with his or her regular duties (secouriste)

*first-aid kit* means an approved container with approved first-aid supplies (trousse de premiers soins);

*first-aid room* means a room provided by a department or agency to be used exclusively for purposes of administering first-aid (salle de premiers soins);

*first-aid station* means a place, other than a first-aid room, where a first-aid kit and equipment are located (poste de premiers soins);

*hazardous substance* includes a controlled product and a chemical, biological or physical agent that by reason of a property that the agent possesses is hazardous to the safety or health of a person exposed to it (substance dangereuse);

*health unit* means a space leased by Health Canada to provide occupational health nursing services (infirmerie);

*high voltage* means a line voltage of 751 volts or more between any two conductors or between any conductor and ground (haute tension);

*medical treatment facility* means a hospital, medical clinic or physician's office (installation de traitement médical);

*motor vehicle* means any motorized vehicle that is used for the transportation of material or passengers (véhicule à moteur).

3. Unless otherwise indicated in this standard, all references to Health Canada are to Occupational and Environmental Health Directorate, Medical Services Branch, Health Canada.

## Departmental responsibilities

4. Departments and agencies are responsible to provide first-aid services to employees in accordance with the requirements of this standard. Health Canada Occupational Health nurses will respond to emergency situations depending on their availability.

Where an employee's normal work is located beyond departmental premises, the department concerned shall ensure that first-aid services are available to such an employee or that employees have access to first-aid treatment.

Where an employee's normal work is located beyond departmental premises, departments shall in consultation with the safety and health committee or representative establish procedures respecting the availability of first-aid services.

## First-aid treatment and reporting

5. Any employee, upon sustaining an injury or sudden illness while at work, shall, where possible, report to a first-aid attendant for treatment and, as soon as practicable thereafter, supply all pertinent information relative to the injury or illness to the person in charge of the work. Where it appears that a physician's attention may be required, the employee shall be promptly referred to a medical treatment facility, and the department shall ensure that suitable transportation and escort, if required, are arranged. For information on transportation costs refer to the *Treasury Board Manual*, Employee Services Volume.

6. Departments and agencies shall maintain, at each place of employment, a written record of every injury or illness which requires first-aid treatment, and such record shall be maintained for ten years following treatment. Treatment records shall include:

(1) the full name or the person receiving treatment;
(2) the date, time and location of the occurrence of the injury or illness;
(3) the date and time that the injury or illness was reported;
(4) a brief description of the nature of the injury or illness; and
(5) a brief description of the treatment rendered and any arrangements made relating to the person treated.

7. Each record of entry shall be signed by the first-aid attendant or person rendering first-aid. Such records shall be maintained in a first-aid attendant's treatment record book. Records of treatment shall be inspected by a responsible departmental official and the safety and health committee/representative for the workplace at three-month intervals to verify their proper maintenance.

## First-aid attendants

8. Departments shall ensure that an adequate number of qualified first-aid attendants are available to render first-aid to employees during working hours:

(1)   at least one attendant shall be available at all times during each shift or working period at a location;

(2)   for field parties, at least two first-aid attendants shall be included in each main party, and at least one attendant among the members of each branch party;

(3)   where an employee is working on live high voltage electrical equipment:

    (a)   a first-aid attendant shall be readily available; or

    (b)   at least one employee shall have the training necessary to provide resuscitation by mouth-to-mouth resuscitation, cardio-pulmonary resuscitation or other direct method;

(4)   the selection, location and training of first-aid attendants shall be determined in consultation with the workplace safety and health committee or representative.

9.   Each first-aid attendant shall hold a valid St. John Ambulance Standard First-Aid Certificate, or an equivalent level of first-aid certification which is acceptable to Health Canada. Approved resuscitation methods shall be taught to the attendants as required. Departments shall maintain a record of each attendant's first-aid qualification and ensure that certification is kept up-to-date as required by the issuing organization. Lists containing the names, certification status and the location of first-aid attendants shall be maintained in an up-to-date status.

10.   Departments shall ensure that all first-aid attendants are made aware of the contents of the policy entitled "Indemnification of Servants of the Crown" (Part III, chapter 2 of the "Materiel, Services and Risk Management" volume, *Treasury Board Manual*).

11.   Notwithstanding additional responsibilities that may be assigned by a department, first-aid attendants shall be responsible for:

(1)   ensuring the day-to-day good order and maintenance of first-aid kits, supplies and related equipment and records;

(2)   providing first-aid treatment within the scope of their competence; and,

(3)   referring an employee to a medical treatment facility, where required.

## First-aid training

12.   Departments shall normally make arrangements for the first-aid training and certification of employees through the appropriate officer of the St. John Ambulance Association or any other organization acceptable to Health Canada. Where consultation with the safety and health committee or representative has determined that cardio-pulmonary resuscitation training is required, the appropriate level of such training shall be provided.

13.   First-aid training shall be provided for employees who have been designated by the department as requiring such training, and where the incumbents voluntarily agree to act as first-aid attendants. The positions so identified will depend upon several factors, including the number and location of first-aid stations, first-aid rooms and health units, and the degree of hazard of the work.

14. Where unusual and variable occupational hazards may exist, such as those found in laboratories or during field operations in isolated areas, Health Canada shall be consulted in regard to specialized first-aid training and/or equipment which may be required.

15. For field operations in isolated areas, specialized first-aid training will include "Field Party Advanced First-Aid, Level 1" as offered by the St. John Ambulance Association or any other organization acceptable to Health Canada.

**First-aid supplies and equipment**

16. Except where a first-aid room is operating reasonably close to the work, departments shall provide and maintain at least one first-aid station at each place of employment. First-aid kits for placement at these fixed locations shall contain the items detailed in appendix A, such items to be obtained from Public Works and Government Services Canada.

17. First-aid kits are to be provided in accordance with the following scale:

(1)   1 to 5 employees — one type "A" first-aid kit;
(2)   6 to 19 employees — one type "B" first-aid kit;
(3)   20 or more employees — one type "C" first-aid kit.

18. Where necessary the kits shall include supplies for protection against infectious disease.

19. Where a hazard from a hazardous substance exists in the workplace, shower facilities to wash the skin and eye-wash facilities to irrigate the eyes shall be provided for immediate use by employees.

Where it is not practicable to comply with the above, portable equipment that may be used in place of the facilities referred to above shall be provided.

The design and installation of emergency eye-wash and shower facilities shall comply, as a minimum, with the American National Standards Institute (ANSI) Standard ANSI Z358.1-1981.

20. A motor vehicle used by a department or agency shall be equipped with a type "A" first-aid kit. In the case of smaller vehicles such as snowmobiles, pocket first-aid kits referred to in paragraph 21(3) shall be provided.

In addition, for operations in the field, auto emergency kits shall be provided.

21. For operations in the field, parties shall be equipped with first-aid kits as detailed in appendix B, such kits to be provided in accordance with the following requirements:

(1)   main party — one standard first-aid kit;
(2)   each party detached from the main party — one intermediate first-aid kit; and,

(3)    individual members who are isolated during operations — one pocket first-aid kit.

## First-aid rooms/stations

22. A first-aid room shall be provided to serve a location where there are 200 or more employees working at any one time. A first-aid room may be provided to serve a lesser number of employees, if justified according to the types of operations and the injury hazard experience at the location. A first-aid room is not required where a health unit or a similar emergency treatment facility is conveniently available to provide first-aid services.

**Note**: First-aid attendants may have use of the health unit premises in an emergency situation in the absence of the nurse. Entry access to a Health Canada facility by a first-aid attendant must be controlled by a responsible officer who shall ensure that material and equipment which must be exclusively used by a health professional, and medical files and health protected documents, cannot be accessed.

23. Where, at a location, the total number of employees of more than one department substantiates the need for a first-aid room, a common first-aid room may be established under coordinated control as agreed upon locally between the departments concerned. Should a common first-aid room prove impracticable, first-aid stations as required by paragraph 14 shall be established by the individual departments.

24. First-aid rooms shall be maintained in a neat and sanitary condition, be located within easy access of male and female toilet facilities, and be situated convenient to the main working areas.

25. A first-aid room shall have a minimum floor area of 15 square metres and shall be provided with:

(1)    adequate lighting, heating and ventilation;
(2)    a sink and hot and cold running water;
(3)    liquid soap and dispenser;
(4)    a separate cubicle or curtained-off area with a cot or bed;
(5)    a cabinet or cupboard space with a lock, suitable for the storage of first-aid supplies;
(6)    a suitable table and several chairs;
(7)    paper towels and dispenser;
(8)    paper cups and dispenser;
(9)    a telephone, or continuous access to an adjacent telephone;
(10)  a type "A" first-aid kit and flashlight for use at the scene of an accident; and
(11)  first-aid supplies in accordance with appendix C.

## Emergency communications

26. All appropriate names, work locations (addresses) and telephone numbers which may be required in respect of any emergency shall be conspicuously

posted at each first-aid station and first-aid room, and such numbers shall, as a minimum, include the following:

(1) first-aid attendant;
(2) emergency transportation (including taxis);
(3) medical treatment facility;
(4) fire department;
(5) police department; and
(6) poison control centre.

27. Communication by land-line or radio shall be established between field parties and those facilities which can provide emergency medical advice, assistance or rescue services, including those operated by the Medical Services Branch, Health Canada. Communications shall also be maintained between main camps and parties working out of such camps, whenever possible.

## Location of first-aid facilities

28. The direction to, and location of, each first-aid station and first-aid room shall be indicated by symbols in accordance with requirements specified in the *Federal Identity Program Manual*.

## Field operations

29. Before proceeding on field operations, the person in charge of a field party shall:

(1) ensure that the required number of first-aid attendants is available;
(2) obtain the required first-aid kits and other first-aid supplies required under this standard; and,
(3) contact the medical treatment facility nearest the intended work area to arrange for emergency services. Normally, the appropriate regional Medical Services Branch office of Health Canada shall be contacted for this purpose.

30. When parties will be operating under conditions which may require special supplies beyond those considered as normal first-aid requirements, departments shall obtain the approval of Health Canada before such supplies are acquired.

31. Whenever a camp is to be established as a base for field operations, the person in charge of the party shall ensure that arrangements have been made for emergency evacuation of casualties and for the communication procedures required to obtain medical advice and/or assistance, and that all members of the party have been advised of such arrangements.

## Other first-aid matters

32. Departments shall, where necessary, consult with Health Canada concerning approval and direction respecting:

(1)    first-aid matters not specifically covered by this standard;
(2)    the interpretation and application of existing first-aid requirements; and,
(3)    the provision of specific first-aid supplies and equipment not detailed in
       this standard.


## CHAPTER 2-5 - APPENDIX A - GENERAL PURPOSE FIRST-AID KITS

| DESCRIPTION | QUANTITY | STOCK NUMBER (PWGSC) |
|---|---|---|
| *First-Aid Kit, general purpose, Type "A", complete | 1 | 6545-21-852-9432 |
| First-Aid Kit, general purpose, Type "B" | 1 | 6545-21-852-9433 |
| First-Aid Kit, general purpose, Type "C" | 1 | 6545-21-852-9434 |

| KIT CONTENTS | A | B | C | |
|---|---|---|---|---|
| Adhesive tape, surgical, 7.5 cm x 4.6 m | 1 | 1 | 2 | 6510-00-203-5000 |
| Applicator, disposable, 25s | 1 | 1 | 1 | 6515-21-852-9428 |
| Bandage, adhesive, 100s | 1 | 1 | 1 | 6510-21-845-2239 |
| Bandage, felt, orthopaedic | — | 2 | 2 | 6510-21-116-0170 |
| Bandage, gauze, 5.0 cm x 4.6 m | — | 6 | 8 | 6510-21-116-0174 |
| Bandage, self-adhering roller, 7.5 cm x 4.6 m | 4 | 6 | 8 | 6510-21-845 2200 |
| Bandage, self-adhering roller, 10 cm x 4.6 m | — | 6 | 8 | 6510-21-845-2201 |
| Bandage, triangular, 2s | 1 | 3 | 6 | 6510-21-880-9702 |
| **Basin, wash | — | — | 2 | 6530-21-846-9260 |
| **Blanket, bed, grey | — | — | 2 | 7210-21-849-9452 |
| Book, Pocket Guide to First-Aid | 1 | 1 | 1 | 7610-21-843-6190 |
| Book, Record | 1 | 1 | 1 | 7530-21-852-9254 |
| Case, First-Aid Kit | 1 | — | — | 6545-21-852-9431 |
| Case, First-Aid Kit | — | 1 | — | 6545-21-852-9429 |
| Case, First-Aid Kit | — | — | 1 | 6545-21-852-9430 |
| Cotton, purified, 28.0 g | 2 | 4 | 16 | 6510-21-116-0197 |
| Depressor, tongue, 25s | 1 | — | 1 | 6515-31-852-9427 |
| Dressing, first-aid, field | 2 | 2 | 3 | 6510-21-102-7867 |
| Dressing, surgical, combination | — | 2 | 3 | 6510-21-849-9539 |
| Gloves, disposable (pair) | 2 | 4 | 6 | 6515-21-845-9482 |
| Forceps, splinter | 1 | 1 | 1 | 6515-00-337-2400 |
| **Litter, folding | — | — | 1 | 6530-21-848-4908 |
| Mask, resuscitation with disposable one-way valve | 1 | 1 | 1 | 6515-21-904-7049 |
| Pad, cotton, eye | 1 | 2 | 4 | 6510-21-845-2189 |

| Description | | | | Stock Number |
|---|---|---|---|---|
| Pin, safety, 9s | 1 | 1 | 2 | 8315-21-843-6856 |
| ***Povidone Iodine swabs (10s) | 1 | 3 | 3 | 6510-21-877-0825 |
| Scissors, bandage | 1 | 1 | 1 | 6515-21-846-0206 |
| Shield, eye, surgical | 1 | 2 | 4 | 6515-21-116-3164 |
| Splint set, wood | — | 1 | 1 | 6545-21-116-2912 |
| Sponge, surgical, 5.0 cm x 5.0 cm, 2s | 3 | 6 | 12 | 6510-21-845-2171 |
| Sponge, surgical, 10.0 cm x 10.0 cm, 2s | 3 | 6 | 12 | 6510-21-845-2440 |
| | | | | |
| * Additional for motor vehicles: Dressing, first-aid, field | 2 | — | — | 6510-21-102-7867 |

\*\* Item is not included in kit,
and must be ordered separately.

\*\*\* Shelf-life of 18 months maximum
from date received for kit.

## CHAPTER 2-5 - APPENDIX B - FIELD PARTY FIRST-AID KITS

| DESCRIPTION | ABBREVIATION |
|---|---|
| *Standard First-Aid Kit | S |
| *Intermediate First-Aid Kit | I |
| *Pocket First-Aid Kit | P |

| KITS CONTENTS | QUANTITY | | | STOCK NUMBER (PWGSC) |
|---|---|---|---|---|
| Acetaminophen tablets, 100s | 1 | 1 | — | 6505-21-870-6175 |
| Adhesive tape, surgical, 7.5 cm x 4.6 m | 1 | 1 | — | 6510-00-203-5000 |
| Aluminum hydroxide and magnesium carbonate gel tablets, 50s | 2 | 2 | — | 6505-21-857-6473 |
| Applicator, disposable, 100s | 1 | — | — | 6515-21-844-5204 |
| Bandage, adhesive, butterfly closure, 100s | 1 | 1 | — | 6510-21-845-2182 |
| Bandage, adhesive, 25s | — | 1 | 1 | 6510-21-845-2238 |
| Bandage, adhesive, 100s | 1 | — | — | 6510-21-845-2239 |
| Bandage, cotton, elastic, 7.5 cm | 4 | 2 | — | 6510-21-845-2449 |
| Bandage, self-adhering roller 7.5 cm x 4.6 m | 6 | 4 | 2 | 6510-21-845-2200 |
| Bandage, triangular, 2s | 8 | 4 | 1 | 6510-21-880-9702 |
| Bath, eye | 1 | — | — | 6515-21-844-5214 |
| Blanket, emergency, pocket | 2 | 2 | 1 | 7210-21-870-6172 |
| Book, First-Aid, English | 1 | — | — | 7610-21-848-3664 |
| Book, First-Aid, French | 1 | — | — | 7610-21-848-3665 |
| Book, Pocket Guide to First-Aid | — | 1 | 1 | 7610-21-843-6190 |
| Brush, scrub, nail | 1 | — | — | 7920-21-116-2811 |

| Description | | | | Stock Number |
|---|---|---|---|---|
| Calamine lotion | — | 1 | — | 6505-21-870-6173 |
| Case, First-Aid Kit | 1 | — | — | 6545-21-852-9429 |
| Case, thermometer | 1 | — | — | 6515-21-857-9045 |
| Cotton, purified, 28.0 g | 6 | 6 | — | 6510-21-116-0197 |
| Depressor, tongue, 25s | 4 | — | — | 6515-31-852-9427 |
| Dressing, first-aid, field | 12 | 6 | 1 | 6510-21-102-7867 |
| Forceps, hemostatic | 1 | — | — | 6515-21-116-4057 |
| Forceps, splinter | 1 | 1 | — | 6515-00-337-2400 |
| Form, Medical Field Card | 20 | — | — | 7530-21-870-5029 |
| Gloves, disposable (pair) | 4 | 2 | — | 6515-21-845-9482 |
| Litter, folding | 1 | — | — | 6530-21-851-3111 |
| Mask, resuscitation with disposable one-way valve | 1 | 1 | — | 6515-21-904-7049 |
| Masks, surgical disposable | 2 | 2 | — | |
| Pad, cotton, eye | 12 | 6 | — | 6510-21-845-2189 |
| Pad, nonadherent, 200s | 1 | — | — | 6510-21-845-2194 |
| Pin, safety, 9s | 4 | — | — | 8315-21-843-6856 |
| **Povidone Iodine swabs 10s | 2 | 2 | 1 | 6510-21-877-0825 |
| Scissors, bandage | 1 | 1 | — | 6515-21-846-0206 |
| Scissors, super shears | 1 | — | — | |
| Shield, eye, surgical | 1 | — | — | 6515-21-116-3164 |
| Soap, surgical | 1 | — | — | 6505-21-855-2230 |
| Splint set, wood | 1 | 1 | — | 6545-21-116-2912 |
| Sponge, surgical, 10.0 cm x 10.0 cm, 2s | 12 | 4 | — | 6510-21-845-2440 |
| Thermometer, clinical | 1 | — | — | 6515-21-857-6956 |
| Waste, matted yarns | 1 | 1 | — | 8305-21-116-0248 |
| Water purification tablets 100s | 2 | 1 | — | 6850-21-894-4444 |

\* Not stocked as a complete kit.
  Kit contents must be ordered separately.

\*\* Shelf-life of 18 months maximum from date
  received for kit.

## CHAPTER 2-5 - APPENDIX C - SUPPLIES FOR FIRST-AID ROOMS

| DESCRIPTION | QUANTITY | STOCK NUMBER (PWGSC) |
|---|---|---|
| Adhesive tape, surgical, 7.5 cm x 4.6 m | 3 | 6510-00-203-5000 |
| Applicator, disposable, 25s | 4 | 6515-21-852-9428 |
| Bag, ice, throat | 1 | 6530-00-770-7500 |
| Bag, hot water | 1 | 6530-21-116-0396 |
| Bandage, adhesive, 100s | 1 | 6510-21-845-2239 |
| Bandage, felt, orthopaedic | 2 | 6510-21-116-0170 |
| Bandage, gauze, 5.0 cm x 4.6 m | 12 | 6510-21-116-0174 |
| Bandage, gauze, 7.5 cm x 4.6 m | 12 | 6510-21-116-0175 |
| Bandage, gauze, 10.0 cm x 5.46 m | 12 | 6510-21-849-9537 |
| Bandage, triangular, 2s | 6 | 6510-21-880-9702 |

| | | |
|---|---|---|
| Basin, wash | 2 | 6530-21-846-9260 |
| Benzalkonium Chloride Tincture | 6 | 6505-21-852-9421 |
| Blanket, bed, grey | 2 | 7210-21-849-9452 |
| Brush, scrub, nail | 1 | 7920-21-116-2811 |
| Cloth, coated (rubber sheeting) | 1 | 8305-21-846-0459 |
| Cotton, non-sterile, 0.454 kg | 1 | 6510-21-116-0194 |
| Cup, paper | 20 | 7350-21-845-1041 |
| Depressors, tongue, 25s | 4 | 6515-21-852-9427 |
| Disposable bedding, (sheets and pillow case) | 12 | |
| Dressing, first-aid, field | 6 | 6510-21-102-7867 |
| Dressing, surgical, combination | 6 | 6510-21-849-9539 |
| Eye irrigation solution, 200 ml | 4 | |
| Forceps, hemostatic | 1 | 6515-21-116-4058 |
| Forceps, splinter | 1 | 6515-00-337-2400 |
| Gloves, rubber, domestic | 1 | 8415-21-845-9584 |
| Gloves, surgeon | 1 | 6515-21-877-0841 |
| Isopropyl alcohol | 1 | 6505-21-852-9420 |
| Litter, folding | 1 | 6530-21-848-4908 |
| Mask resuscitation with disposable one-way valve | 1 | 6515-21-904-7049 |
| Manual, first-aid, English | 1 | 7610-21-848-3664 |
| Manual, first-aid, French | 1 | 7610-21-848-3665 |
| Masks, surgical disposable | 2 | |
| Medicine cup, graduated | 1 | 6530-21-846-9540 |
| Pad, cotton, eye | 4 | 6510-21-845-2189 |
| Pin, safety, 9s | 2 | 8315-21-843-6856 |
| *Povidone Iodine swabs | 10 | 6510-21-877-0825 |
| Scissors, bandage | 1 | 6515-21-846-0206 |
| Shield, eye, surgical | 4 | 6515-21-116 3164 |
| Soap, surgical | 1 | 6505-21-855-2230 |
| Splint set, wood | 1 | 6545-21-116-2912 |
| Splint set, speed type, multi-purpose | 1 | |
| Sponge, surgical, 5.0 cm x 5.0 cm, 2s | 100 | 6510-21-845-2171 |
| Sponge, surgical, 10.0 cm x 10.0 cm, 2s | 100 | 6510-21-845-2440 |
| Tray, instrument | 1 | 6530-21-846-9818 |

*Shelf-life 18 months maximum from date
received for kit.

## CHAPTER 2-7 - HAZARDOUS CONFINED SPACES DIRECTIVE

### General

### Collective agreement

This directive is deemed to be part of collective agreements between the parties to the National Joint Council (NJC). Employees are to be afforded ready access to this directive.

### Grievance procedure

In cases of alleged misinterpretation or misapplication arising out of this directive, the grievance procedure, for all represented employees within the meaning of the *Public Service Staff Relations Act*, will be in accordance with section 7.0 of the *National Joint Council By-Laws*. For unrepresented employees, the departmental or agency grievance procedure applies.

### Effective date

This directive was effective on November 1, 1993.

### Application

This directive incorporates the minimum requirements of the *Canada Labour Code,* Part II, and applicable regulations issued pursuant to that legislation, and applies to all departments and other portions of the Public Service, as defined in Part 1 of Schedule 1 of the *Public Service Staff Relations Act*.

### Definitions

In this directive:

***class of confined spaces*** (*catégorie d'espaces clos*) - a group of at least two confined spaces that are likely, by reason of their similarity, to present the same hazards to persons entering, exiting or occupying them.

***confined space*** (*espace clos*) - an enclosed or partially enclosed space that

(a)  is not designed or intended for human occupancy except for the purpose of performing work,
(b)  has restricted means of access and egress, and
(c)  may become hazardous to an employee entering it due to
   (i)    its design, construction, location or atmosphere,
   (ii)   the materials or substances in it, or
   (iii)  any other conditions relating to it, and

includes but is not limited to: a tank, silo, storage bin, process vessel or other enclosure not designed or intended for human occupancy, in respect of which special precautions are necessary when an employee is required to enter therein to protect the employee from a dangerous atmosphere, prevent the employee from becoming entrapped in stored material, or otherwise ensure the employee's safety.

**confined space ship repair** (*espace clos dans un navire en réparation*) - where the confined space relates to ships or vessels in repair, maintenance or refit, confined space means a storage tank, ballast tank, pump room, coffer dam or other enclosure, other than a hold, not designed or intended for human occupancy, except for the purpose of performing work,

(a)   that has poor ventilation,
(b)   where there may be an oxygen deficient atmosphere, or
(c)   in which there may be an airborne dangerous substance.

**hot work** (*travail à chaud*) - any work where flame is used or a source of ignition may be produced.

**qualified person** (*personne qualifiée*) - a person who, because of knowledge, training and experience, is qualified to perform safely and properly the duties specified under the directive in the following areas:

-   hazard assessment
-   entry procedures
-   emergency procedures
-   emergency requirement and equipment maintenance
-   entry permits
-   hot work

performance of these duties may be assigned to different qualified persons.

**ventilation equipment** (*équipement d'aération*) - where the confined space relates to ships or vessels in repair, maintenance or refit means a fan, blower, induced draft or other ventilation device used to force a supply of fresh, respirable atmospheric air into an enclosed space or to remove ambient air from such space.

## Requirements

### 7.1 Hazard assessment

7.1.1   Where a hazard assessment required by this section identifies existing or potential hazards, each confined space or class of confined spaces shall be identified as "permit entry only" and no employee shall enter the confined space unless the appropriate entry permit has been issued and signed by the qualified person and, prior to entry, explained to, understood by and signed by the employee

7.1.2   Where it is likely that a person will, in order to perform work for an employer, enter a confined space and an assessment pursuant to this subsection has not been carried out in respect of the confined space, or in respect of the class of confined spaces to which it belongs, the employer shall appoint a qualified person.

(a)   to carry out an assessment of the physical and chemical hazards to which the person is likely to be exposed in the confined space or the class of confined spaces; and

235

(b)    to specify the tests that are necessary to determine whether the person
       would be likely to be exposed to any of the hazards identified.

7.1.3  The qualified person shall, in a signed and dated report to the employer,
record the findings of the assessment carried
out.

7.1.4  The employer shall make a copy of any report available to the safety and
health committee or the safety and health representative, if either exists.

7.1.5  The report shall be reviewed by a qualified person at least once every
three years to ensure that its assessment of the hazards with which it is
concerned is still accurate.

7.1.6  If a confined space has not been entered in the three years preceding
the time when the report should have been reviewed and no entry is scheduled,
the report need not be reviewed until it becomes likely that a person will, in
order to perform work for an employer, enter the confined space.

## 7.2 Entry procedures

7.2.1  For the purposes of this section, any procedures developed by a
department shall include an entry permit system which shall include a check list
of entry requirements to be given to and signed by the employee or employees.
Employees have the right to a full explanation of the implications of entry into
the confined space prior to signing the entry permit.

7.2.2  Departments shall ensure that the procedures developed in consultation
with the safety and health committee include the procedures to be followed by
the qualified persons responsible for the inspection, maintenance and testing of
all monitoring equipment, personal protective equipment, ventilating equipment,
safety harnesses and any other entry, protective and rescue equipment used in
conjunction with entry into a confined space.

7.2.3  Following consideration of the report made pursuant to 7.1.3, every
employer shall, in consultation with the safety and health committee or the
safety and health representative, if either exists, establish procedures that are
to be followed by a person entering, exiting or occupying a confined space
assessed pursuant to section 7.1, or a confined space that belongs to a class of
confined spaces assessed pursuant to the section.

7.2.4  The procedures shall specify the date on which they are established.

7.2.5  The procedures shall establish an entry permit system that provides for

(a)    specifying, in each case, the length of time for which an entry permit is
       valid, and
(b)    recording the name of the person entering the confined space, and the
       time of entry and the anticipated time of exit.

7.2.6 The employer shall specify the protection equipment referred to in the Personal Protective Equipment directive that is to be used by every person who is granted access to the confined space by the employer.

7.2.7 The employer shall specify any insulated protection equipment and tools referred to in the Tools and Machinery directive that a person may need in the confined space.

7.2.8 The employer shall specify the protection equipment and emergency equipment to be used by a person who takes part in the rescue of a person from the confined space or in responding to other emergency situations in the confined space.

## 7.3 Confined space entry

7.3.1 The employer shall, where a person is about to enter a confined space pursuant to an entry permit system, appoint a qualified person, who could be the same person, to verify by means of tests, that compliance with the following specifications can be achieved during the period of time that the person will be in the confined space, namely,

(a) the concentration of any chemical agent, or combination of chemical agents, in the confined space to which the person is likely to be exposed will not result in the exposure of the person
    (i) to a concentration of an airborne chemical agent or combination of chemical agents in excess of the value for that chemical agent adopted by the American Conference of Governmental Industrial Hygienists in its publication entitled *Threshold Limit Values and Biological Exposure Indices for 1991-92*, as amended from time to time, or
    (ii) to a concentration of an airborne chemical agent or combination of chemical agents that is more than 50 per cent of the lower explosive limit of the chemical agent or combination of chemical agents, or
    (iii) where a source of ignition may ignite the concentration of an airborne agent or combination of chemical agents, to a concentration that is more than 10 per cent of the lower explosive limit of the chemical agent or combination of chemical agents.
(b) the concentration of airborne hazardous substances, other than chemical agents, in the confined space is not hazardous to the safety or health of the person, and
(c) the percentage of oxygen in the air in the confined space is not less than 19.5 per cent by volume and not more than 23 per cent by volume, at normal atmospheric pressure.

7.3.2 The person appointed pursuant to this section shall also verify that

(a) any liquid in which the person could drown has been removed from the confined space,
(b) any free-flowing solid in which the person may become entrapped has been removed from the confined space,

(c) the entry of any liquid, free-flowing solid or hazardous substance into the confined space has been prevented by a secure means of disconnection or the fitting of blank flanges,

(d) all electrical and mechanical equipment that may present a hazard to the person has been disconnected from its power source, real or residual, and has been locked out, and

(e) the opening for entry into and exit from the confined space is sufficient to allow the safe passage of a person using protection equipment.

7.3.3  The qualified person shall, subject to paragraph 7.4.1, verify that the specifications set out in this section are complied with at all times that a person is in the confined space.

7.3.4  The qualified person shall, in a signed and dated report to the employer, set out the results of the verification carried out, including the test methods, the test results and a list of the test equipment used.

7.3.5  Where the report indicates that a person who has entered the confined space has been in danger, the employer shall send the report to the safety and health committee or the safety and health representative, if either exists.

7.3.6  In all other cases, the employer shall make a written copy or a machine-readable version of the report available to the safety and health committee or the safety and health representative, if either exists.

## 7.4  Repair of ships and vessels

7.4.1  For confined spaces related to ships or vessels in repair, maintenance or refit, the employer may meet the requirements of paragraph 7.3.1(a) by forced ventilation from the lowest point in the confined space as long as the requirements of paragraph 7.3.1(b) have been met. When conditions in the confined space are maintainable in this state, the employer may establish an entry permit system in accordance with section 7.2 which will be valid for multiple entries into the confined space during the period for which the entry permit is valid. The length of time for permit validity shall be to a maximum of a shift for uncoated fuel tanks and to a maximum of 24 hours in all other situations.

## 7.5  Emergency procedures and equipment

7.5.1  Where conditions in a confined space or the nature of the work to be performed in a confined space is such that paragraph 7.3.3 cannot be complied with, the employer shall in consultation with the safety and health committee or the safety and health representative, if either exists, establish emergency procedures to be followed in the event of an accident or other emergency in or near the confined space.

7.5.2  The procedures shall specify the date on which they are established and provide for the immediate evacuation of the confined space when an alarm is activated, or there is any significant change in a concentration or percentage referred to in paragraph 7.3.1 that would adversely affect the safety or health of a person in the confined space.

7.5.3 The employer shall provide the protection equipment referred to in paragraphs 7.2.6, 7.2.7 and 7.2.8 for each person who is about to enter the confined space.

7.5.4 The employer shall ensure that a qualified person trained in the entry and emergency procedures established pursuant to this directive is in attendance outside the confined space, and in communication with the person inside the confined space.

7.5.5 The employer shall provide the qualified person with a suitable alarm device for summoning assistance, and ensure that two or more persons are in the immediate vicinity of the confined space to assist in the event of an accident or other emergency.

7.5.6 One of the persons referred to above shall

(a) be trained in emergency procedures;
(b) be the holder of a basic first aid certificate; and
(c) be provided with the required protection equipment and emergency equipment.

7.5.7 The employer shall ensure that every person entering, exiting or occupying the confined space wears an appropriate safety harness that is securely attached to a lifeline.

7.5.8 The lifeline shall be attached to a secure anchor outside the confined space and be controlled by the qualified person.

7.5.9 The lifeline shall protect the person from the hazard for which it is provided and shall not in itself create a hazard.

7.5.10 The lifeline shall, where reasonably practicable, be equipped with a mechanical lifting device.

## 7.6 Record of emergency procedures and equipment

7.6.1 When a person enters a confined space under circumstances such that 7.3.1 cannot be complied with, the qualified person shall, in a signed and dated report to the employer,

(a) specify the emergency procedures that are to be followed and the protection equipment, insulated protection equipment and tools and the emergency equipment that are to be used; and
(b) specify any other procedures to be followed and any other equipment that could be needed.

7.6.2 The report and any procedures specified therein shall be explained by the qualified person to every employee who is about to enter a confined space, and a copy of the report shall be signed and dated by any employee to whom the report and the procedures have been so explained, acknowledging by signature the reading of the report and the explanation thereof.

## 7.7 Provision and use of equipment

7.7.1 The employer shall provide each person who is granted access to a confined space with the protection equipment specified.

7.7.2 The employer shall provide each person who is to undertake rescue operations with the protection equipment and emergency equipment specified.

7.7.3 The employer shall ensure that every person who enters, exits or occupies a confined space follows the procedures established and uses the protection equipment specified in this directive.

## 7.8 Precaution

7.8.1 No person shall close off a confined space until a qualified person has verified that no person is inside it.

## 7.9 Hot work

7.9.1 Unless a qualified person has determined that the work can be performed safely, hot work shall not be performed in a confined space that contains

(a) an explosive or flammable hazardous substance in a concentration in excess of 10 per cent of its lower explosive limit; or
(b) oxygen in a concentration in excess of 23 per cent.

7.9.2 Where hot work is to be performed in a confined space that contains hazardous concentrations of flammable or explosive materials, a qualified person shall patrol the area surrounding the confined space and maintain a fire-protection watch in that area until all fire hazard has passed.

7.9.3 Fire extinguishers specified as emergency equipment shall be provided in the area.

7.9.4 Where an airborne hazardous substance may be produced by hot work in a confined space, no person shall enter or occupy the confined space unless

(a) section 7.10 is complied with; or
(b) the person uses a respiratory protective device that meets the requirements of the Personal Protective Equipment directive.

## 7.10 Ventilation equipment

7.10.1 Where ventilation equipment is used to maintain the concentration of a chemical agent or combination of chemical agents in a confined space at or below the concentration referred to in paragraph 7.3.1(a), or to maintain the percentage of oxygen in the air of a confined space within the limits referred to in paragraph 7.3.1(c), the employer shall not grant access to the confined space to any person unless the requirements of this section are met.

7.10.2 The ventilation equipment shall be equipped with an alarm that will, if the equipment fails, be activated automatically and be audible or visible to every person in the confined space, or be monitored by an employee who is in constant attendance of the equipment and who is in communication with the person or persons in the confined space.

7.10.3 If the ventilation equipment fails to operate properly, the employee shall immediately inform the person or persons in the confined space of the failure of the equipment.

7.10.4 In the event of failure of the ventilation equipment, sufficient time will be available for the person to escape from the confined space before

(a) the concentration of the chemical agent or combination of chemical agents in the confined space exceeds the concentrations referred to in paragraph 7.3.1(a), or

(b) the percentage of oxygen in the air ceases to remain within the limits referred to in paragraph 7.3.1(c).

## 7.11 Training

7.11.1 The employer shall provide every employee who is likely to enter a confined space with instruction and training in entry and emergency procedures.

7.11.2 The employer shall also provide every such employee with instruction and training on all protection and emergency equipment that may be used.

7.11.3 The employer shall ensure that no person enters a confined space unless the person is instructed in entry and emergency procedures.

7.11.4 The employer shall also ensure that every such person is instructed in the use of protection and emergency equipment that may be used.

7.11.5 Employees likely to enter a confined space shall be provided with information on the hazard assessment related to the confined space or class of confined spaces required by section 7.1, the entry permit system required by section 7.2 and the verification system required by section 7.3.

## 7.12 Record keeping

7.12.1 The employer shall, at the employer's place of business nearest to the workplace in which the confined space is located, keep a written copy or a machine-readable version of the reports specified in this section.

7.12.2 Any report made pursuant to paragraph 7.1.3 and the procedures established pursuant to paragraphs 7.2.3 and 7.5.1 for a period of 10 years after the date on which the qualified person signed the report or the procedures were established.

7.12.3 Any report made pursuant to paragraph .3.4 for a period of 10 years after the date on which the qualified person signed the report where the verification procedures undertaken pursuant to 7.3.1 and 7.3.2 indicate that the specifications set out in the above-noted subsections were not complied with.

7.12.4  Any report, in every other case, for a period of two years after the date on which the qualified person signed the report.

## 7.13  **Resolving "qualified person" disputes**

7.13.1  Where there is a dispute regarding the term "qualified person" for purposes of an occupational safety and health standard, the following procedure shall be implemented:

(a)  the employee shall raise the matter directly with the person-in-charge;
(b)  the person-in-charge shall review the employee's qualifications and decide upon the employee's status as a qualified person;
(c)  if the employee is dissatisfied with the decision, the matter shall be referred to the safety and health committee established for the employee's workplace;
(d)  the safety and health committee shall review the matter and make appropriate recommendations to the person-in-charge;
(e)  if the safety and health committee does not consider itself competent to deal with the case, it shall recommend an acceptable third party to the person-in-charge;
(f)  the person-in-charge shall, pursuant to (d) or (e), take the recommendations into consideration, render a final management decision and undertake the appropriate action.

If the employee does not agree with the final decision which has been rendered, a grievance may be initiated pursuant to the NJC redress procedure.

## CHAPTER 2-8 - OCCUPATIONAL HEALTH UNIT STANDARD

### Application

1.  This standard applies to all departments and other portions of the Public Service, as defined in Part I of Schedule I of the *Public Service Staff Relations Act*.

### Consultation

2.  The attention of departments is drawn to the consultation provisions of the relevant collective agreements when applying this standard.

### Definitions

3.  In this standard:

(1)  ***director*** means a Regional or Zone Director of Medical Services Branch, Health Canada, or another official of Health Canada authorized to act on the Director's behalf (directeur);
(2)  ***occupational health nurse*** means a health professional who is qualified to do such work through specialized training, knowledge and experience,

and whose general responsibilities are listed in appendix A (infirmière en santé du travail);

(3) ***occupational health unit*** means a facility designed to provide an operational base for provision of occupational health nursing services (service de santé au travail).

## General requirements

4. Consideration shall be given, subject to paragraph 5, to the establishment of an occupational health unit where 750 or more public servants are normally employed on any shift at a specific geographic location.

5. The requirements of paragraph 4 shall be modified in accordance with the following variable factors:

(a) the distance (taking into consideration time, convenience and weather factors) to other similar health services facilities; and

(b) the occupational risks and hazards of the work, including the local requirements for special services such as periodic health evaluations.

## Responsibilities

6. Health Canada shall, in accordance with the requirements of this standard, be responsible through its Medical Services Branch for:

(1) responding to departmental requests for establishing occupational health units;

(2) assessing the need for occupational health units;

(3) recommending to the Treasury Board the requirement for a new occupational health unit, indicating the location, space required, host department(s), number of employees by department to be served, and other pertinent data;

(4) advising and consulting with host department(s) and the department having administration and control of the accommodation to be occupied, concerning the location, space and facilities required in a new occupational health unit which has been approved by the Treasury Board, or in respect of an existing installation, any major renovations or additions of special equipment or facilities;

(5) providing or arranging for the staffing of occupational health units;

(6) developing arrangements for the provision of other occupational health services where a local occupational health unit is not or will not be available, and advising departments of such arrangements.

7. Departments shall:

(1) communicate requests for an occupational health unit to the director, stating the number and location of employees that will be served by the unit;

(2) consult with Health Canada and the department having administration and control of the accommodation to be occupied in respect of:
  (a) the selection of a mutually suitable location,
  (b) the amount of space required,

| | |
|---|---|
| (c) | the fit-up requirements, |
| (d) | the services necessary (heat, light, water, electrical), |
| (e) | any special equipment or facilities, and |
| (f) | any renovations, additional special equipment or facilities in existing installations. |
| (3) | ensure that, where a health unit is not available, all employees are continuously informed of the alternative health services which have been arranged by Health Canada. |

## Location

8. The direction to, and location of, an occupational health unit shall be indicated by graphic symbols in accordance with the Federal Identity Program Guide.

An occupational health unit shall be located:

(1) in a building conveniently situated and accessible to the greatest number of employees;
(2) within that building, in an area permitting the most convenient access to all employees; and
(3) in an area as free as possible of dust, odours, noise and vibration and, where possible, with convenient access to washroom facilities near the health unit.

## Design and facilities

9. The host department(s), the department having administration and control of the accommodation to be occupied, and Health Canada shall, before construction or renovation begins, consult on the proposed space, design and facilities for an approved occupational health unit installation.

10. Access to rooms in the unit shall be wide enough to accommodate stretchers and wheelchairs. Rooms to be provided shall include:

(1) **A waiting room** - The waiting room shall contain adequate waiting space and furnishings for at least four persons and a display rack for health promotion literature.
(2) **A treatment room** - Each treatment room shall have a minimum floor area of 15 m. The treatment room shall have counter space, lockable cupboards, and hot and cold running water. Where this room may also be used as a first-aid room, it shall conform to the requirements of the First-aid directive, chapter 2-5.
(3) **A quiet room** - A minimum of one quiet room, with space for two cots, shall be constructed in each occupational health unit. Curtains or other suitable arrangements should be installed so as to provide each cot area with visual privacy.
(4) **A counselling office** - A completely enclosed nurse's office, suitable for private counselling, shall be provided in each occupational health unit. Where the unit is routinely staffed by more than one nurse, an additional office or offices may be provided, as required. Offices should be provided

with telephone communication, and be directly accessible to the waiting room.

(5) **An examining room** - A separate room shall be provided which is suitable for testing and examination purposes where this is justified by the work load.

(6) **A storage room** - Sufficient lockable storage space shall be provided to store medical and office supplies and equipment. Lockable filing cabinets shall be supplied and may be placed in this room.

(7) **Toilets and washrooms** - These facilities, including hot and cold running water, shall be available in each occupational health unit and be designed to accommodate handicapped employees including those in wheelchairs.

(8) **Additional space** - Where it is convenient to use space in a health unit for clerical staff or environmental health officers, the provision of additional space and furnishings should be considered.

## Supplies and equipment

11. Supplies and equipment for occupational health units shall be provided as per appendix B.

## Occupational health unit staff

12. Where an occupational health unit serves employees in separate buildings, the occupational health nurse(s) from the central unit shall, in consultation with the department(s) to be served, provide special occupational health nursing services in such buildings. In this regard, a first-aid room or a soundproof private office may be used. For other occupational health services such as special testing, treatment, file storage, etc., the occupational health unit shall be used.

13. In areas where a fixed occupational health unit is not practicable, mobile occupational health services for employees may be provided through the appropriate regional offices of Medical Services Branch.

## Records

14. The provision, maintenance, use, interpretation and confidentiality of all records, forms and procedures as required in the operation of occupational health units by occupational health nurses shall be the responsibility of Health Canada.

## CHAPTER 2-8 - APPENDIX A - THE ROLE OF THE OCCUPATIONAL HEALTH NURSE

The occupational health nurse fulfils a key role in the operation of the Public Service Health Program. The nurse's prime function is to provide a preventive health care program designed to promote and maintain within the nurse's role, an optimum level of health in the employee, and to control or minimize any possible adverse effects of the work or work environment upon the employee's health. .

The occupational health nurse meets this objective through the following activities:

## 1. Health education, promotion and prevention programs

The occupational health nurse encourages the employee's increase of knowledge and the development of attitudes required to maintain a healthy lifestyle; educates the employee concerning sound, safe and hygienic practices, particularly those related to the work environment; offers and provides appropriate health monitoring, information, health education and promotion programs to individual employees or groups of employees.

## 2. Liaison with community medical resources — Health and social agencies and services

The occupational health nurse maintains an up-to-date knowledge of the community resources available, and maintains an awareness of the appropriate referral and follow-up procedures. Where indicated in a Public Service health or safety standard or procedure, employees may be referred to the local medical officer of Medical Services Branch, Health Canada, or the designated physician in the community.

## 3. Environmental health

The occupational health nurse develops a thorough knowledge of the work environment, the type of occupational hazards encountered, an appreciation of the potential occupational injuries or illnesses, and remains alert to the various effects of the work environment upon the employee's health. The occupational health nurse assists departments to identify those employees who require occupational health evaluations; performs special health monitoring and testing (e.g. audiometric tests); communicates recommendations concerning the need for environmental health surveys or investigations, and the procedures to be followed in requesting such surveys to the responsible manager in charge of the work.

## 4. Health care

The occupational health nurse provides emergency nursing treatment in cases of illness or injury which occur at work and, where necessary, will suggest referral to the employee's personal physician. The occupational health nurse performs periodic health evaluations, health monitoring and, on occasion, may perform special nursing treatments and care to employees according to approved Health Canada procedures and the signed approval or authorization of the employee's treating physician.

## 5. Counselling

The occupational health nurse provides confidential counselling to those employees voluntarily seeking help and advice because of health problems, stress-related illnesses and symptoms, and the effects of alcohol and drug abuse. In certain situations where agreed to by the department concerned, the occupational health nurse may fulfill the role of an Employee Assistance Program Counsellor and will, at the supervisor's request and with the cooperation of the employee, provide counselling in those cases where poor health may be the cause of deteriorating work performance.

## 6. Administrative activities

The occupational health nurse is responsible for the operation of the health unit, the provision of occupational health services according to the established procedures of Health Canada, and for the preparation and compilation of all records and other data required in the operation of the Public Service Health Program.

## 7. Summary

The role of the occupational health nurse constantly evolves and therefore the foregoing headings and activity descriptions are not intended to encompass all the duties that may be undertaken by the nurse. The occupational health nurse continuously adapts the occupational health services to respond to those changes occasioned by new technology, products and procedures and new occupational health and safety standards, policies and procedures approved for application in the Public Service.

## CHAPTER 2-8 - APPENDIX B - LIST OF EQUIPMENT FOR OCCUPATIONAL HEALTH UNITS

### Counselling office

Desk, double pedestal (150 cm x 75 cm)
Chairs (2) with arm rests, upholstered
Chair (1), straight, with arm rests
Costumer (1)
Bookcase
Electric fan

### Waiting room

Table(s), end
Chairs, with arm rests, upholstered
Costumer(s)
Rubber mats for outer footwear
Pamphlet rack(s)
Weigh scale (metric)

### Quiet room

Chairs, straight without arm rests
Mirrors, wall
Costumer
Cots (specifications available through Medical Services Branch, Health Canada)
Curtains or screens around cots
Electric fan

**Treatment room**

Chairs, metal, without arm rests
Refrigeration unit
Table(s), metal, (specifications available from Medical Services Branch, Health
    Canada)
Special magnifying glass with lamp
Electric fan

**Examining room**

Cot
Bedside table
Desk
Examining table
Straight back chair
Mirror
Various medical equipment as specified by Health Canada

**Storage room/File room**

Shelves
Portable filing cabinets

**\* Supplies, equipment and furnishings**

\* All medical supplies and equipment, office furniture, and furnishings shall be
determined and provided by Health Canada.

## CHAPTER 2-9 - TOOLS AND MACHINERY DIRECTIVE

### Grievance procedure

In cases of alleged misinterpretation or misapplication arising out of this
standard, the grievance procedure, for all represented employees within the
meaning of the *Public Service Staff Relations Act*, will be in accordance with
Section 7.0 of the *National Joint Council By-Laws*. For unrepresented
employees, the departmental grievance procedure applies.

This standard is deemed to be part of collective agreements between the
parties to the National Joint Council and employees are to be afforded ready
access to this standard.

### Application

1. This standard incorporates the minimum requirements of the *Canada Labour
Code*, Part II, and application regulations issued pursuant to that legislation,
and applies to all departments and other portions of the Public Service, as
defined in Part I of Schedule I of the *Public Service Staff Relations Act*.

## Definitions

2. In this standard:

(1) **explosive actuated fastening tool** means a tool that, by means of an explosive force, propels or discharges a fastener for the purpose of impinging it on, affixing it to or causing it to penetrate another object or material (pistolet de scellement à cartouches explosives);

(2) **fire hazard area** means an area that contains explosive or flammable concentrations of dangerous substances (endroit présentant un risque d'incendie);

(3) **locked out** means, in respect of any equipment, machine or device, that the equipment, machine or device has been rendered inoperative and cannot be operated or energized without the consent of the person who rendered it inoperative (verrouillé);

(4) **machine guard** means a device that is installed on a machine to prevent a person, or any part of his body or clothing, from becoming engaged in any rotating, moving, electrically charged, hot or other dangerous part of a machine, or the material that the machine is processing, transporting or handling. It also means a device that makes the machine inoperative if a person or any part of his clothing is in or near a part of the machine that can cause injury (dispositif de protection);

(5) **person in charge** means a qualified person who supervises employees performing work in order to ensure the safe and proper conduct of an operation or of the work of employees (responsable);

(6) **qualified person** means, in respect of a specified duty, a person who, because of knowledge, training and experience, is qualified to perform that duty safely and properly (see paragraph 48) (personne qualifiée);

(7) **safety officer** means a person so designated by the Minister of Human Resources Development pursuant to Part II of the *Canada Labour Code* (agent de sécurité).

3. Departments shall ensure that requirements specified in the *Canada Occupational Safety and Health Regulations*, Part XIII, *Tools and Machinery Regulation*, issued pursuant to the *Canada Labour Code*, Part II, are applied at every workplace occupied by employees.

## Design, construction, operation and use

4. To the extent that it is practicable, departments shall ensure that all tools have been designed and constructed so as to be safe under all conditions of intended use.

5. Where there is a hazard that an explosive or flammable atmosphere in a workplace is likely to be ignited by sparks, the exterior surface of any tool used by an employee shall be made of non-sparking material.

6. All portable electric tools used by employees shall meet the standards set out in CSA Standard C22.2 No. 71.1-M1985, *Portable Electric Tools*, dated March, 1985.

7. Where a portable electric tool is used in a hazardous location, it shall be of a type that complies with the appropriate recommendation of CSA Standard C22.1-1986, *Canadian Electrical Code*, Part I, *Safety Standards for Electrical Installations*.

8. Subject to paragraph 9, all portable electric tools used by employees shall be grounded in accordance with CSA Standard C22.2 No. 71.1-M1985.

9. Paragraph 8 does not apply to tools that:

(1)  are powered by a self-contained battery;
(2)  have a protective system of double insulation; or
(3)  are used in a location where reliable grounding cannot be obtained if the tools are supplied from a double insulated portable ground fault circuit interrupter of the class A type that meets the standards set out in CSA Standard C22.2 No. 144-1977, *Ground Fault Circuit Interrupters*, dated March, 1977.

10. No person shall ground an electrical portable power tool that has a protective system of double insulation.

11. All portable electric tools used by employees in a fire hazard area shall be marked as appropriate for use or designed for use in the area of that hazard.

12. Where an air hose is connected to a portable air-powered tool used by an employee, a restraining device shall be attached:

(1)  to the tool, where an employee may be injured by the tool falling; and
(2)  to all hose connections, in order to prevent injury to an employee in the event of an accidental disconnection of a hose.

13. Employees shall ensure that the tool end of any flexible shaft portable power tool is secured in a manner that will prevent the flexible shaft from whipping when the motor is started.

**Operation and use**

14. Employees shall not operate tools or machinery unless they are wearing the appropriate personal protective equipment pursuant to the Personal protective equipment safety directive (chapter 2-14).

15. Determination of the protective clothing and equipment to be used shall be in accordance with the Personal protective equipment directive, chapter 2-14 and the Clothing directive as outlined in chapter 1-5.

16. All explosive actuated fastening tools used by employees shall meet the standards set out in CSA Standard Z166-1975, *Explosive Actuated Fastening Tools*, dated June, 1975.

17. No employee is to use an explosive actuated portable power tool without the approval of the person in charge and unless the employee possesses an

operator's certificate issued by the manufacturer, or he or she has been trained in the safe use of the tool.

18. Every employee who operates an explosive actuated fastening tool shall operate it in accordance with CSA Standard Z166-1975, *Explosive Actuated Fastening Tools*, dated June, 1975.

19. An employee shall not be permitted to use a tool or machine unless he or she is qualified by knowledge, training and experience, and is authorized to do so.

20. Where it is necessary to remove or change an attachment, or make any adjustment or repair to a power tool, such work shall not proceed unless the tool is disconnected from its power source in a manner that ensures that it cannot be inadvertently reconnected.

21. Employees who use a pneumatic portable power tool shall shut off the air supply to that tool and bleed the air line before disconnecting it from the tool, unless the air line is equipped with a quick disconnect coupling that makes such precautions unnecessary.

22. No person is to use a pneumatic portable power tool or air hose in such a manner that an air stream might be directed forcibly against his or her body, or the body of any other person.

23. All chain saws used by employees shall meet the standards set out in CSA Standard CAN3-Z62.1-M85, *Chain Saws*, dated February, 1985.

24. Departments shall ensure, to the extent that is practicable, that exposure to continuous vibration from tools and machinery is minimized.

**Inspection and maintenance**

25. Departments shall ensure that all hand tools and portable power tools used are inspected at regular intervals and maintained in a safe working condition.

26. An inspection and maintenance plan for tools and machinery shall be instituted by departments and a record kept of all inspections and maintenance work performed in accordance with such plan.

27. Each tool and machine shall be checked by employees before use to ensure that there is no visible defect.

28. All hand tools and portable power tools shall be transported and stored in a safe manner.

**Defective tools and machinery**

29. Where an employee finds any defect in a tool or machine that may render it unsafe for use, he or she shall report the defect to the person in charge as soon as possible.

30. Every department shall mark or tag as unsafe and remove from service any tool or machinery used by employees that has a defect that may render it unsafe to use.

## Instructions and training

31. Every employee shall be instructed and trained by a qualified person appointed by the department in the safe and proper inspection, maintenance and use of all tools and machinery that he or she is required to use.

32. Every department shall maintain a manual of operating instructions for each type of portable electric tool, portable air-powered tool, explosive actuated fastening tool and machine used by employees.

33. A manual referred to in paragraph 32 shall be kept by the department readily available for examination by an employee who is required to use the tool or machine to which the manual refers.

## General requirements for machine guards

34. Every machine that has exposed, moving, rotating, electrically charged or hot parts or that processes, transports or handles material that constitutes a hazard to an employee shall be equipped with a machine guard that:

(1)    prevents the employee or any part of his or her body from coming into contact with the parts or material;
(2)    prevents access by the employee to the area of exposure to the hazard during the operation of the machine; or
(3)    makes the machine inoperative if the employee or any part of his or her clothing is in or near a part of the machine that is likely to cause injury.

35. Machine guards shall be installed on any machine or part of a machine that constitutes a source of danger to employees. The guard shall be designed and placed in such a manner that it does not in itself create a hazard. Such machine guards shall be maintained by a qualified person.

## Use, operation, repair and maintenance of machine guards

36. Machine guards shall be operated, maintained and repaired by a qualified person.

37. Subject to paragraph 38, where a machine guard is installed on a machine, no person shall use or operate the machine unless the machine guard is in its proper position.

38. A machine may be operated when the machine guard is not in its proper position in order to permit the removal of an injured person from the machine.

39. Subject to paragraph 40, where it is necessary to remove a machine guard from a machine in order to perform repair or perform maintenance work on the machine, no person shall perform the repair or maintenance work unless the machine has been locked out.

40. Where it is not practicable to lock out a machine referred to in paragraph 39 in order to perform repair or maintenance work on the machine, the work may be performed if:

(1)  the person performing the work follows written instructions provided by the person in charge that will ensure that any hazard to that person is not significantly greater than it would be if the machine had been locked out;
(2)  the person performing the work:
  (a)  obtains a written authorization from the person in charge each time the work is performed; and
  (b)  performs the work under the direct supervision of a qualified person;
(3)  such work is performed in the presence of and under the direct supervision of the person in charge, or a qualified person authorized by the person in charge.

41. Departments shall ensure that a copy of every written procedure referred to in paragraph 40 is readily available to persons who repair and maintain machines.

## Abrasive wheels

42. Abrasive wheels shall be:

(1)  used only on machines equipped with machine guards;
(2)  mounted between flanges; and
(3)  operated in accordance with sections 4 to 6 of CSA Standard B173.5-1979 *Safety Requirements for the Use, Care and Protection of Abrasive Wheels*, dated February, 1979.

43. A bench grinder shall be equipped with a work rest or other device that:

(1)  prevents the work piece from jamming between the abrasive wheel and the wheel guard; and
(2)  does not make contact with the abrasive wheel at any time.

## Mechanical power transmission apparatus

44. Equipment used in the mechanical transmission of power shall be guarded in accordance with sections 7 to 10 of ANSI Standards ANSI B15.1-1972, *Safety Standard for Mechanical Power Transmission Apparatus*, dated July, 1972.

## Woodworking machinery

45. Woodworking machinery shall be guarded in accordance with clause 3.3 of CSA standard Z114-M1977, *Safety Code for the Woodworking Industry*, dated March, 1977.

## Punch presses

46. Punch presses shall meet the standards set out in CSA Standard Z142-1976, *Code for the Guarding of Punch Presses at Point of Operation*, dated February, 1976.

## Robot systems

47. Departments shall ensure, to the extent that is practicable, that guarding of a robot machine or a robot machine system conforms, as a minimum, to the *American National Standard for Industrial Robots and Robot Systems — Safety Requirements*, ANSI/RIA RIS.06-1986 as amended from time to time.

## Resolving qualified person disputes

48. Where there is a dispute regarding the term "qualified person" for purposes of an occupational safety and health standard, the following procedure shall be implemented:

(a)   The employee shall raise the matter directly with the person in charge.
(b)   The person in charge shall review the employee's qualifications and decide upon the employee's status as a qualified person.
(c)   If the employee is dissatisfied with the decision, the matter shall be referred to the safety and health committee established for the employee's workplace.
(d)   The safety and health committee shall review the matter and make appropriate recommendations to the person in charge.
(e)   If the safety and health committee does not consider itself competent to deal with the case, it shall recommend an acceptable third party to the person in charge.
(f)   The person in charge shall, pursuant to (d) or (e), take the recommendations into consideration, render a final management decision and undertake the appropriate action.

If the employee does not agree with the final decision which has been rendered, a grievance may be initiated pursuant to the NJC redress procedure.

## CHAPTER 2-10 - MATERIALS HANDLING SAFETY DIRECTIVE

### General

### Grievance procedure

In cases of alleged misinterpretation or misapplication arising out of this standard, the grievance procedure, for all represented employees, within the meaning of the *Public Service Staff Relations Act*, will be in accordance with Section 7.0 of the *National Joint Council By-Laws*. For unrepresented employees, the departmental grievance procedure applies.

This standard is deemed to be part of collective agreements between the parties to the National Joint Council and employees are to be afforded ready access to this standard.

### Application

1. This standard applies to all departments and other portions of the Public Service, as defined in Part I of Schedule I of the *Public Service Staff Relations Act*.

**Definitions**

2. In this standard:

(1) ***materials handling equipment*** means any machine, equipment or mechanical device used when transporting, lifting, moving or positioning any materials, goods, articles, persons or things and includes any crane, derrick, loading tower, powered industrial truck, hand truck, conveyor, hoist, earth-moving equipment, rope, chain, sling, dock, ramp, storage rack, container, pallet and skid; but does not include elevating devices that are subject to chapter 2-4, Elevating devices directive, or tools that are subject to chapter 2-9, Tools and machinery directive (appareil de manutention des matériaux);

(2) ***maximum safe load***, with respect to any materials handling equipment or any floor, dock or other structure used in handling materials, means the lesser of:
   (a) the maximum load that such equipment or structure was designed or constructed to handle or support safely, or
   (b) the maximum load that such equipment or structure is guaranteed in writing by the manufacturer to handle or support safely (charge maximale admissible);

(3) ***mobile equipment*** means any materials handling equipment that is self-propelled or in respect of which mobility is the predominant characteristic (appareil mobile);

(4) ***motor vehicle*** means a truck, tractor, trailer, semi-trailer, automobile, bus or other similar self-propelled vehicle used primarily for transporting materials, persons or things by road; (véhicule automobile);

(5) ***operator*** means a person who has the qualifications described in paragraph 6 and who has been designated to operate or assist in the operation of materials handling equipment (conducteur);

(6) ***person in charge*** means a qualified person appointed by management to supervise the safe and proper conduct of an operation or of the work of employees (responsable);

(7) ***qualified person*** means a person who, because of knowledge, training and experience, is qualified to perform safely and properly a specified job (see paragraph 145) (personne qualifiée);

(8) ***safety officer*** means a person designated as a safety officer by the Minister of Human Resources Development pursuant to Part II of the *Canada Labour Code* (agent de sécurité);

(9) ***signaller*** means a qualified person designated by a person in charge to direct, by means of hand, voice or other signals, the safe movement or operation of materials handling equipment (signaleur);

(10) ***standard code of signals*** means a code of signals that:
   (a) is adopted by a department for use by all signallers in directing the safe movement or operation of materials handling equipment; and
   (b) complies with the code of signals recommended by the National Safety Council, the American National Standards Institute or any other standard approved by Human Resources Development Canada (code uniforme de signalisation).

## Applicability

3. This standard does not apply to the operation and use of motor vehicles on public roads.

## General responsibility of departments

4. Departments shall ensure that all materials handling equipment, floors, docks and other structures that are operated or used for handling materials:

(1)   are suitable and safe for the purposes for which they are operated or used;
(2)   are maintained in a safe operating condition; and
(3)   comply with this standard.

5. The operator of any materials handling equipment shall have ready access to such operating manuals as may be necessary for the safe and proper operation and maintenance of the materials handling equipment.

6. Every operator shall be:

(1)   appropriately trained, instructed and tested in the safe and proper use of the applicable materials handling equipment, and be familiar with departmental safety standards; and
(2)   familiar with the highway vehicle laws of every province and territory within which mobile equipment is operated during the course of employment.

7. A record shall be maintained of any training and instruction provided pursuant to paragraph 6(1).

8. Subject to paragraph 76, no employee shall be permitted to operate or assist in the operation of any materials handling equipment unless the employee has either been authorized as a qualified operator, or is an employee engaged in maintaining or repairing materials handling equipment and is required under the authority of the person in charge to operate such equipment for purposes of testing, ascertaining fault or verifying repairs.

9. Where, in the opinion of a Human Resources Development Canada Regional Director, a code, procedure or practice referred to in this standard, or utilized by a department, does not provide a sufficient degree of safety or may be otherwise inappropriate, the Director may, in accordance with the procedures outlined in the Occupational safety policy for the Public Service, make recommendations to departments concerning the specific safety procedures or codes to be followed in the circumstances. Departments may obtain information and/or advice concerning good industrial safety practice or applicable safety codes or procedures by contacting the appropriate Regional Office of Human Resources Development Canada.

**General responsibility of employees**

10. Every operator shall operate any assigned materials handling equipment in the manner in which the operator was trained and instructed as stipulated in paragraph 6(1).

11. No operator shall operate any materials handling equipment in a careless or reckless manner or otherwise endanger his or her safety or that of other employees.

12. No operator shall operate or use any materials handling equipment from which a machine guard or other safety device has been removed or rendered ineffective except in accordance with chapter 2-9, Tools and machinery safety and health directive.

13. No employee shall interfere with the safe operation of any materials handling equipment.

14. No employee shall remove or render ineffective a machine guard or other safety device with which any materials handling equipment is fitted except with the express approval of the person in charge or a safety officer.

**Design and construction of materials handling equipment**

**General**

15. Materials handling equipment required to be operated or used by an employee shall be designed and constructed in such a way that:

(1)   the equipment will perform safely under the severest conditions of its operation and use that are likely to be encountered;
(2)   to the extent practicable, all parts of the equipment that are subject to failure are so designed that, if a failure occurs, it will not result in a loss of control of the equipment or otherwise create an unsafe condition;
(3)   all glass in doors, windows and other parts of the equipment is safety glass that will not shatter into sharp and dangerous pieces under impact;
(4)   any equipment employed directly in the loading and unloading of ships complies with the *Tackle Regulations* made under the *Canada Shipping Act*; and
(5)   it conforms to the applicable requirements of the *Canada Motor Vehicle Safety Standards* prescribed by Transport Canada.

**Protection from falling objects**

16. Where materials handling equipment is used under such circumstances that the operator may be struck by a falling object or shifting load, the materials handling equipment shall be provided with a protective cab, roof, screen, bulkhead or guard of a design, construction and strength that will prevent, under all foreseeable conditions, the penetration of the object or load into the area occupied by the operator.

17. A protective device referred to in paragraph 16 shall be:

(1) fabricated from non-combustible or fire resistant material; and
(2) designed to permit quick exit from the materials handling equipment in an emergency.

18. Paragraph 16 does not apply where a protective device referred to therein would interfere with the effective operation of the materials handling equipment and a procedure or method is used that, in the opinion of a safety officer, will protect the operator from a falling object or shifting load.

## Protection from turn over

19. Where mobile equipment is likely to turn over under any circumstances of its use, it is to be fitted with roll-over bars or a similar protective device acceptable to a safety officer that will prevent the operator of the mobile equipment from being trapped or crushed under the equipment if it does turn over.

## Fuel tanks

20. Any fuel tank, compressed gas cylinder or similar container of a dangerous substance that is mounted on any materials handling equipment is to be:

(1) located or protected so that, under all conditions, it constitutes a minimal hazard to any employee who is required to operate or ride on that equipment and, in the case of a fuel tank, is separated from the operator by an adequate protective shield or partition;
(2) connected to fuel overflow and vent pipes that are located so that fuel spills and vapours cannot be ignited by hot exhaust pipes or other hot or sparking parts, or otherwise endanger the safety or health of any employee who is required to operate or ride on that equipment; and
(3) labelled on servicing caps or covers as to the contents of each tank.

## Protection from elements

21. The operator of any materials handling equipment regularly used out-of-doors is to be protected from exposure to any weather condition that might endanger the operator's safety or health.

22. Where the temperature in the operator's compartment or position on any materials handling equipment is, as a result of heat coming from or associated with the equipment, normally and consistently above 27°C, the position shall be protected from the heat by an insulated barrier or some other suitable means.

## Vibration

23. All materials handling equipment operated by any employee is to be so designed and constructed that the operator will not be injured, or control of the equipment be impaired, by any vibrations, jolting or uneven movements of the equipment.

24. Any protection provided in accordance with paragraph 23 shall be an intrinsic part of the design and construction of the materials handling equipment.

## Controls

25. The arrangement and design of dial displays and controls, and the general layout and design of the operator's compartment or position on all materials handling equipment shall:

(1) contribute to the safe operation of the materials handling equipment; and
(2) not hinder or prevent the operator from operating the materials handling equipment in accordance with good industrial safety practice.

## Fire extinguishers

26. Fire extinguishers are to be provided in accordance with fire protection engineering standards and practices published by the Fire Commissioner of Canada.

## Means of entry or exit

27. All materials handling equipment operated or maintained by an employee is to be provided with a safe means of entry into and exit from:

(1) the compartment or position of the operator; and
(2) any place on the equipment in which an employee must be in order to service the equipment.

## Tool boxes

28. Tools, materials or parts carried on any materials handling equipment shall be stored in a tool box or other secure and safe place where they will not endanger the operator or any employee.

29. Where any mobile equipment is operated or used by an employee at night, or in areas where the illumination level within the area of operation of the equipment is less than 1 dalx, the mobile equipment is to be:

(1) fitted on the front and rear thereof with warning lights that are visible at night from a distance of not less than 100 m; and
(2) provided with general illumination sufficient to ensure the safe operation of the equipment under all conditions of use.

30. Where the general illumination referred to in paragraph 29(2) is provided by lighting facilities on the mobile equipment, the lighting facilities shall comply with paragraphs 31 and 32.

31. Notwithstanding paragraph 29, no operator shall operate any mobile equipment at night on a road that is used by other vehicles unless it is equipped with such lighting facilities for mobile equipment as are prescribed by the laws of the province or territory in which the equipment is operated.

32. Where lighting facilities for mobile equipment are not prescribed by a law of the province or territory in which the equipment is operated, the lighting facilities on that equipment shall comply with CSA Standard D106.1-1977, *Vehicle Lighting Equipment*, or such other standard as is acceptable to Human Resources Development Canada.

## Slow moving vehicles

33. Mobile equipment operated at a rate of speed that is more than 30 km below the posted speed for a road or area shall be equipped with a slow moving vehicle warning device as prescribed by the laws of the province or territory in which the equipment is operated.

34. Where the laws of the province or territory in which the mobile equipment is operated do not prescribe a slow moving vehicle warning device, such mobile equipment shall be equipped with a warning device in accordance with CSA Standard D198-M1977, *Slow Moving Identification Vehicle Emblem*.

## Safe loads

35. Materials handling equipment shall not be operated or used with a load that is in excess of its maximum safe load.

36. Where the failure of any materials handling equipment due to overloading would create an unsafe work condition:

(1)   the maximum safe load of that equipment shall be clearly marked on the equipment or on a label securely attached to a permanent part of the equipment in a position where the mark or label can be easily read by the operator, and such mark or label shall be maintained in a legible condition; and
(2)   where appropriate, equipment shall be provided with a diagram securely attached to a permanent part of the equipment, showing the lift capacity at each attitude and length of the boom or other lifting member.

37. Notwithstanding paragraph 36, the labelling of chains, ropes and slings may be waived by a safety officer if some other means of determining their safe loads is readily available to all employees required to use them, and procedures are adopted that are acceptable to a safety officer and that will ensure that the safe loads are not exceeded.

## Control systems

38. All mobile equipment that is operated or used by an employee shall be fitted with braking, steering and other control systems that:

(1)   are capable of safely controlling and stopping the movement of the mobile equipment and any hoist, bucket or other part thereof; and
(2)   respond positively, reliably and quickly to moderate effort on the part of the operator.

39. When directed in writing by a safety officer:

(1)    power-assisted systems shall be provided for the braking, steering or other control systems of the mobile equipment; and

(2)    an alternate power source for braking and steering shall be provided on equipment that cannot be controlled safely by an operator in the event of engine failure.

40. Any mobile equipment normally used for transporting employees as passengers from place to place on a work site shall be equipped with a mechanical parking brake as well as a hydraulic or pneumatic braking system.

41. Where any materials handling equipment has a moving part with a limit as to safe operating speed or safe travelling distance, an automatic control shall be provided for that part, where practicable, to prevent its speed or distance of travel, as the case may be, from exceeding that limit.

## Starting devices

42. Where it is practicable to do so, all mobile equipment that is operated or used shall be fitted with a power operated starting device.

43. No operator of any mobile equipment that is fitted with a power operated starting device shall use a hand crank to start the equipment unless the starting device fails.

44. When the starting device referred to in paragraph 42 fails, it shall be repaired or replaced.

## Warnings

45. Any mobile equipment shall be fitted with a horn or similar audible warning device, having a distinctive sound that can be clearly heard above the noise of the equipment and any surrounding noise. Where audible warning devices do not provide adequate warning, visual flashing signals such as strobe lights shall be used.

## Seat belts

46. Any mobile equipment, operated or used under conditions where safety seat belts or shoulder type restraining devices are likely to contribute to the safety of the operator or passengers, shall be fitted with such seat belts or devices.

47. Every employee, while travelling on any mobile equipment that has been fitted with safety seat belts or a shoulder type restraining device, shall use or wear such seat belts or device.

## Rear view mirror

48. Where a motor vehicle or any other mobile equipment cannot be operated safely unless the operator has a clear view of the area behind the vehicle or equipment, such vehicle or equipment is to be equipped with sufficient mirrors to provide a clear view of the rear. Where an operator does not have a clear view around a vehicle or equipment, a signaller shall act as a guide.

## Electrical equipment

49. Any electrically powered materials handling equipment shall be so designed and constructed that the operator or any other employee will be protected from electrical shock or injury by means of securely fastened protective guards, screens or panels.

## Automatic equipment

50. Any mobile equipment, controlled or operated by a remote or automatic system, and used in any place where it may make dangerous contact with any employee, is to be prevented from making such contact.

51. Where it is not practicable to comply with paragraph 50, at every dangerous intersection or place along the roadway or path of travel of any mobile equipment referred to therein, the safety of employees is to be protected by an alarm system, an emergency stop system or protective barriers or other safety devices, systems or procedures that are approved by a safety officer.

## Docks and ramps

52. Every loading and unloading dock, platform and ramp used by any employee shall be:

(1)   of sufficient strength to support, without failure, the maximum load to which it will be subjected; and
(2)   generally free of surface irregularities that may interfere with the safe control of mobile equipment; and
(3)   fitted, around any of its sides that are not used for loading or unloading, with side rails, bumpers or rolled edges of sufficient height and strength to prevent mobile equipment from running over the edge; or
(4)   used in a manner that will prevent mobile equipment from running over the edge;
(5)   constructed to provide protection against injury by moving edges.

53. Every portable ramp or dock plate used by an employee shall be:

(1)   clearly marked or tagged to indicate its maximum safe load; and
(2)   provided with a means of attaching it firmly and securely in place, except where it is so designed that it cannot slide, move or otherwise be displaced under the load that it is required to support.

54. Paragraph 53(1) does not apply where:

(1)   some means of establishing the safe load of a portable ramp or dock plate, other than that described in that paragraph, is available to all employees required to use such ramp or plate; and
(2)   procedures are adopted by employees which will prevent the maximum safe load of the portable ramp or dock from being exceeded.

55. No operator shall operate or be permitted to operate any mobile equipment on a ramp with a gradient in excess of:

(1)    the gradient that is recommended as safe for that type of ramp by the manufacturer of the mobile equipment, either loaded or unloaded, as applicable;

(2)    such lesser gradient as is safe in the opinion of the person in charge, with regard to the mechanical condition of the mobile equipment and its load and traction; or

(3)    such gradient as may be directed in writing by a safety officer.

## Conveyors

56. Each conveyor, cableway or other similar materials handling equipment that is operated, used or serviced by any employee is to be designed, constructed, operated and maintained in accordance with American Society of Mechanical Engineers standard B20.1(1957), or with a standard recommended by Human Resources Development Canada.

57. Where a conveyor, cableway or other similar materials handling equipment crosses a roadway or walkway at ground or floor level, and such equipment could come into contact with an employee or a vehicle, a safe passageway for such employee or vehicle, as the case may be, shall be provided.

58. Where a conveyor, cableway or other similar materials handling equipment crosses a roadway, walkway or work area used by employees, the equipment is to be so guarded that material from the equipment cannot fall on any employee or vehicle passing underneath it.

59. Where a "go-slow" sign, restricted clearance sign or other warning sign is placed at a conveyor cross-over for the safety of employees, the approaches to such a cross-over shall be clearly marked at a safe distance from the cross-over with an appropriate warning sign.

## Clearances

60. Subject to paragraphs 63 and 64, the person in charge shall ensure, on any route that is regularly travelled by mobile equipment operated or used by any employee, that clearances are provided that comply with paragraph 61.

61. Every clearance referred to in paragraph 60 shall:

(1)    in the case of an overhead clearance, be at least 150 mm above:

    (a)    that part of the mobile equipment or its load that is highest when the mobile equipment is in its highest normal operating position at the point of clearance; and

    (b)    the top of the head of every employee authorized to ride on the mobile equipment, when the employee is occupying the highest normal position at the point of clearance; and

(2)    in the case of a side clearance, be adequate to permit the mobile equipment and its load to be manoeuvred safely by an operator, but in no case less than 150 mm on each side measured from the farthest projecting part of the equipment or its load, when the equipment is being operated in a normal manner.

62. If an overhead clearance measured in accordance with paragraph 61(1)(a) and (b) is less than 300 mm, the top of the doorway or object that restricts the clearance shall be marked with a distinguishing colour or mark; and the height of the passageway in metres shall be shown near the top of the passageway in letters that are not less than 50 mm in height and are on a contrasting background.

63. Paragraphs 61(1)(a) and 62 do not apply to:

(1) any mobile equipment whose course of travel is controlled by fixed rails or guides;
(2) that portion of the route of any mobile equipment that is inside a railway car, truck or trailer truck, including the doorway of the car, truck or trailer truck and the warehouse doorway leading directly thereto; or
(3) a load that is larger than that normally transported by mobile equipment, if such a load is transported infrequently and special precautions are taken to prevent contact with objects that might restrict the movement of the equipment.

64. Where it is not practicable to provide a clearance prescribed by paragraph 61 and, in the opinion of a safety officer, a lesser clearance would not be dangerous, such lesser clearance may be approved in writing by the safety officer.

**Aisles and corridors**

65. Where an aisle, corridor or other course of travel which exceeds 15 m in length is a principal traffic route for pedestrians and mobile equipment, a clearly marked walkway not less than 750 mm wide shall be provided along one side of the route for the use of pedestrians only.

66. Paragraph 65 does not apply where measures other than those described therein are adopted for the purpose of controlling traffic and protecting pedestrians and such measures comply with good industrial safety practice.

67. Where an aisle, corridor or other course of travel that is a principal traffic route intersects another such aisle, corridor or course of travel and, in the opinion of a responsible departmental officer or a safety officer, such intersection is dangerous:

(1) warning signs marked with the words "Dangerous Intersection — Croisement dangereux" or similar words in letters not less than 50 mm in height on a contrasting background are to be posted along the approaches to the intersection; and,
(2) to the extent that it is practicable, every blind corner is to be provided with mirrors in such manner that an operator of any mobile equipment that is approaching the corner along one course of travel can see a pedestrian or vehicle approaching the intersection along the other intersecting course of travel.

## Ropes, chains and slings

68. Subject to this standard, and to any direction in writing that may be given by a safety officer, the design and construction of any rope, chain or sling and of any fittings and attachments thereon that are used by any employee shall comply with the recommendations contained in the *Accident Prevention Manual of the National Safety Council.*

69. Notwithstanding paragraph 68, any steel wire rope intended for use as materials handling equipment by any employee shall comply with:

(1)   the recommendations of the manual referred to in paragraph 68;
(2)   the recommendations of CSA standard G4-1976, *Steel Wire Rope for General Purpose and for Mine Hoisting and Mine Haulage*; or
(3)   any other standard acceptable to Human Resources Development Canada.

## Operation, use and maintenance of materials handling equipment

### Inspections and checks

70. Before any materials handling equipment is operated or used for the first time, it is to be inspected and tested by a qualified person in accordance with the requirements of the department and the operating and maintenance manuals for that equipment, to determine whether it is in a safe operating condition and is suited to the purpose for which it is to be used.

71. Following the inspection and test referred to in paragraph 70, a safety and maintenance check of all materials handling equipment is to be made by a qualified person as frequently as is necessary to ensure the safe operation of the equipment, or as may be directed in writing by a safety officer.

72. Where required by good industrial safety practice, or where directed in writing by a safety officer, a safety and maintenance check schedule is to be provided and maintained for all materials handling equipment, which schedule shall show:

(1)   the equipment checked;
(2)   the date of the check;
(3)   the nature of the check; and
(4)   the maintenance work performed on the equipment.

73. A copy of each schedule referred to in paragraph 72 shall be retained on file for at least one year at the location where the materials handling equipment is maintained and shall be readily available for examination upon request by a safety officer.

74. Every operator who has been assigned any materials handling equipment shall, immediately before placing that equipment in operation for the first time on a shift, make visual inspection of that equipment and such other inspection as may be directed by the person in charge in order to ensure, to the extent that is possible from such inspections, that the equipment is safe for operation.

75. Every operator referred to in paragraph 74 shall report and document in writing to the person in charge, any defect or condition in any materials handling equipment that is believed will affect the safe operation of the equipment that is required to be operated and if, in the operator's opinion, any such defect or condition is unsafe, the operator shall not operate the materials handling equipment until it has been examined by the person in charge or a safety officer and declared to be safe to operate.

**Operators**

76. For the purpose of training, an employee who is not an operator may be permitted to operate materials handling equipment if that employee is accompanied by an operator who can take over control of the equipment in the case of an emergency.

77. An operator who, in the opinion of the person in charge or a safety officer, appears to be suffering from a physical condition that may suddenly incapacitate him or her, or appears to have some other physical disability that may affect the ability to safely steer or otherwise safely operate the materials handling equipment that has been assigned, shall not be permitted to operate the materials handling equipment until it has been determined through a medical assessment arranged in accordance with the Occupational health evaluation standard (chapter 2-13), that the operator is free of any such condition or disability or is incapable or operating that equipment safely.

78. If a law of the province or territory requires that the operator of a certain type of materials handling equipment possess an operator's licence, no operator shall operate or be permitted to operate that type of materials handling equipment unless he or she possesses the operator's licence required by that law.

79. Paragraph 78 does not apply in respect to an operator who has successfully passed an appropriate competency test, conducted by or on behalf of a department and is in possession of a valid permit or authority issued by the department to operate the equipment for which he or she was tested, providing such operation is restricted to the department's premises.

80. Subject to paragraph 79, no operator shall operate any mobile equipment from other than the operator's regular position, or another position designed specifically for that purpose.

81. Mobile equipment may be operated from a position other than one referred to in paragraph 80, where the control of the equipment and the view of the work area from the position is at least as good and as safe as from the operator's regular position on that equipment, and such position is approved by the person in charge.

82. Operators shall not operate any mobile equipment unless they:

(1) have a clear and unobstructed view of the work area and the course to be travelled; or

(2) are under the direction of a signaller and have the approval of the person in charge.

## Repairs

83. For the purposes of paragraphs 84 and 85, "integrity" means the ability of any materials handling equipment, part or fixture thereof to retain all of its safe and reliable performance.

84. Any repair, modification or replacement of a part of any materials handling equipment that is operated or used by any employee shall not decrease the safety factor and integrity of the equipment or part.

85. If a part of lesser strength or quality than the original part is used in the repair, modification or replacement of a part of any materials handling equipment, the use of the equipment is to be restricted to such loading and use as will ensure the retention of the original safety factor and integrity of the equipment or part.

## Combination of equipment

86. Mobile equipment shall not be operated or used by any employee in an assembly or a combination with other materials handling equipment unless the safety of that assembly or combination is at least equal to that required by this standard for the separate parts of the assembly or combination in respect of braking, steering and general operating controls and safety.

## Passengers

87. Subject to paragraph 88, unless authorized by the person in charge, no employee other than the operator and his or her assistants shall ride or be permitted to ride on any mobile equipment or any part thereof, or on any material transported thereon, unless the mobile equipment is specifically designed for the transport of such additional passengers.

88. A trainee operator or a person inspecting or testing any mobile equipment may accompany the operator if a secure seat or other safe place is provided on the mobile equipment.

## Loading and maintenance while in motion

89. No employee shall pick up from, or place upon, any mobile equipment any materials or supplies while the mobile equipment is in motion unless the mobile equipment is specifically designed for that purpose.

90. Except in the case of an emergency, no employee shall get on or off any mobile equipment while it is in motion.

91. No employee shall perform any repairs, maintenance or cleaning work on any materials handling equipment while it is being operated, except on those fixed parts of the equipment that are so isolated or protected that the operation of the equipment does not affect the safety of the employee performing that work.

## Starting precautions

92.  No employee shall start the power unit of any materials handling equipment until all drive clutches have been disengaged, all brakes set and the operator is assured that no person will be endangered by the starting of the power unit.

93.  Where the power unit of any materials handling equipment operated or used by any employee cannot be started from the operator's position, specific procedures or safeguards are to be employed that will prevent the accidental movement of the equipment during the starting of the power unit.

94.  Subject to paragraphs 95 and 96, all mobile equipment that is operated by any employee is to be shut down during any period that it is unattended.

95.  Subject to paragraph 96, where it is not reasonably practicable to shut down any mobile equipment while it is unattended, the operator of the mobile equipment shall secure it against accidental movement by placing the transmission in neutral, and setting a parking or mechanical brake, blocking the wheels and/or tracks, or by using other measures acceptable to a safety officer.

96.  Where, in any circumstances described in paragraph 95, any mobile equipment is left unattended on an incline, it shall be secured against accidental movement by setting the parking or mechanical brake and blocking the wheels.

97.  No operator who is operating a crane, hoist or similar materials handling equipment shall leave any such equipment unattended other than in a condition of maximum stability unless some other equally safe measure approved by the person in charge is taken to prevent the equipment from tilting or otherwise accidentally moving.

## Positioning and securing load

98.  Where mobile equipment is travelling with a raised or suspended load, the operator shall ensure that the load is carried as close to the ground or floor level as good industrial safety practice and local conditions permit, and in no case shall the load be carried at a point above the centre of gravity of the loaded mobile equipment or the point at which the loaded mobile equipment becomes unstable.

99.  No operator shall operate or be permitted to operate mobile equipment that is loaded in such a manner as to obstruct the view in the direction of travel.

100.  No operator shall operate or be permitted to operate any materials handling equipment, unless the load that it is carrying is so secured that it cannot slide or move to a dangerous extent or be toppled or dislodged from the equipment under any normal condition of operation, including a sudden swerve or an emergency stop at the maximum speed at which the equipment is permitted to operate.

## Housekeeping

101. The floor, cab and other occupied parts of any materials handling equipment that is operated or used by any employee is to be kept free of any grease, oil, materials, tools or equipment that might cause a fire hazard or an employee to slip or trip, or might otherwise interfere with the safe operation of the equipment.

## Parking

102. No operator shall park any mobile equipment in a corridor, aisle, doorway or other place where that equipment might interfere with the safe movement of other equipment, materials or persons.

## Hazard area

103. In paragraphs 104 to 108, "hazard area" means any area within which a crane, hoist, shovel or other readily mobile materials handling equipment or equipment with wide swinging booms or other similar parts is operating and might injure any person.

104. The main approaches to any hazard area are to be posted with suitable warning signs or shall be under the control of a signaller while operations are in progress.

105. No employee or other person shall enter or be permitted to enter a danger area while operations are in progress unless that person is a safety officer, or an employee whose presence in the hazard area is essential to the conduct, supervision or safety of the operations, or a person who has been authorized by the person in charge to be in the hazard area.

106. If any person, other than a person referred to in paragraph 105, enters a hazard area while operations are in progress, the person in charge shall ensure that operations in that area are immediately discontinued and are not resumed until that person has left the area.

107. Subject to paragraph 104, any materials handling equipment or part thereof that is operated or used by any employee shall not, because of the wide swing of its booms or overhead loads, or for any other reason, extend into any adjacent travelled or other occupied areas outside the hazard area.

108. Where an operator cannot avoid the extension of any materials handling equipment or part thereof into areas outside the hazard area, the person in charge shall ensure that barricades, overhead protection or other barriers are erected to prevent any such extension, or a signaller is provided to warn the operator when the operation is hazardous.

## Overhead loading

109. No operator shall occupy the operator's position on any motor vehicle or mobile equipment if, during the overhead loading or unloading thereof, the load must pass over the operator's position, unless that position is protected by an

overhead shield or guard of sufficient strength to prevent injury to the operator in the event that the load accidentally falls onto the motor vehicle or mobile equipment.

110. No operator shall begin, or be permitted to begin, the operation of any materials handling equipment in an area where there is a possibility that it might contact an electrical cable, gas pipeline or other overhead or underground hazard unless the operator has been:

(1) warned of the presence of every such known hazard;
(2) instructed, in accordance with the best information available, concerning the exact location of every overhead or underground electrical cable, gas pipeline or other hazard in the immediate vicinity of the operation; and
(3) informed of the specified safety clearances that must be maintained with respect to any overhead or underground hazard in order to avoid contact with it.

111. Where the location of a hazard referred to in paragraph 110 cannot be determined with certainty, or the person in charge is unable to provide the safety clearances referred to in paragraph 110(3), every electrical cable is to be de-energized and every pipeline containing a dangerous substance is to be shut down and drained before a digging or other operation involving the use of materials handling equipment commences within the area of possible contact with such a hazard.

**Bumping blocks**

112. Where rear dumping mobile equipment is required to discharge its load at the edge of a sudden drop in grade level that is of sufficient depth to cause tipping of the mobile equipment, a substantial bumping block, a signaller or other means that is acceptable to a safety officer is to be provided to prevent the mobile equipment from backing over that edge.

**Fuelling**

113. Materials handling equipment using flammable fuels shall be fuelled in accordance with the requirements or standards prescribed by the Fire Commissioner of Canada.

114. No operator shall operate or be permitted to operate any mobile equipment on a gradient in excess of:

(1) the gradient that is recommended as safe by the manufacturer of the mobile equipment; or
(2) such lesser gradient that is safe, having regard to the mechanical condition of the mobile equipment, the weight of the load it is transporting and the condition of the roadway.

## Signals

115. When signals are used to direct the safe movement of any materials handling equipment that is operated or used by employees, the signals are to be given only by a signaller and only a standard code of signals is to be used.

116. No employee, other than a signaller referred to in paragraph 115, shall give signals to direct the movement or operation of any materials handling equipment; however, any person may cause a stop signal to be given in an emergency and any such signal shall be obeyed by an operator.

117. No signaller shall direct the movement or operation of any materials handling equipment, except in accordance with a standard code of signals.

118. No signaller shall be employed or occupied otherwise than as a signaller during the time that any mobile equipment under his or her direction is in motion or operation in an area where signals are required to be given.

119. A standard code of signals shall be provided to each signaller, operator and to other persons who are required to understand such signals, and such employees are to be instructed, trained and tested in the use of the code.

120. A copy of the standard code of signals shall be filed and be readily available for examination by a safety officer.

121. Where, in any work area, signals are required for the safe direction of any mobile equipment, and it is not practicable to use visual signals, the person in charge shall ensure that a telephone, radio or other signalling device acceptable to a safety officer is used.

122. No person shall use radio transmitting equipment for the purpose of transmitting signals in any area occupied by employees when such use might activate electric blasting equipment in that area.

123. Before using a radio for the purpose of transmitting signals, the person in charge shall ensure that another transmitting device within the vicinity will not interfere with reliable transmission of signals.

124. Every wire and cable used in a signalling system is to be protected against damage that is likely to interfere with the transmission of signals.

125. Any signalling device referred to in paragraph 121 that functions unreliably or improperly shall be immediately removed from service and shall not be returned to service until it has been examined, repaired and tested by a qualified person, and found to be functioning properly.

126. Where a signalling device referred to in paragraph 121 functions unreliably or improperly and the operation of any mobile equipment cannot be safely directed by another means of signalling, the mobile equipment is to be shut down until the signalling device is functioning properly.

127. Every employee operating a defective signalling device shall report the defect to the person in charge.

128. Where the safety of any person is likely to be endangered by the unexpected movement of any materials handling equipment that is controlled by a signal, the signaller shall not give the signal to move until that person is properly warned or protected.

129. Where the operator of any materials handling equipment does not clearly understand a signal referred to in paragraphs 121 to 128, he or she shall regard that signal as a stop signal.

**Ropes, slings and chains**

130. Any rope or sling, or any attachment or fitting thereon that is used by any employee shall be used and maintained in accordance with the recommendations contained in the *Accident Prevention Manual of the National Safety Council*, or with a standard that conforms to good industrial safety practice and is acceptable to a safety officer.

131. Any chain that is used by an employee shall be used and maintained in accordance with recommendations contained in CSA Standard B-75 (R-1964), *Code of Practice for the Use and Care of Chains*, or with a standard that conforms to good industrial safety practice and is acceptable to a safety officer.

**Manual handling of materials**

132. Where, because of the weight, size, shape, toxicity or other characteristic of a material or object, the manual handling of that material or object may endanger the safety or health of an employee, the person in charge shall ensure that the material or object is not so handled.

133. Employees who are regularly required to manually lift or carry loads in excess of 10 kg shall be instructed and trained in a safe method of lifting and carrying such loads, and in a work procedure appropriate to the conditions of that work and the employee's physical condition.

134. No employee shall manually lift or carry loads in excess of 10 kg except in conformity with the method and work procedure referred to in paragraph 133.

135. Where weight, shape, size or other characteristic of a material or object to be moved are excessive for one employee, the department shall determine an alternate method for handling the material or object.

136. Each method and work procedure adopted pursuant to paragraph 133 for the manual lifting and carrying of loads in excess of 45 kg shall be set out in writing and that record shall be readily available to any safety officer and to any employee to whom it applies.

137. If a safety officer is of the opinion that a method or work procedure prescribed by a department for the manual lifting and carrying of loads in

excess of 10 kg is not sufficiently safe, the safety officer may direct in writing that the method or procedure be modified.

## Storage of materials

138. Subject to this standard and chapter 2-2, Dangerous substances directive, all materials are to be stored in accordance with the recommendations contained in the *Accident Prevention Manual of the National Safety Council*, or with any standard that conforms to good industrial safety practice. They shall also be stored in accordance with the fire safety requirements of the Fire Commissioner of Canada.

139. All materials are to be placed and stored in such a manner that the maximum safe load-carrying capacity of the floor and any other supporting structure is not exceeded.

140. Unless otherwise approved in writing by a safety officer, any materials that are stacked in piles shall be stacked in such a manner that the piles do not:

(1)  create a hazard by interfering with the distribution of light;
(2)  obstruct or encroach upon passage-ways, traffic lanes or exits;
(3)  impede the safe operation of materials handling equipment;
(4)  obstruct the ready access to, and the use and free operation of, fire fighting equipment;
(5)  interfere with the proper operation of sprinklers and other fixed fire protection and prevention equipment and devices;
(6)  endanger the safety and health of any employee; and
(7)  conceal any warning signs or symbols.

141. All materials are to be stored in such a manner that they will not collapse, fall, slip, topple or otherwise endanger employees who are stacking or removing materials or working in the storage area.

142. Where a person in charge, or other responsible departmental official, is informed in writing by a Human Resources Development Canada Regional Director that any materials handling equipment, floor, dock or any other structure that is used by employees may not be adequate because of deterioration, age, misuse, abuse or for any other reason, the person in charge, or official, shall ensure that the maximum safe load or other operating restrictions in respect to the equipment, floor, dock or other structure is determined and reported by a qualified person who is acceptable to the Human Resources Development Canada Regional Director.

143. Any safety restriction or revised safe load limits shall continue in force until the original safety of the equipment or other facilities is restored.

144. One copy of the report of the determination by the qualified person referred to in paragraph 142 shall be submitted to the Human Resources Development Canada Regional Director.

**Resolving "qualified person" disputes**

145.  Where there is a dispute regarding the term "qualified person" for purposes of an occupational safety and health standard, the following procedure shall be implemented:

(a)  The employee shall raise the matter directly with the person in charge.
(b)  The person in charge shall review the employee's qualifications and decide upon the employee's status as a qualified person.
(c)  If the employee is dissatisfied with the decision, the matter shall be referred to the safety and health committee established for the employee's workplace.
(d)  The safety and health committee shall review the matter and make appropriate recommendations to the person in charge.
(e)  If the safety and health committee does not consider itself competent to deal with the case, it shall recommend an acceptable third party to the person in charge.
(f)  The person in charge shall, pursuant to (d) or (e), take the recommendations into consideration, render a final management decision and undertake the appropriate action.

If the employee does not agree with the final decision which has been rendered, a grievance may be initiated pursuant to the NJC redress procedure.

## CHAPTER 2-11 - MOTOR VEHICLE OPERATIONS DIRECTIVE

### Grievance procedure

In cases of alleged misinterpretation or misapplication arising out of this standard, the grievance procedure, for all represented employees, within the meaning of the *Public Service Staff Relations Act*, will be in accordance with Section 7.0 of the *National Joint Council By-Laws*. For unrepresented employees, the departmental grievance procedure applies.

This standard is deemed to be part of collective agreements between the parties to the National Joint Council and employees are to be afforded ready access to this standard.

### Application

1.  This standard applies to all departments and other portions of the Public Service, as defined in Part I of Schedule I of the *Public Service Staff Relations Act*.

### Purpose

2.  This standard outlines the requirements for the safe operation of motor vehicles owned or leased by Public Service departments, to ensure the safety and health of employees and the public, and to avoid property or equipment damage.

## Definitions

3. In this standard:

(1) **motor vehicle** means a truck, tractor, trailer, semi-trailer, automobile, bus, all-terrain vehicle, snowmobile or other similar self-propelled vehicle used primarily for transporting personnel and/or material (véhicule à moteur);

(2) **motor vehicle accident** means an event involving the operation of a motor vehicle which results in injury to persons and/or damage to equipment or property (accident);

(3) **motor vehicle operator** means any employee who is required to operate a motor vehicle in the performance of the employee's duties (conducteur);

(4) **qualified person** means, in respect of a specified duty, a person who, because of knowledge, training and experience, is qualified to perform that duty safely and properly (See paragraph 35) (personne qualifiée).

## General responsibilities

4. Departments are responsible for:

(1) developing accurate departmental rules and procedures for the safe operation of motor vehicles, in accordance with the general principles set forth in this standard;

(2) analyzing and evaluating motor vehicle accident reports and statistics, determining the causes of accidents and utilizing this information to prevent additional accidents from similar causes;

(3) ensuring that every motor vehicle is maintained in a safe operating condition;

(4) ensuring that motor vehicle operators are qualified in all respects to operate the vehicles to which they are assigned;

(5) enforcing safe driving rules and traffic regulations on premises and in operations under their control;

(6) cooperating with civil authorities in the enforcement of traffic laws and the observance of safe practices; and

(7) ensuring that employees are fully informed of the correct procedures to be followed in the event of an accident.

## Safe operation of motor vehicles

5. The operation of an unsafe motor vehicle is prohibited. A motor vehicle is unsafe when any defect exists which, in the judgment of the responsible supervisor in consultation with a qualified motor vehicle mechanic, could contribute to an accident. A motor vehicle operator shall not be required to operate a mechanically unsafe vehicle or a vehicle loaded in a hazardous manner.

6. All motor vehicles, including emergency motor vehicles such as ambulances, shall be operated in a prudent manner and at speeds compatible with road, traffic, weather and visibility conditions, and in compliance with the appropriate federal, provincial, territorial or municipal laws.

### Hazardous movement

7. Prior to the movement of oversize or overweight motor vehicles, or those carrying dangerous articles or equipment over public highways, notification of the route and the utilization of public bridges, tunnels and/or highways is to be given to appropriate civil officials. The movement of dangerous substances by motor vehicle shall be subject to requirements specified in chapter 2-2, Dangerous substances directive.

8. Motor vehicles that are regularly operated in remote or isolated areas shall be equipped with appropriate communications devices for emergency purposes.

### Medical examinations

9. Employees who are required to operate buses, ambulances, emergency vehicles and heavy mechanical or mobile equipment shall undergo health evaluations pursuant to chapter 2-13, Occupational health evaluation standard.

### Qualification of motor vehicle operators

10. Every motor vehicle operator shall possess a valid licence to operate the motor vehicle to which the operator is assigned in accordance with the appropriate provincial or territorial law, or as may be otherwise required by regulations or statutes applicable to the Public Service.

11. In addition, motor vehicle operators may be required to demonstrate their competence to operate assigned motor vehicles and, in this regard, appropriate records shall be maintained.

### Training

12. Departments shall, where appropriate, institute or participate in motor vehicle operator training programs designed to provide:

(1) refresher training to acquaint personnel with changes in equipment or operating conditions; and
(2) remedial training to offset specific weaknesses indicated by accident records, traffic rule violations or other instances of inadequate operating performance.

13. Departments shall ensure that a record of the training required by paragraph 12 is maintained for each employee.

### Investigation of accidents

14. Every motor vehicle accident is to be investigated, the cause or causes determined and appropriate corrective action applied. Additionally, a Supervisor's accident investigation report is to be completed pursuant to chapter 4-1, Accident investigation and reporting.

15. Departments shall maintain a record of vehicle repairs or replacement as a result of accidents.

## Servicing and inspection

16. Each department is responsible for ensuring that the servicing and inspection of its motor vehicles meet normal preventive maintenance and safety requirements commensurate with the use of motor vehicles, but in no case shall the level of maintenance be less than the requirements outlined in the appropriate manufacturer's user manual.

## Safe transportation of personnel

17. At the start of each shift, each operator is to be responsible for carrying out a brief inspection of the motor vehicle assigned. Defects are to be reported promptly to the responsible supervisor.

18. To the extent possible, personnel are to be transported in passenger type motor vehicles such as sedans, station wagons and buses. The following safety rules shall apply:

(1)   only authorized personnel shall be permitted to ride in motor vehicles;
(2)   the number of persons permitted to ride in a passenger motor vehicle must not exceed the seating capacity of that motor vehicle except when being transported locally for short distances in buses provided with handholds;
(3)   personnel shall not be permitted to ride with any part of their person extended outside the motor vehicle, or on a running board, fender, cab, side or tailgate of a motor vehicle;
(4)   personnel shall not board or alight from a motor vehicle while it is in motion; and
(5)   tools, equipment and cargo shall be properly stowed and secured to prevent shifting in transit.

19. When it is not possible or practicable to use passenger motor vehicles to transport personnel, truck type motor vehicles may be used. In such cases the additional safety measures listed below shall apply:

(1)   fixed seating is to be provided and sideboards or stakes and tailgates fitted;
(2)   the number of personnel to be transported may not exceed that for which fixed seating is provided;
(3)   a suitable cover should be provided for protection from the elements;
(4)   tools, equipment and cargo shall be properly stowed and secured to prevent shifting in transit;
(5)   a motor vehicle operator shall:
(a)   brief personnel on safety requirements and appoint a person in charge of passenger conduct;
(b)   release and lower the tailgate prior to the loading and unloading of passengers; and
(c)   operate the motor vehicle with special caution relating to speed, road conditions, starting, stopping and turning.

20. Under special conditions, trucks without fixed seating may be used for transporting small groups (less than ten) for short distances on departmental property. Passengers are to be in a secure position within the body of the truck, and the vehicle driven with extreme caution. If the use of a dump truck is authorized for such a purpose, the hoist controls are to be positively secured to prevent inadvertent operation.

## Fire prevention

21. No motor vehicle shall be operated unless it is entirely free of fuel leaks.

22. Motor vehicles shall be equipped with portable fire extinguishers conforming to FC Standard No. 401, *Fire Extinguishers*, published by the Fire Commissioner of Canada.

## Motor vehicle fuelling and operations

23. The following safety procedures and any other applicable procedures specified by the Fire Commissioner of Canada, or that office's authorized representative, shall be followed during the fuelling of motor vehicles:

(1)   motor vehicles are not to be fuelled indoors;
(2)   only a qualified person shall be permitted to fuel a motor vehicle;
(3)   open flame, spark producing devices or smoking are not to be allowed within 7.5 m of fuelling operations or areas;
(4)   during fuelling, the engine of the motor vehicle must be stopped, the ignition and lights turned off, the parking or emergency brake applied, and the nozzle of the fuel hose kept in contact with the fuel intake pipe to prevent electrical arcing;
(5)   when reserve supplies of fuel are to be carried on motor vehicles, they shall be carried in approved containers adequately secured and protected.

24. Tank trucks shall be loaded and unloaded in authorized areas by qualified personnel and under controlled procedures, in accordance with the *National Fire Code of Canada*, 1985.

25. Fire safety operations for industrial trucks shall conform to FC Standard No. 304, *Industrial Trucks*, published by the Fire Commissioner of Canada.

## Propane vehicles

26. Requirements specified by the Fire Commissioner of Canada shall be followed concerning the operation and storage of propane vehicles.

27. Each employee who is required to fuel a propane vehicle shall be tested and licensed by those provinces where such licensing is required pursuant to provincial statute.

28. In provinces where licenses are not required, departments shall certify employees for propane fuelling through internal departmental training and such training shall be equal to, or better than, that provided in a province where licenses are required.

29. Each employee who is certified pursuant to paragraph 28 shall:

(1)  be familiar with the specific safety precautions and operating procedures applicable to vehicle fuelling;
(2)  be able to identify and understand the functions and components of vehicle fuel supply systems;
(3)  be able to identify all components of a fuel dispenser and demonstrate capability in safely fuelling a vehicle; and
(4)  pass a written examination on the fuelling procedures applicable to the fuel handled.

## Safety measures against asphyxiation

30. The concentration of toxic exhaust fumes to which the operator and other persons are exposed when working on or near motor vehicles shall not exceed the levels as are prescribed pursuant to chapter 2-2, Dangerous substances directive.

## Motor vehicle safety belts

31. Operators of, and passengers in, motor vehicles which are equipped with safety belts shall be required to fasten such safety belts in the approved manner at all times when the vehicle is in motion.

## Highway warning devices

32. Motor vehicles operated on roads or in areas at speeds of more than 30 km per hour below the posted speed for the road or area, shall be equipped with a warning device as prescribed by the statutes of the province or territory in which the vehicle is operated, or in the absence of such requirements, in accordance with CSA Standard D198-1977, *Slow Moving Identification Vehicle Emblem*.

33. In the event that a motor vehicle becomes disabled on or adjacent to the highway, advance warning devices such as flares or reflectors shall be placed in accordance with the statutes of the province or territory in which the vehicle is disabled.

## First-aid kits

34. Motor vehicles shall be equipped with first-aid kits in accordance with the requirements of chapter 2-5, First-aid directive.

## Resolving "qualified person" disputes

35. Where there is a dispute regarding the term "qualified person" for purposes of an occupational safety and health standard, the following procedure shall be implemented:

(a)  The employee shall raise the matter directly with the person in charge.
(b)  The person in charge shall review the employee's qualifications and decide upon the employee's status as a qualified person.

(c) If the employee is dissatisfied with the decision, the matter shall be referred to the safety and health committee established for the employee's workplace.

(d) The safety and health committee shall review the matter and make appropriate recommendations to the person in charge.

(e) If the safety and health committee does not consider itself competent to deal with the case, it shall recommend an acceptable third party to the person in charge.

(f) The person in charge shall, pursuant to (d) or (e), take the recommendations into consideration, render a final management decision and undertake the appropriate action.

If the employee does not agree with the final decision which has been rendered, a grievance may be initiated pursuant to the NJC redress procedure.

## CHAPTER 2-12 - NOISE CONTROL AND HEARING CONSERVATION DIRECTIVE

### Grievance procedure

In cases of alleged misinterpretation or misapplication arising out of this standard, the grievance procedure, for all represented employees, within the meaning of the *Public Service Staff Relations Act*, will be in accordance with Section 7.0 of the *National Joint Council By-Laws*. For unrepresented employees, the departmental grievance procedure applies.

This standard is deemed to be part of collective agreements between the parties to the National Joint Council and employees are to be afforded ready access to this standard.

### Application

1. This standard incorporates the minimum requirements of the *Canada Labour Code*, Part II, and applicable regulations issued pursuant to that legislation, and applies to all departments and other portions of the Public Service, as defined in Part I of Schedule I of the *Public Service Staff Relations Act*.

### Definitions

2. In this standard:

(1) **A-weighted sound pressure level** means a sound pressure level indicated by a measurement system that includes an A-weighting filter as specified by the International Electrotechnical Commission (IEC) Standard 651-1979 "Sound Level Meters", as amended from time to time (niveau de pression acoustique pondéré A);

(2) **dBA** means "decibel A-weighted" and is a unit of A-weighted sound pressure level (dBa);

(3) **motor vehicle operator** means an employee operating a motor vehicle with a gross vehicle weight of more than 4,500 kg (10,000 lb) (conducteur de véhicules automobiles);

(4)     *noise exposure level* (LEX,8) means ten times the logarithm to the base 10 of the time integral over any 24-hour period of the squared, A-weighted sound pressure divided by 8, the reference pressure being 20 micropascals. (niveau d'ambiance sonore (LEX,8));

(5)     *sound level meter* means a device for measuring sound pressure level, the specifications of which meet the performance requirements for a Type 2 instrument as specified by the International Electrotechnical Commission (IEC) Standard 651-1979, *Sound Level Meters*, as amended from time to time (sonomètre);

(6)     **sound pressure level** means 20 times the logarithm to the base 10 of the ratio of the root mean square pressure of a sound to the reference pressure of 20 micropascals expressed in decibels (niveau de pression acoustique).

## Monitoring and exposure

3.  Environmental health officers of Health Canada will monitor noise levels and exposure to noise and, where necessary, give appropriate direction to departments and agencies in accordance with the requirements of this standard and Treasury Board procedure 4-2: Occupational health investigations.

## Hazard investigation

4.  Where an employee, other than a motor vehicle operator, is or may be exposed in a workplace to an A-weighted sound pressure level equal to or greater than 84 dBA for a period that is likely to endanger his or her hearing, the department shall, without delay:

(a)     appoint a qualified person to carry out an investigation of the degree of exposure; and

(b)     notify the safety and health committee or safety and health representative, if any, of the proposed investigation and the name of the qualified person appointed to carry out that investigation.

## Measurement of exposure

5.(1)  The exposure of an employee to sound shall be measured using an instrument appropriate for that measurement that meets the standard set out for such an instrument in clause 4, Instrumentation, of the Canadian Standards Association Standard CAN/CSA-Z107.56, the English version of which is dated December 1986 and the French version of which is dated September 1987, as amended from time to time.

(2)  The procedure for measuring the exposure of an employee to sound shall be appropriate for that measurement and meet the standards set out for that measurement in clauses 5, 6.4.1, 6.4.4, 6.5.2, 6.5.4, 6.6.2, and 6.6.4 of the Canadian Standards Association Standard CAN/CSA-Z107.56-M86, the English and French versions of which are dated December 1986, as amended from time to time.

(3) The A-weighted sound pressure level referred to in paragraph 4 shall be measured instantaneously using the slow response setting of a sound level meter at any time during normal operating conditions.

6. In the investigation referred to in paragraph 4, the following criteria shall be taken into consideration:

(a)    the levels of sound to which the employee is or may be exposed;
(b)    the duration of exposure to such levels of sound;
(c)    the sources of sound;
(d)    the control methods used to eliminate or reduce the exposure; and
(e)    whether the exposure of the employee to sound is likely to exceed the limits prescribed in paragraph 11.

7.(1) On completion of the investigation referred to in paragraph 4 and after consultation with the safety and health committee or the safety and health representative, if any, has taken place, the qualified person shall set out in a signed written report:

(a)    observations respecting the criteria considered under paragraph 6;
(b)    recommendations respecting the manner of compliance with paragraphs 4 and 6;
(c)    recommendations respecting the use of hearing protectors by employees who are exposed to a noise exposure level equal to or greater than 84 dBA but less than 87 dBA; and
(d)    an indication of which employees are likely to be exposed to a noise exposure level equal to or greater than 84 dBA.

(2) The signed report referred to in this paragraph shall be kept readily identifiable to the workplace to which it applies for a period of ten years and a copy shall be given to the appropriate safety and health committee.

8. Where the investigation referred to in paragraph 4 reveals that the employee is or is likely to be exposed to a noise exposure level equal to or greater than 84 dBA, the department shall:

(a)    post and keep posted a copy of the report referred to in paragraph 7 in a conspicuous place in the workplace to which it applies; and
(b)    provide information and instruction to the employee that outlines the hazards associated with routine exposure to high sound levels.

9.(1) Departments shall ensure that employees exposed to a noise exposure level equal to or greater than 84 dBA have their hearing level tested, including the required audiograms, in accordance with the requirements outlined in the Occupational health evaluation standard, chapter 2-13.

(2) Audiogram test results shall be permanently retained on employees' medical files.

**Engineering controls**

10. As far as is reasonably practicable, every department shall, through engineering controls, reduce the exposure of every employee, other than a motor vehicle operator, to or below a noise exposure level of 87 dBA whether or not the employee is using a hearing protector.

**Limits of exposure**

11.(1) Notwithstanding paragraph 10, no employee other than a motor vehicle operator shall, in any 24-hour period, be exposed to:

(a)  an A-weighted sound pressure level listed in column I of Appendix A for a duration of exposure that exceeds the number listed in column II of that appendix for that A-weighted sound pressure level; or

(b)  a noise exposure level that exceeds 87 dBA.

(2) No motor vehicle operator shall, in any 24-hour period, be exposed to an A-weighted sound pressure level listed in column I of Appendix B for a duration of exposure that exceeds the number listed in column II of that A-weighted sound pressure level.

**Report to regional safety officer (Human Resources Development Canada)**

12. Where it is not reasonably practicable, without providing hearing protectors, for a department to maintain the exposure to sound of an employee, other than a motor vehicle operator, at or below the limits prescribed in paragraph 11, departments shall:

(a)  make a report in writing to the regional safety officer at the regional office setting out the reasons why the exposure cannot be so maintained; and

(b)  provide a copy of the report referred to above to the safety and health committee or the safety and health representative, if any.

13.(1) Where a department has made a report referred to in paragraph 12(a), the department shall provide every employee, other than a motor vehicle operator, who is exposed to sound in excess of a limit prescribed in paragraph 11 with a hearing protector that:

(a)  meets the standards set out in CSA Standard Z94.2-M1984, *Hearing Protectors*, the English version of which is dated June 1984 and the French version of which is dated February 1985, as amended from time to time; and

(b)  prevents the exposure of the employee using the hearing protector from exceeding the limits prescribed in paragraph 11.

(2) Where the department has provided a hearing protector to an employee, the department shall:

(a)  in consultation with the safety and health committee or the safety and health representative, if any, formulate a program to train the employee provided with the hearing protector in its fit, care and use; and

(b) implement the program referred to above.

(3) Every person, other than an employee, granted access to a workplace whose exposure to sound is likely to exceed a limit prescribed in paragraph 11 shall use a hearing protector that meets the standard referred to in paragraph 13.(1)(a).

**Warning signs**

14. Departments shall post and keep posted at conspicuous locations at any workplace where an employee, other then a motor vehicle operator, is or may be exposed to an A-weighted sound pressure level equal to or greater than 87 dBA, warning signs that there exists a potentially hazardous level of sound in the workplace.

**CHAPTER 2-12 - APPENDIX A - MAXIMUM PERMITTED DURATION OF EXPOSURE TO A-WEIGHTED SOUND PRESSURE LEVEL AT Workplace**

| Column I | Column II |
|---|---|
| A-weighted sound pressure level dBA | Maximum duration of exposure in hours per employee, per 24-hour period |
| 80 | 40 |
| 81 | 32 |
| 82 | 25 |
| 83 | 20 |
| 84 | 16 |
| 85 | 13 |
| 86 | 10 |
| **87** | **8.0** |
| 88 | 6.4 |
| 89 | 5.0 |
| 90 | 4.0 |
| 91 | 3.2 |
| 92 | 2.5 |
| 93 | 2.0 |
| 94 | 1.6 |
| 95 | 1.3 |
| 96 | 1.0 |
| 97 | 0.80 |
| 98 | 0.64 |
| 99 | 0.50 |
| 100 | 0.40 |
| 101 | 0.32 |
| More than 102 | 0.00 |

**Note 1** - While exposure durations in excess of 24 hours per day are not possible, they have been listed in this table to allow calculation of the noise exposure.

**Note 2** - This table reflects an exchange rate of 3 dB.

# CHAPTER 2-12 - APPENDIX B - MAXIMUM PERMITTED DURATION OF EXPOSURE TO A-WEIGHTED SOUND PRESSURE LEVEL FOR MOTOR VEHICLE OPERATORS

| Column I | Column II |
|---|---|
| A-weighted sound pressure level dBA | Maximum duration of exposure in hours per employee, per 24-hour period |
| 90 | 8 |
| 92 | 6 |
| 95 | 4 |
| 97 | 3 |
| 100 | 2 |
| 102 | 1.5 |
| 105 | 1 |
| 110 | 0.5 |
| 115 | 0.25 |
| More than 115 | 0 |

# CHAPTER 2-13 - OCCUPATIONAL HEALTH EVALUATION STANDARD

## Application

1. This standard applies to all Public Service departments and agencies, as defined in Part I of Schedule I of the *Public Service Staff Relations Act*.

## Definition

2. In this standard, the term *health evaluation* (examen de santé) means any specific screening, assessment or examination of an employee which is carried out by a health professional to determine or monitor the employee's occupational health status, and includes simple interventions such as immunizations.

## Purpose

3. The principal objectives of health evaluations are: to act as a means of preventing illness and disability arising out of, or aggravated by, conditions of work; to establish that individuals are able to continue working without detriment to their health or safety or that of others; and to establish the conditions under which certain individuals with illnesses or disabilities are able to continue working.

4. Health evaluations are generally provided in respect of specific occupations which have an inherent element of risk to the health or safety of an employee; where an employee's actions could result in a threat to the health and safety of another; before certain postings; and where a Public Service standard, policy directive or guideline provides that such evaluations may be requested at the discretion of departmental management or Health Canada.

## Specific application and exclusions

5.  Health evaluations shall be required for employees engaged in the occupations/activities listed in Appendix A, Health Evaluation Schedule. Occupations referred to in Orders-in-Council governing physical standards for Civil Aviation Personnel are excluded from this standard.

## Procedures

6.  All health evaluations are to be arranged by the employing departments and agencies through the appropriate Zone or Regional Office of Medical Services Branch, Health Canada, using the forms and procedures prescribed by that Department.

7.  Health evaluations will be initiated in accordance with the frequencies specified herein. However, in individual cases, Health Canada may recommend investigations or evaluations at intervals more frequent than those prescribed in Appendix A. Health evaluations shall be carried out wherever practicable during normal working hours.

8.  The medical costs associated with health evaluations, except those subject to Foreign Service directives, are the responsibility of Health Canada. Where an employee is required to undertake travel for a health evaluation, the employee will be considered in travel status, and reimbursement of expenses incurred shall be governed by the Treasury Board Travel directive or other applicable authority.

9.  Following any health evaluation, a report indicating the employee's capability to perform the required work will be forwarded by the appropriate Medical Services Branch office of Health Canada to the employing department.. This report will not contain any medical or psychological diagnosis or provide any reasons for the conclusions drawn in the report.

10.  Where work limitations are identified, the statement will incorporate advice to management concerning the adaptation or selection of work, or the placement or reassignment of the employee, and the estimated duration of such limitations. In this regard, departments shall make every effort to provide work for which the employee is physically qualified and is, or can be, trained to perform.

11.  Each employee will be advised of the results of a health evaluation by Health Canada. In the event of a health problem being discovered, the employee should be referred to his or her own physician for advice.

## Special evaluations

12.  Subject to the requirements of paragraphs 6 and 9, departments should arrange for special health evaluations where:

(1)  on the advice of Health Canada, an employee has been exposed to an occupational health hazard;

(2) a health problem may be the cause of impaired work performance which has been identified through an Employee Assistance Program;

(3) in the judgement of the person in charge, the work performance of an employee during the employee's probationary period appears to be consistently impaired owing to factors which may be related to health or physical condition. In this case, the health evaluation should be directed towards confirming the employee's physical suitability for continued employment in that position.

## Health evaluation categories

13. The three health evaluation categories are as follows:

**Category 1**: A confidential general health questionnaire completed by the employee and screened by a nurse. Unusual clinical histories will be brought to the attention of a physician, who will determine whether follow-up action is necessary. (This category is not applicable at this time.)

**Category 2**: A confidential general health questionnaire administered by a nurse, who may also perform certain basic investigations depending on the type of work and particular hazards involved. A qualified technician may carry out some of these procedures, including the completion of special reports, but does not administer the general health questionnaire.

**Category 3**: A confidential health questionnaire, administered by a nurse or physician, followed by a full clinical history and physical examination, and special investigations as required.

## Physical examination standards

14. The establishment of all physical examination standards is the responsibility of Health Canada, in consultation with appropriate specialists.

## Medical confidentiality

15. All medical information, forms and records transmitted or used in connection with these health evaluations will be maintained in a medical confidential status, and retained within the medical community as authorized by Health Canada.

## CHAPTER 2-13 - APPENDIX A - HEALTH EVALUATION SCHEDULE

| GROUP | OCCUPATION/ACTIVITY | CATEGORY | FREQUENCY |
|---|---|---|---|
| 1 | Not allocated. | | |
| 2 | Hospital and health service employees. | 3 | Every 2 years. |
| 3 | a. Ship's personnel, including all other persons required to accompany a ship on cruise. | 3 | Every 2 years to age 40 and annually thereafter; dental exams before every voyage longer a than 4 months; audiograms every 6 months for engine room personnel and radio operators. |
| | | 2 | Before each voyage where the ship will be more than 24 hours cruising from a hospital. |
| | b. Marine surveyors. | 3 | Every 3 years to age 40 and annually thereafter (audiograms annually). |
| 4 | Personnel operating buses, ambulances, emergency vehicles, heavy mechanical or mobile equipment (See Note 1) and school buses. | 3 | Every 3 years to age 45, and annually thereafter; annual Tuberculin tests for school bus drivers if clinically indicated. |
| 5 | Not allocated. | | |
| 6 | a. Isolated and/or remote posting areas. | 3 | Annually, before proceeding to remote areas (includes dental examination). |
| | b. Ice observers. | | Every 2 years to age 40 and annually thereafter (dental exams before each voyage longer than 4 months). |
| 7 | Not allocated. | | |

| 8 | Animal keepers, veterinarians, primary products inspectors (health of animals). | As required | As determined by Health and Welfare Canada. |
|---|---|---|---|
| 9 | Firefighters (part time and full time). | 3 | Every 2 years to age 35 and annually thereafter. |
| 10 | Flight service specialists, Marine traffic regulators, Coast guard radio operators and Marine search and rescue controllers. | 3 | Every 2 years to age 40 and annually thereafter (audiograms every 6 months). |
| 11 | Personnel exposed to ionizing or non-ionizing radiation. See Note 2. | As required | As required by Radiation Protection Bureau. |
| 12 | Personnel exposed to excessive noise levels, as defined in the Noise control and hearing conservation standard, chapter 2-12. | 2 | Every 2 years to age 40 and annually thereafter (audiograms every 6 months). |
| 13 | Personnel exposed to chemical or biological hazards. See Note 3. | As required | As determined by Health Canada |
| 14 | Lighthouse keepers and dependants at remote locations. | 3 | Annually. |
| 15 | Personnel engaged in underwater diving, whether or not this is their primary duty. | 3 | Annually. |
| 16 | Personnel (including dependants) posted to isolated posts, as defined in the Isolated Posts directives. | 3 | Before each posting. |
| 17 | Personnel (including dependants) serving abroad, subject to | 3 | Before each posting and cross-posting, and upon return to Canada. |

|  |  |  |  |
|---|---|---|---|
|  | Foreign Service directives. |  |  |
| 18 | Correctional services personnel (COF, LUF and members of emergency response teams). | 3 | Every 3 years to age 40; every 2 years to age 50 and annually thereafter (ECGs annually over age 40). |
| 19 | Personnel liable to consistent arduous physical effort or exposure outdoors, as determined by Health Canada and in consultation with Treasury Board. See Note 4. | 3 | Every 2 years to age 40 annually thereafter. |
| 20 | Not allocated. |  |  |

**Notes:**

**Note 1:**  Heavy mechanical or mobile equipment includes trucks over 10,800 kg gross vehicle weight and hoists, cranes, excavators, graders, loaders, heavy tractors, trenchers, snow blowers and sweepers.

**Note 2:**  Personnel in this group include those who are required to wear film badges for the detection of ionizing radiation, and personnel exposed to microwave radiation above the permitted energy level.

**Note 3:**  Chemical hazards include liquids, gases, dusts, fumes, mists and vapours.  Biological hazards include insects, mites, nematodes, molds, yeasts, fungi, viruses and bacteria.

**Note 4:**  This group may include certain employees of Parks Canada, avalanche forecasters and observers, and individuals in similar circumstances.

## CHAPTER 2-14 - PERSONAL PROTECTIVE EQUIPMENT DIRECTIVE

### Grievance procedure

In cases of alleged misinterpretation or misapplication arising out of this standard, the grievance procedure, for all represented employees, within the meaning of the *Public Service Staff Relations Act*, will be in accordance with Section 7.0 of the *National Joint Council By-Laws*. For unrepresented employees, the departmental grievance procedure applies.

This standard is deemed to be part of collective agreements between the parties to the National Joint Council and employees are to be afforded ready access to this standard.

## Application

1. This standard applies to all departments and other portions of the Public Service, as defined in Part I of Schedule I of the *Public Service Staff Relations Act*.

## Definitions

2. In this standard:

(1)    *integrity* means, in respect of any device or equipment, the ability of that device or equipment to retain all of the qualities essential to its safe, reliable and adequate performance (intégrité);

(2)    *person in charge* means a qualified person appointed by management to ensure the safe and proper conduct of an operation or of the work of employees (responsable);

(3)    *personal protective equipment* means any clothing, equipment or device worn or used by a person to protect that person from injury or illness (équipement de protection individuelle);

(4)    *qualified person* means a person who, because of knowledge, training and experience, is qualified to perform safely and properly a specified job (see paragraph 50) (personne qualifiée);

(5)    *safety officer* means a person designated as a safety officer by the Minister of Human Resources Development pursuant to the *Canada Labour Code*, Part II (agent de sécurité);

(6)    *safety restraining device* means any safety belt, safety harness, seat, rope, belt, strap or life-line designed to be used by an employee to protect that employee from injury due to falling, and includes every fitting, fastening or accessory thereto (dispositif protecteur de soutien).

## General responsibility of departments

3. Where an employment hazard (refer to appendix A) cannot be eliminated or controlled within safe limits, and the wearing or use of personal protective equipment by an employee may prevent an injury or reduce its severity, departments shall ensure that each employee who is exposed to an employment hazard wears or uses that equipment as prescribed by this standard.

4. The provision of personal protective equipment shall be in accordance with the Clothing directive as outlined in chapter 1-3.

5. Employees shall be instructed and trained by a person who is qualified in the proper and safe operation, use and care of all personal protective equipment that they are required by this standard to wear or use.

6. All personal protective equipment worn or used by any employee shall be adequate in all respects to protect the employee from the hazards of employment, be otherwise suitable for use by the employee, and have been so designed that it does not in itself create an employment hazard.

7. All personal protective equipment shall be stored, maintained, inspected and tested by a qualified person for the purpose of ensuring that it is in a safe and fully effective condition at all times.

8. A record of personal protective equipment shall be maintained in accordance with good industrial safety practice, or as recommended by a Human Resources Development Canada Regional Director, and shall be readily available for examination. The record should contain the following information:

(1)  a description of the equipment and the date of its purchase or acquisition;
(2)  the date and result of each inspection and test of the equipment; and
(3)  the date and nature of any maintenance work performed on the equipment since its purchase or acquisition.

9. Every record required to be kept pursuant to paragraph 8 shall be retained by the department for a period of two years in a place where it is readily available for examination by a safety officer.

### General responsibility of employees

10. No employee shall commence a work assignment or enter a work area where any kind of personal protective equipment is required by this standard to be worn or used unless the employee:

(1)  is wearing or using that kind of personal protective equipment in the manner prescribed in this standard;
(2)  has been instructed and trained in the proper and safe operation and use of that personal protective equipment pursuant to paragraph 5; and
(3)  has visually inspected that personal protective equipment to ensure it will protect against the hazards of employment.

11. Every employee shall care for all personal protective equipment that is assigned, in accordance with the instructions and training given as outlined in paragraph 5.

12. Every employee shall immediately report to the person in charge any personal protective equipment that, in the opinion of the employee, no longer adequately provides protection from the hazards of employment.

### Head protection

13. Where, in accordance with paragraph 3, an employee is required to wear a safety hat, the safety hat shall comply with the recommendations of Canadian Standards Association Standard Z94.1-M1977, *Industrial Protective Headwear*, or with a standard acceptable to Human Resources Development Canada.

14. Where, in accordance with paragraph 3, an employee is required to wear a form of head protection other than a safety hat, such head protection shall comply with good industrial safety practice or with a standard acceptable to Human Resources Development Canada.

## Eye and face protection

15. Where, in accordance with paragraph 3, an employee is required to wear eye or face protection, such eye or face protection shall comply with CSA Standard Z94.3-1982, *Industrial Eye and Face Protectors*, or with a standard acceptable to Human Resources Development Canada.

## Foot and leg protection

16. Where, in accordance with paragraph 3, an employee is required to wear protective footwear or purpose-designed footwear, such footwear shall have soles and heels of a material that will minimize slipping under all conditions of their normal use, and in all other respects they shall comply with CSA Standard Z195-M1981, *Protective Footwear*, or with a standard acceptable to Human Resources Development Canada.

17. Where, in accordance with paragraph 3, an employee is required to wear leg protection or foot protection other than protective footwear or purpose-designed footwear, such leg protection or foot protection shall comply with the appropriate Canadian Standards Association standard, or with a standard acceptable to Human Resources Development Canada.

18. In an industrial fabricating, processing, maintenance, repair or storage area or in any other workplace designated by a safety officer, employees shall not wear, or be permitted to wear:

(1) footwear constructed of canvas, fabric, soft rubber or other lightweight material that provides insufficient protection from impact or puncture;
(2) sandals or other footwear that have open toes; or
(3) any other type of footwear the wearing of which is prohibited by a safety officer.

## Skin protection

19. Where, in accordance with paragraph 3, an employee is required to wear personal protective equipment or a barrier cream for skin protection:

(1) such personal protective equipment or barrier cream shall be adequate to protect the skin of the employee during the entire period during which the skin is exposed to any hazard; and
(2) if such personal protective equipment is not disposable, it shall be maintained in a clean and sanitary condition.

With respect to the hazards of ultraviolet radiation (UVR) associated with sunlight:

(1) exposure to UVR must be reduced as much as possible, and where such exposure cannot be avoided employees' skin must be protected against the adverse effects of UVR;
(2) departments, in consultation with local occupational safety and health committees, shall carefully review the various situations where employees

are required to work outdoors and take all reasonable practicable measures to reduce exposure to the harmful effects of the sun;

(3)    where such consultations identify potential health risks, departments shall supply an appropriate broad-spectrum sunscreen with a minimum sun protective factor of 15 as recommended by Health Canada.

It should be noted that this section does not apply to sunglasses.

## Respiratory protection

20.  Where, in accordance with paragraph 3, an employee is required to wear respiratory equipment, such respiratory equipment shall be of a type approved for its intended use by the United States Bureau of Mines, the United States National Institute of Occupational Safety and Health or by a person or agency acceptable to Health Canada or Human Resources Development Canada.

21.  Where air or oxygen is provided in connection with any respiratory equipment referred to in paragraph 20, the air or oxygen shall comply with CSA Standard Z.180.1-M1978, *Compressed Breathing Air*, or with a standard approved in writing by Human Resources Development Canada.

## Safety restraining devices

22.  Unless an employee is wearing a safety restraining device that complies with this standard, the employee shall not be required or permitted to work while standing on or supported by:

(1)    any unenclosed or unguarded work structure that is:
    (a)    more than 2.4 m directly above the nearest permanent safe level,
    (b)    above an operating machine, device, structure or obstruction that could cause injury to the employee upon contact, or
    (c)    above any open-top tank, pit or vat;
(2)    any scaffold or other similar elevated work structure that is more than 6 m above a permanent safe level and from which the employee may fall if the structure tips or fails;
(3)    any ladder at a height more than 2.4 m directly above the nearest permanent safe level if, because of the nature of the work, one hand cannot be used to hold onto the ladder; or
(4)    any other elevated work structure in respect of which a safety officer directs in writing that a safety restraining device shall be used.

23.  Notwithstanding paragraph 22, the use of a safety restraining device is not required where:

(1)    such use is, in the circumstances, unsafe or not practicable as determined by a qualified person; and
(2)    other safety measures approved in writing by a safety officer are employed.

24.  Every ladder from which an employee is working, as described in paragraph 22(3), shall be secured in such a manner that it cannot be accidentally or inadvertently dislodged from its position.

25. Every safety restraining device used by employees shall be of sufficient strength, at all times and under all conditions of its normal use, to support, without failure or loss of integrity, the maximum load to which it will be subjected and, in any case:

(1) a static load of not less than 450 kg; and
(2) a load of not less than 180 kg that is applied suddenly at the end of a 1.2 m vertical drop or such greater distance as the safety restraining device may permit the load to fall.

26. Each type of safety restraining device required by this standard to be worn or used by employees shall, prior to being worn or used, be tested as prescribed in paragraphs 25 to 42, for the purpose of determining whether the design and fabrication of that type of safety restraining device satisfies the test requirements stated therein.

27. The test referred to in paragraph 26 shall be conducted by the manufacturer, distributor or seller of the safety restraining device, or by a person acceptable to Human Resources Development Canada.

28. At least one representative sample of each type of safety restraining device produced by each manufacturer shall, after it is assembled, be subjected for test purposes to:

(1) loads that are one and one-half times the loads prescribed in paragraphs 25(1) and 25(2); or
(2) any other load or stress approved in writing by Human Resources Development Canada.

29. If the sample restraining device referred to in paragraph 28 is unable to support without failure of any kind the test loads prescribed therein, departments shall ensure that none of the restraining devices of which it is representative is used by their employees.

30. The sample restraining device tested pursuant to paragraph 28 shall:

(1) not be placed in service after being subjected to the test loads prescribed by that paragraph;
(2) be marked or tagged to indicate that it is not to be placed in service;
(3) be marked or tagged with the date of the test and the name and position of the person who conducted the test; and
(4) be readily available for examination by a safety officer.

31. The test referred to in paragraph 28 shall, with respect to each manufacturer of safety restraining devices, be conducted before distribution. Where there is a change in the design, method of fabrication or the kind or quality of material used in the fabrication of the safety restraining device, the test shall be conducted after the change is made and, in any event, before the safety restraining device incorporating or resulting from that change is distributed.

32. Where a written guarantee or warranty is given by the manufacturer, distributor or seller of a type of safety restraining device representing that the type of safety restraining device in question has been tested and complies with the requirements of this standard, that type of safety restraining device may be deemed to have been tested and to comply with the requirements of paragraphs 26 and 28.

33. Any body safety belt is deemed to satisfy the requirements of paragraph 28(2) if it complies in all respects with the requirements of CSA Standard Z259.3-M1978, *Lineman's Body Belt and Lineman's Safety Strap*.

34. Each fitting, anchor and accessory used in connection with a safety restraining device shall comply with the recommendations contained in CSA Standard Z91-M1980, *Safety Code for Window Cleaning Operations*, or with a standard acceptable to Human Resources Development Canada.

35. To the extent that it is practicable, every safety restraining device shall be worn or used in such a manner that the person wearing or using it cannot fall freely for more than 1.2 m.

36. Paragraph 35 does not apply to a safety restraining device that incorporates a shock absorbing mechanism that limits the effect of the fall to that produced by a free fall of 1.2 m or less.

37. Not more than one person shall use one lifeline at the same time.

38. All safety restraining devices shall be inspected and serviced by a qualified person at intervals appropriate to their use, and safety restraining devices that are used once a week or more often shall be inspected and serviced by such a person at least once each month. A record of such inspection and servicing shall be maintained in accordance with good industrial safety practice.

39. Where a safety officer is of the opinion that an inspection made pursuant to paragraph 38 is not sufficient to determine the strength or integrity of a safety restraining device, or where its strength or integrity is likely to be decreased because of its age or use, the safety officer may direct in writing that a representative sample of the safety restraining device be subjected to test loads as prescribed in paragraphs 25 to 32 or such lesser loads as he or she considers appropriate.

40. Where a safety restraining device fails to meet the requirements of the test referred to in paragraph 39, two additional representative samples of the same type of safety restraining device shall be tested and, if either of such samples fails to meet those requirements, all safety restraining devices of which the samples are representative shall be removed from service.

41. Every safety restraining device that has been subjected to a load exceeding its maximum safe working load while being tested shall not be returned to service.

42. For the purpose of paragraph 41, the maximum safe working load is the quotient obtained when the minimum load in kilograms necessary to break the weakest part of the assembled safety restraining device is divided by five.

## Drowning hazards

43. No employee shall work or be permitted to work at any work location where there is a risk to that employee of drowning unless:

(1)   the employee is wearing an approved life jacket or other buoyancy device of a type described in paragraph 44 or the employee is prevented at all times during such work from falling into the water by a safety net, platform or a safety restraining device; and

(2)   the employee is accompanied by at least one other person who can provide assistance, if required.

44. The life jacket or buoyancy device referred to in paragraph 43(1) shall be a jacket or device capable of supporting a person with the head above water in a face-up position, without effort on the person's part, until rescue can be effected.

45. Where there is a risk of hypothermia due to falling into icy water, suitable insulated protective clothing shall be provided to employees who are so exposed.

46. Where circumstances of work over or near water are such that, in the opinion of the person in charge or where recommended by a safety officer, a rescue boat is required, a suitable boat shall be provided and:

(1)   where required according to local conditions, be equipped with a suitable motor maintained in operational readiness;

(2)   be operated by a qualified person and fitted with appropriate rescue equipment; and

(3)   be held in readiness at a location enabling quick rescue during periods that such rescue services are required.

## Loose clothing

47. Where an employee is wearing loose clothing, long hair, dangling accessories, rings or other jewellery that might become entangled with a machine or any rotating or moving part of that machine, or the metallic part of which might come into contact with energized electrical equipment, the employee shall not enter or be permitted to enter a work area where any such machine or equipment is operating unless the clothing, hair, accessories, rings or other jewellery is so tied, fitted, covered or otherwise secured as to prevent such entanglement or contact.

## Traffic hazards

48. Any employee who is assigned to give traffic signals or direction or who is otherwise exposed to a possible hazard from vehicular traffic during work, shall:

(1) wear a high visibility vest or other similar clothing; or
(2) be protected by a high visibility barricade.

49. The high visibility vest and barricade referred to in paragraph 48 shall be readily noticeable or distinguishable all of the time and under all of the conditions that the employee is exposed to vehicular traffic.

## Resolving "qualified person" disputes

50. Where there is a dispute regarding the term "qualified person" for purposes of an occupational safety and health standard, the following procedure shall be implemented:

(a) The employee shall raise the matter directly with the person in charge.
(b) The person in charge shall review the employee's qualifications and decide upon the employee's status as a qualified person.
(c) If the employee is dissatisfied with the decision, the matter shall be referred to the safety and health committee established for the employee's workplace.
(d) The safety and health committee shall review the matter and make appropriate recommendations to the person in charge.
(e) If the safety and health committee does not consider itself competent to deal with the case, it shall recommend an acceptable third party to the person in charge.
(f) The person in charge shall, pursuant to (d) or (e), take the recommendations into consideration, render a final management decision and undertake the appropriate action.

If the employee does not agree with the final decision which has been rendered, a grievance may be initiated pursuant to the NJC redress procedure.

## CHAPTER 2-14 - APPENDIX A - EMPLOYMENT HAZARDS

The following is only a representative listing of occupational safety and health hazards where an employee may require the protection provided by personal protective equipment.

### Safety hazard sources

– animals, birds, reptiles
– heavy boxes, crates, packages
– tanks, bins, excavations
– confined spaces
– buildings and structures
– dangerous substances
– mechanical transmission equipment
– electrical apparatus
– fire
– glass
– hand and power tools
– hoisting apparatus
– machines

- metal processing
- minerals and mineral processing
- paper and pulp processing
- plants, trees, vegetation
- plastics processing
- scrap, debris, waste materials
- steam
- textile processing
- wood processing

## Health hazard sources

Chemical:

- liquids
- gases
- dusts
- fumes
- mists
- vapours

Physical:

- ionizing and non-ionizing radiation
- noise
- vibration
- sanitation
- ventilation
- extremes of temperatures and pressure

Biological:

- insects
- mites
- molds
- yeasts
- fungi
- viruses
- bacteria

## CHAPTER 2-15 - PESTICIDES DIRECTIVE
## General

### Collective agreement

This directive is deemed to be part of collective agreements between the parties to the National Joint Council and employees are to be afforded ready access to this directive.

### Grievance procedure

In cases of alleged misinterpretation or misapplication arising out of this directive, the grievance procedure, for all represented employees within the meaning of the *Public Service Staff Relations Act*, will be in accordance with section 7.0 of the *National Joint Council By-Laws*. For unrepresented employees the departmental grievance procedure applies.

### Application

This directive incorporates the minimum requirements of the *Canada Labour Code,* Part II, and applicable regulations issued pursuant to that legislation, and applies to all departments and other portions of the Public Service, as defined in Part 1 of Schedule 1 of the *Public Service Staff Relations Act*.

Definitions

In this directive:

*pest (parasite)* - any injurious, noxious or troublesome insect, fungus, bacterial organism, virus, weed, rodent or other plant or animal pest, and includes any injurious, noxious or troublesome organic function of a plant or animal;

*pesticide (pesticide)* - a product registered and listed under the *Pest Control Products Act* (PCPA) and its regulations intended to prevent, destroy or manage a pest; this includes antimicrobial agents such as disinfectants and sanitizers listed in the PCP Regulations;

*qualified person (personne qualifiée)* - a person who, because of knowledge, training and experience, is licensed or certified in accordance with a provincial or national program.

### Requirements

#### 15.1 Integrated Pest Management (IPM)

15.1.1 Departments shall develop pest management programs that incorporate integrated pest management (IPM) principles and practices to reduce the use of broad-spectrum pesticides.

15.1.2 When a decision is made to use pesticides within the context of an IPM program, the department must ensure that employees are not exposed to health hazards from pesticides.

15.1.3 The goal of IPM is to manage pests effectively, safely and economically, by:

(a)   reducing the use of broad-spectrum pesticides, and using more pest/target specific control products;
(b)   reducing the level of toxicity of products used;
(c)   using alternate control methods; and
(d)   improving and perfecting on methodology used.

15.1.4 IPM is an approach to pest management that integrates all pest management practices and control methods into one pest management program. IPM does not usually try to eliminate all pests, but tries to reduce the pest population to an acceptable level. In IPM, the use of pesticides is advocated as a last resort only.

15.1.5 IPM involves:

(a) identifying pests;
(b) determining the cause and source of the pest;
(c) knowing the pest's life cycle, behaviour and effects on its host, and the most vulnerable period in its life cycle;
(d) monitoring pest activities and effectiveness of control or management methods.

15.1.6 IPM requires knowing and using available methods, such as:

(a) approved biological controls including:
    (i) parasitic and predatory insects; and
    (ii) host-specific pathogens,
(b) maximizing a plant's health and minimizing its susceptibility to pest infestations by:
    (i) crop rotation;
    (ii) moisture control;
    (iii) planting techniques; and
    (iv) sanitation.
(c) genetic selection, i.e., choosing resistant species and varieties of plants;
(d) mechanical controls; e.g., trapping, cultivating, physical barriers;
(e) the use of pesticides which are of relatively low toxicity to human and animal populations, and of low persistency in the environment; e.g., insecticidal soaps;
(f) the use of conventional pesticides in a prescribed manner.

## 15.2 **Work procedures**

15.2.1 Each department in which pesticides are used, handled, stored or disposed of shall ensure the manufacturer's instructions as detailed on the pesticide label, on a material safety data sheet (MSDS) or other manufacturer's literature is readily available in the workplace and is followed.

15.2.2 Where, for research purposes or otherwise, deviations from the manufacturer's instructions are required, an application for a research permit under the requirements of the PCPA shall be obtained before proceeding with such use.

15.2.3 Detailed written procedures regarding the safe use, handling, storage, transportation and disposal of such pesticides, including circumstances where the employee may be required to work alone, are developed in consultation with the Health and Safety Committee, prominently displayed in the workplace, and explained to all employees concerned.

15.2.4  Pesticides shall be used, handled, mixed and disposed of by qualified persons.

15.2.5  When pest control is contracted out, contractors shall be certified or licensed in accordance with the applicable provincial requirements; the provisions of the IPM program shall apply.

15.2.6  A spill contingency plan appropriate to the scale of operations shall be in place prior to any application of pesticides.

## 15.3  Substitution

15.3.1  The least toxic of pesticides recommended for control of a pest or an alternate acceptable control method shall be used whenever control or management of a pest is required.  Pesticides known or suspected of being human carcinogens shall not be utilized except under restricted procedures, and such use shall be conducted by a qualified person.

## 15.4  Isolation

15.4.1  To the extent possible, potentially hazardous pesticide operations should be either isolated from the worker or the worker isolated from the operation.  Isolation techniques that should be considered include but are not limited to the following:

(a)  positive pressure tractor cabs with filtered air supply;
(b)  conducting pesticide operations when the fewest number of employees are in the area;
(c)  isolation chambers for research application of high concentrations of toxic pesticides; and
(d)  enclosing pesticide transfer points in handling facilities and automated application of pesticides.

## 15.5  Protective equipment and clothing

15.5.1  Where total isolation from exposure to pesticides is not feasible, approved respiratory protective devices, eye protection, and personal protective clothing and equipment  appropriate to the potential hazard as identified on the pesticide label or MSDS shall be provided and worn whenever pesticides are handled or used.  Personal protective equipment (including first aid supplies and portable eye wash stations) shall not be kept in the same storage room as pesticides, to avoid contamination.  Departments shall provide personal protective equipment and clothing in accordance with the manufacturer's recommendations and Directive 2-14 (Personal Protective Equipment) and the Clothing directive, as a minimum requirement.

## 15.6  Storage

15.6.1  To the extent possible, quantities of pesticides purchased and stored shall not exceed the needs of one season in accordance with a pest management program.  To the extent possible, pesticides shall be kept in their original containers with labels intact and separately stored in locked cabinets.

Storage cabinets and rooms shall be vented to the outside with controlled access to avoid unauthorized use. Shelving shall be secure and impervious; and no higher than 150 cm, unless specifically designed for safe access above eye level. Appropriate warning signs shall be prominently displayed to identify such locations. Spill-control material appropriate to the pesticides in storage shall be maintained at the storage site.

## 15.7 Disposal

15.7.1 During disposal procedures, all possible precautions shall be taken to ensure that persons and the environment cannot be subsequently contaminated. Waste disposal shall be conducted in accordance with the manufacturer's directions on labels, MSDS and Environment Canada's Code of Good Practice for Handling, Storage, Use and Disposal of Pesticides at Federal Facilities or with other codes or requirements authorized by Environment Canada for this purpose.

## 15.8 Mixing, loading and application equipment

15.8.1 Before mixing and using pesticides, the work procedures developed under Section 15.2 shall be read for special instructions for personal protection and special procedures.

15.8.2 Measuring, mixing, and loading pesticides is the most hazardous stage of pesticide use because of the possibility of contact with the concentrated product.

15.8.3 In addition to appropriate protective clothing and safety equipment identified on the label or MSDS, a liquid-proof apron, to cover the body from chest to knees, should be worn.

15.8.4 Scales, measuring cups, mixing pails, and other equipment used in these operations shall be used only for pesticides. Equipment shall be cleaned and returned to locked storage when not in use.

15.8.5 Application equipment shall be selected, calibrated, operated, and maintained in accordance with established procedures to ensure employee safety and uniform application of the pesticide only to the desired target area at the correct rate and to prevent contamination of non-target areas.

## Application of pesticides

### 15.9 General

15.9.1 Departments shall ensure that decisions related to pesticide application programs and subsequent re-entry shall be developed in consultation with the workplace Occupational Safety and Health (OSH) Committee. To the extent possible, all pesticide applications shall be carried out when employees are not present.

## 15.10 **Indoors**

15.10.1  Five days prior to the application, employees shall be informed of the intended pesticide application by way of posted signs and a notice.  Both of them shall include:

(a)  name of the product to be used;
(b)  PCP registration number;
(c)  reason for application;
(d)  date of application;
(e)  telephone number to contact for information;
(f)  time for safe re-entry into the treated area.

15.10.2  Signs shall remain posted for at least 48 hours after application, unless a longer time is specified for safe re-entry.

15.10.3  The time for safe re-entry into the treated area shall be determined by referring to the product label, the MSDS, or the Occupational and Environmental Health Services Directorate of Health Canada.

## 15.11 **Outdoors**

15.11.1  Warning signs shall be posted 24 hours prior to application.  However, it is recognized that under certain unforeseen weather conditions, spraying operations may have to be initiated on short notice; under these circumstances, the 24-hour pre-application posting requirement may not be possible, but signs must nonetheless be posted prior to pesticide application.

15.11.2  Signs shall remain posted for at least 48 hours after application, unless a longer time is specified for safe re-entry.

15.11.3  Signs must be made of weather-resistant material.  They should be approximately 50 cm high by 40 cm wide.

15.11.4  The sign shall contain the following wording:

**WARNING - PESTICIDES USED/
ATTENTION - PESTICIDES UTILISÉS**

15.11.5  The sign shall also contain a warning pictogram that alerts the public not to touch or walk on treated plants or areas.

15.11.6  The sign shall also indicate the following:

(a)  date of application;
(b)  name of pesticide used;
(c)  PCP registration number;
(d)  reason for application;
(e)  telephone number for information; and
(f)  safe re-entry date.

### 15.12 Greenhouses, barns, etc.

15.12.1 Application requirements are the same as for outdoors, except that signs shall be posted 24 hours prior to application.

### 15.13 Personal hygiene

15.13.1 After handling pesticides and before attending to personal needs, employees should wash thoroughly, with special attention to the face, hands, hair, and under fingernails.

15.13.2 Departments shall ensure that protective clothing and equipment are cleaned after every use in accordance with Directive 14, Personal Protective Equipment.

### 15.14 Pesticide emergencies

15.14.1 If a spill or leak of pesticides occurs, the spill contingency plan prepared in accordance with paragraph 15.16.1 shall be implemented.

## Transportation and decontamination

### 15.15 Transportation

15.15.1 Procedures developed for transporting pesticides, as outlined in article 5(2) shall meet the requirements of the *Transportation of Dangerous Goods Act* (TDGA) concerning preparation and packaging for transportation, and transporting pesticides. This includes documentation, placarding, and labelling requirements of pesticides being transported, as well as training requirements and responsibilities of employees involved in these operations.

15.15.2 Certain small quantities of pesticides may be exempted from the requirements of the TDGA. This can be determined by referring to the appropriate sections of the Transportation of Dangerous Goods Regulations (TDGR).

15.15.3 Pesticides shall be transported in a separate compartment from the driver and passengers, and shall not be transported in the same compartment containing animals, food, animal feed, clothing, household furnishings, or other personal items.

15.15.4 All pesticides being transported shall be inspected to ensure the integrity of containers, and be placed in the vehicle in a safe manner to avoid tipping, spilling, or leaking.

15.15.5 All pesticide containers shall have the original label intact. A list of pesticides being transported, with a copy of the labels, shall be kept by the driver.

15.15.6 Spill clean-up equipment appropriate to the quantities of pesticides being transported shall accompany the shipment.

15.15.7 Vehicles used for transportation of pesticides shall be posted with a warning sign as follows:

## WARNING - PESTICIDES - ATTENTION

15.15.8 Vehicles used for transportation of pesticides shall also be:

(a)   decontaminated before being used for any other purpose;
(b)   equipped with safety locks and locked when unattended.

15.15.9 Vehicles used for the occasional transportation of pesticides shall meet the above requirements to the extent practicable.

### 15.16  Decontamination

15.16.1 Decontamination of a spill site shall be in accordance with a predetermined spill contingency plan and carried out with the latest techniques advocated by emergency organizations.

15.16.2 Decontamination of pesticide spills shall be carried out by a person trained in decontamination of pesticide spills and supervised by a qualified person.

15.16.3 All empty pesticide containers shall be decontaminated, recycled or disposed of in accordance with the Code of Good Practice for Handling, Storage, Use and Disposal of Pesticides at Federal Facilities.

15.16.4 Application equipment shall be decontaminated in accordance with the Code of Good Practice for Handling, Storage, Use and Disposal of Pesticides at Federal Facilities.

### Administration

### 15.17  Inventories

15.17.1 An up-to-date inventory of all pesticides in storage shall be maintained. Containers must be dated when received and, to the extent possible, the shelf-life of the pesticide identified.  The inventory list is to be kept in a separate location and made available to the Occupational Safety and Health (OSH) Committee.

### 15.18  Labelling

15.18.1 As required under the *Pest Control Products Act* (PCPA), all pesticides shall be kept in original containers with the original label intact.

### 15.19  Monitoring

15.19.1 Procedures involving the use of pesticides, either in the laboratory or in general field application, shall be monitored at regular intervals by the responsible authority within the department to ensure that prescribed safety procedures are being followed.  If an independent survey or health investigation is considered advisable at any time, a written request should be submitted to the appropriate Regional Officer, Medical Service Branch of Health Canada in accordance with *Treasury Board Manual*, Chapter 4-2 Procedures for Occupational Health Investigations in the "Occupational Safety and Health" volume of the *Treasury Board Manual*.

15.19.2  The Joint Occupational Safety and Health Committee shall be advised of health and safety investigations prior to their being conducted.  All reports and data from monitoring should be made available to the Occupational Safety and Health Committee or the safety and health representative.

## 15.20  Housekeeping

15.20.1  Appropriate good housekeeping shall be followed in all areas where pesticides are mixed, stored or handled.  This includes the maintenance of absolute cleanliness of the workplace and the use of approved waste disposal facilities and techniques including adherence to the requirements of Safety Directive 18, Sanitation.

## 15.21  Education and training

15.21.1  Departments shall ensure that the qualified persons who use, handle, mix and dispose of pesticides are licensed or certified in accordance with a provincial or national program.

15.21.2  In addition, departments shall, in consultation with the Safety and Health Committee or the safety and health representative, develop and implement a workplace educational program for qualified persons.  This program shall include the concepts and principles of the departmental IPM program and instruction related to specific pesticides used in the workplace, their hazards as outlined on labels, MSDS, and manufacturer's literature, the protection required for qualified persons to perform their duties, and first aid and emergency procedures pertinent to pesticide use.

15.21.3  The workplace educational program referred to above shall be reviewed, in consultation with the Safety and Health Committee or the safety and health representative, at least once a year, whenever new pesticides are about to be introduced in the workplace, and when new hazard information about a pesticide becomes available.

## 15.22  First aid

15.22.1  First aid instructions, and emergency procedures as detailed on the product label, the MSDS, and in manufacturer's literature shall be followed for suspected pesticide poisoning. Procedures shall be displayed prominently in all areas where pesticides are stored, handled, used, and disposed of, and where decontamination is carried out.

15.22.2  Emergency telephone numbers for first aid attendants, for the local poison control centre and for the Occupational and Environmental Health Services Directorate, Health Canada shall be prominently displayed.

### 15.23 Personnel monitoring

15.23.1 All personnel engaged regularly in work involving the handling of pesticides shall be examined in accordance with the provisions of *Treasury Board Manual*, Chapter 2-13, Occupational Health Evaluation Standard in the "Occupational Safety and Health" volume.

## Records

### 15.24 Medical

15.24.1 All medical records obtained during examination of an employee under the requirements of the *Treasury Board Manual*, Chapter 2-13, Occupational Health Evaluation Standard, including detailed employee history of exposure, shall be maintained by the Occupational and Environmental Health Services Directorate, Health Canada. Records shall be made available to an employee's physician upon request.

### 15.25 Pesticide application

15.25.1 Departments shall maintain records on the application of pesticides for a period of 30 years after the application date. The records shall contain the following information as a minimum:

(a)   pesticide applied;
(b)   PCP registration number;
(c)   application rate;
(d)   application site;
(e)   method of application;
(f)   persons applying the pesticide;
(g)   reason for application;
(h)   unusual circumstances which occurred during the application;
(i)   reports of health or safety investigations conducted, including all sampling data and other relevant information.

15.25.2 Copies of the above records shall be placed on the personal file of employees applying pesticides and as a reference on the personal file of other employees who request it.

### 15.26 Environmental monitoring

15.26.1 Departments shall maintain records of all environmental sampling data and reports for a period of 30 years from the date of reporting.

## Resources

### 15.27 Organizations

15.27.1 Information on registered pesticides may be obtained from Agriculture and Agri-food Canada, the department responsible for the regulation of such products. Departments and employees can obtain information from Agriculture Canada's National Pesticides call line at 1-800-267-6315.

15.27.2 Health Canada will provide information on the effects of pesticide exposure, the treatment of exposed persons and advice concerning appropriate training, including emergency first aid.

15.27.3 The Environmental Protection Directorate of Environment Canada will provide advice concerning the disposal of pesticides.

15.27.4 The Canadian Centre for Occupational Health and Safety (CCOHS) maintains databases on Material Safety Data Sheets (MSDS), Pest Management Research Information System, and Regulatory Information on Pesticide Products.

> CCOHS
> 250 Main Street East
> Hamilton, Ontario
> L8N 1H6
> Tel: (416) 572-4400
> 1-800-263-8466
> Fax: (416) 572-4500

15.27.5 The Pest Management Alternatives Office (PMAO) promotes measures that encourage the judicious use of pesticides through integrated crop management strategies.

> PMAO
> Vanguard Building
> 71 Slater Street, Room 701
> Ottawa, Ontario
> K1P 5H7
> Tel: (613) 991-1001
> Fax: (613) 991-0999

15.27.6 The Crop Protection Institute has available publications and videos on pesticides.

> Crop Protection Institute
> 21 Four Seasons Place, Suite 627
> Etobicoke, Ontario
> M9B 6J8
> Tel: (416) 622-0771
> Fax: (416) 622-6764

### 15.28 Materials

15.28.1 The following publications are available at the address indicated.

Code of Good Practice for the Handling, Storage, Use, and Disposal of Pesticides at Federal Facilities in Canada: (sheduled for publication: early 1994)

> Environment Canada
> Conservation and Protection

Environmental Protection Directorate

*Pesticide Safety* (1988)

*Pesticide Handling - A Safety Handbook* (1986)

*DND Pest Management Manual*, 6th edition

The *Pest Control Products Act* and Regulations

> Canada Communication Group
> 45 Sacré-Coeur Blvd.
> Hull, Quebec
> K1A OS9
> Tel: (819) 956-4800
> Fax: (819) 994-1498

## CHAPTER 2-16 - ELEVATED WORK STRUCTURES DIRECTIVE

### Grievance procedure

In cases of alleged misinterpretation or misapplication arising out of this standard, the grievance procedure, for all represented employees within the meaning of the *Public Service Staff Relations Act*, will be in accordance with Section 7.0 of the *National Joint Council By-Laws*. For unrepresented employees, the departmental grievance procedure applies.

This standard is deemed to be part of collective agreements between the parties to the National Joint Council and employees are to be afforded ready access to this standard.

### Application

1. This standard incorporates the minimum requirements of the *Canada Labour Code*, Part II, and applicable regulations issued pursuant to that legislation, and applies to all departments and other portions of the Public Service, as defined in Part I of Schedule I of the *Public Service Staff Relations Act*.

### Scope

2. This standard is applicable to portable ladders, temporary ramps and stairs, temporary elevated work bases used by employees and temporary elevated platforms used for materials.

### Definitions

3. In this standard:

(1)   ***elevated work structure*** means a structure or device that is used as an elevated work base for persons or as an elevated platform for material and includes any scaffold, stage or staging, walk-way, decking, bridge, boatswain's chair, tower, crawling board, temporary floor, any portable

ladder or means of access or egress from any of the foregoing, and any safety net, landing or other device used in connection with such a structure (charpente surélevée);

(2) **mobile elevated work structure** means a vehicle-mounted aerial device, elevating rolling work platform, boom-type elevating work platform or self-propelled elevating work platform (charpente surélevée mobile);

(3) **person in charge** means a qualified person who supervises employees performing work in order to ensure the safe and proper conduct of an operation or of the work of employees (responsable);

(4) **qualified person** means, in respect of a specified duty, a person who, because of knowledge, training and experience, is qualified to perform that duty safely and properly (see paragraph 47) (personne qualifiée).

## Departmental responsibilities

4. Departments shall ensure that requirements specified in the *Canada Occupational Safety and Health Regulations*, Part III, Temporary Structures and Excavations Regulation, issued pursuant to the *Canada Labour Code*, Part II, are applied at every workplace occupied by employees.

5. No department shall permit the use of a temporary structure where it is reasonably practicable to use a permanent structure.

6. Departments shall ensure that each temporary work structure used by an employee is safe for use, and is used in a safe and proper manner.

7. Departments shall ensure that a qualified person visually inspects each temporary structure prior to each work shift to ensure, insofar as possible by such inspection, that it is safe to use and ensure that a record of each inspection is made by the person who carried out the inspection.

8. No employee shall use a temporary structure unless:

(1) authority has been received from the person in charge to use it;
(2) the employee has been trained and instructed in its safe and proper use; and
(3) the employee, or the person in charge, visually inspects the structure prior to each work shift to ensure, insofar as possible by such inspection, that it is safe to use.

9. Where an inspection made in accordance with paragraph 7 reveals a defect or condition that adversely affects the structural integrity of a temporary structure, no employee shall use the temporary structure until the defect or condition is remedied.

10. Every employee shall report to the person in charge, as soon as practicable, any defect or condition in a temporary structure that may, in the opinion of that employee, create a hazard.

11. No employee shall use any temporary structure that has a defect or condition that, in the opinion of that employee, may endanger the employee or

any other employee, until the structure has been examined by a qualified person and declared to be safe.

12. No employee shall work on a temporary structure in rain, snow, hail or an electrical or wind storm that is likely to be hazardous to the safety or health of the employee, except where the work is required to remove a hazard or to rescue an employee.

## Mobile elevated work structures

13. Departments shall ensure that the design, construction, maintenance and use of every mobile elevated work structure shall comply as appropriate, with:

CAN 3 B354.1-M82 *Elevating Rolling Work Platforms.*

CAN 3 B354.2-M82 *Self-Propelled Elevating Work Platforms for Use on Paved/Slab Units.*

CAN 3 B354.3-M82 *Self-Propelled Elevating Work Platforms for Use as Off-Slab Units.*

CAN 3 B354.4-M82 *Boom Type Elevating Work Platforms.*

CSA C225-1976 *Vehicle Mounted Aerial Devices.*

14. Departments shall ensure, to the extent that is practicable, that where it is necessary to use or move a mobile elevated work structure with an employee on such a device, the person in charge ensures that the device is observed until it is no longer in motion.

15. Every platform, hand-rail, guardrail and work area on a temporary structure shall be kept free of accumulations of ice and snow while the temporary structure is in use.

16. The floor of a temporary structure used by an employee shall be kept free of grease, oil or other slippery substance and of any material or object that may cause an employee to trip.

17. Tools, equipment and materials used on a temporary structure shall be arranged or secured in such a manner that they cannot be knocked off the structure accidentally.

## Barricades

18. Where a vehicle or a pedestrian may come into contact with a temporary structure, a person shall be positioned at the base of the temporary structure or a barricade shall be installed around it to prevent any such contact.

## Guardrails and toe boards

19. Guardrails and toe boards shall be installed at every open edge of the platform of a temporary structure.

20. Every guardrail shall consist of:

(1)   a horizontal top rail not less than 900 mm and not more than 1100 mm above the base of the guardrail;
(2)   a horizontal intermediate rail spaced midway between the top rail and the base; and
(3)   supporting posts spaced not more than 3 m apart at their centres.

21. Every guardrail shall be designed to withstand a static load of 890 N applied in any direction at any point on the top rail.

22. Where there is a hazard that tools or other objects may fall from a platform or other raised area onto an employee:

(1)   a toe board that extends from the floor of the platform or other raised area to a height of not less than 125 mm shall be installed; or
(2)   where the tools or other objects are piled to such a height that a toe board referred to in sub-paragraph (1) does not prevent the tools or other objects from falling, a solid or mesh panel shall be installed from the floor to a height of not less than 450 mm.

**Temporary stairs, ramps and platforms**

23. Subject to paragraph 30, temporary stairs, ramps and platforms shall be designed, constructed and maintained to support any load that is likely to be imposed on them and to allow safe passage of persons and equipment on them.

24. Temporary stairs shall have uniform steps in the same flight and:

(1)   a slope not exceeding 1.2 to 1; and
(2)   a hand-rail that is not less than 900 mm and not more than 1100 mm above the stair level on open sides, including landings.

25. Temporary ramps and platforms shall be:

(1)   securely fastened in place;
(2)   braced, if necessary, to ensure their stability; and
(3)   provided with cleats or surfaced in a manner that provides a safe footing for employees.

26. A temporary ramp shall be so constructed that its slope does not exceed:

(1)   where the temporary ramp is installed in the stairwell of a building not exceeding two storeys in height, 1 to 1, if cross cleats are provided at regular intervals not exceeding 300 mm; and
(2)   in any other case, 1 in 3.

**Scaffolds**

27. Departments shall ensure, to the extent that is practicable, that the design, construction and use of scaffolds meet the requirements of CSA Standard S269.2/M87, *Access Scaffolds for Construction Purposes.*

28. The erection, use, dismantling or removal of a scaffold shall be carried out by or under the supervision of a qualified person.

29. The footings and supports of every scaffold shall be capable of carrying, without dangerous settling, all loads that are likely to be imposed on them.

30. Every scaffold shall be capable of supporting at least four times the load that is likely to be imposed on it.

31. The platform of every scaffold shall be at least 480 mm wide and securely fastened in place.

**Portable ladders**

32. Commercially manufactured portable ladders shall meet the standards set out in CSA Standard CAN3-Z11-M81, *Portable Ladders*, the English version of which is dated September 1981, as amended to March, 1983, and the French version of which is dated August, 1982, as amended to June, 1983.

33. Subject to paragraph 34, every portable ladder shall, while being used:

(1) be placed on a firm footing; and
(2) be secured in such a manner that it cannot be dislodged accidentally from its position.

34. Where, because of the nature of the location or of the work being done, a portable ladder cannot be securely fastened in place, it shall, while being used, be sloped so that the base of the ladder is not less than one-quarter and not more than one-third of the length of the ladder from a point directly below the top of the ladder and at the same level as the base.

35. Every portable ladder that provides access from one level to another shall extend at least three rungs above the higher level.

36. Metal or wire-bound portable ladders shall not be used where there is a hazard that they may come into contact with any live electrical circuit or equipment.

37. No employee shall work from any of the three top rungs of any single or extension portable ladder or from either of the two top steps of any portable step ladder.

## Excavation

38. Before the commencement of work on a tunnel, excavation or trench, the department shall mark the location of all underground pipes, cables and conduits in the area where the work is to be done.

39. Where an excavation or trench constitutes a hazard to employees, a barricade shall be installed around it.

40. In a tunnel or in an excavation or trench that is more than 1.4 m deep and whose sides are sloped at an angle of 45° or more to the horizontal:

(1)　the walls of the tunnel, excavation or trench, and
(2)　the roof of the tunnel

shall be supported by shoring and bracing that is installed as the tunnel, trench or excavation is being excavated.

41. Paragraph 40 does not apply in respect of a trench where the department provides an approved system of shoring composed of steel plates and bracing, welded or bolted together, that can support the walls of the trench from the ground level to the trench bottom and can be moved along as work progresses.

42. The installation and removal of the shoring and bracing referred to in paragraph 40 shall be performed or supervised by a qualified person.

43. Tools, machinery, timber, excavated materials or other objects shall not be placed within 1 m from the edge of an excavation or trench.

44. Departments shall ensure that each excavation in which an employee works is safe for use, and is used in a safe and proper manner, and

(1)　prior to a work shift, a qualified person shall undertake a safety inspection of each excavation to be used during that shift;
(2)　a record of each inspection shall be completed by the person who carried out the inspection; and
(3)　every record referred to in paragraph (2):
　　(a)　shall be signed by the person who carried out the inspection, and
　　(b)　shall include:
　　　　(i)　date of the inspection,
　　　　(ii)　identification and location of excavation,
　　　　(iii)　any observation that the person considers relevant to the safety of employees,
　　　　(iv)　declaration that, in the opinion of the person carrying out the inspection, the excavation is safe for its intended use.

## Safety nets

45. Where there is a hazard that tools, equipment or materials may fall onto or from a temporary structure, the department shall provide a protective structure

or safety net to protect from injury any employee on or below the temporary structure.

46. The design, construction and installation of a safety net referred to in paragraph 45 shall meet the standards set out in ANSI Standard ANSI A10.11-1979, *American National Standard for Safety Nets Used During Construction, Repair and Demolition Operations*, dated August 7, 1979.

47. Where there is a dispute regarding the term "qualified person" for purposes of an occupational safety and health standard, the following procedure shall be implemented:

(a)  The employee shall raise the matter directly with the person in charge.
(b)  The person in charge shall review the employee's qualifications and decide upon the employee's status as a qualified person.
(c)  If the employee is dissatisfied with the decision, the matter shall be referred to the safety and health committee established for the employee's workplace.
(d)  The safety and health committee shall review the matter and make appropriate recommendations to the person in charge.
(e)  If the safety and health committee does not consider itself competent to deal with the case, it shall recommend an acceptable third party to the person in charge.
(f)  The person in charge shall, pursuant to (d) or (e), take the recommendations into consideration, render a final management decision and undertake the appropriate action.

If the employee does not agree with the final decision which has been rendered, a grievance may be initiated pursuant to the NJC redress procedure.

# CHAPTER 2-17 - USE AND OCCUPANCY OF BUILDINGS DIRECTIVE

## General

## Grievance procedure

In cases of alleged misinterpretation or misapplication arising out of this standard, the grievance procedure, for all represented employees within the meaning of the *Public Service Staff Relations Act*, will be in accordance with Section 7.0 of the *National Joint Council By-Laws*. For unrepresented employees, the departmental grievance procedure applies.

## Collective agreement

This standard is deemed to be part of collective agreements between the parties to the National Joint Council and employees are to be afforded ready access to this standard.

## Application

This standard incorporates the minimum requirements of the *Canada Labour Code*, Part II, and applicable regulations issued pursuant to that legislation and applies to all departments and other portions of the Public Service, as defined in Part I of Schedule I of the *Public Service Staff Relations Act*.

## Scope

This standard outlines certain safety and health requirements concerning the use and occupancy of buildings occupied by Public Service employees.

## Definitions

*Floor opening* means an opening measuring 300 mm or more in its smallest dimension in a floor, platform, pavement or yard (ouverture dans un plancher);

*person in charge* means a qualified person appointed by management to ensure the safe and proper conduct of an operation or of the work of employees (personne responsable);

*wall opening* means an opening at least 750 mm high and 300 mm wide in a wall or partition (ouverture dans un mur).

## General responsibility

No employee shall use, or be required or permitted to use, any building in a manner likely to endanger the safety or health of that employee or of any other person.

## Requirements

### 17.1 Design and construction

17.1.1 The design and construction of every building shall meet the standards set out in Parts 3 to 9 of the *National Building Code of Canada, 1985*, to the extent that is essential for the safety and health of employees.

### 17.2 Workplace occupancy

17.2.1 Departments shall ensure that requirements specified in the *Canada Occupational Safety and Health Regulations*, Part XVII, Safe Occupancy of the Workplace, issued pursuant to the *Canada Labour Code*, Part II, and the *National Fire Code of Canada*, 1985, are applied at every workplace occupied by employees.

17.2.2 Matters respecting office accommodation, particularly where occupancy of a new or renovated office accommodation is planned, shall be the subject of consultation between management and employees or employee representatives.

### 17.3 Environmental conditions

17.3.1 To the extent practicable, the environmental conditions to be maintained in office buildings shall conform to the requirements specified in the following documents:

(a) ASHRAE Standard 55-1981, *Thermal Environmental Conditions for Human Occupancy*; and

(b) ASHRAE Standard 62-1981, *Ventilation for Acceptable Indoor Air Quality*.

17.3.2 In office accommodation, air (dry bulb) temperatures during working hours should be maintained within the 20°C — 26°C range, which is the ideal temperature operating range. Temperatures between 17°C and 20°C and above 26°C can be uncomfortable, and occupancy should not exceed 3 hours daily or 120 hours annually in each of these extremes. Temperatures above 26°C are deemed to be uncomfortable when the Humidex reading (Appendix A) at a given temperature equals 40 or less. Temperatures shall be measured at desk-top level in those spaces within work stations which would be occupied by employees while they are carrying out the major part of their normal duties.

(a) With regard to the uncomfortable range of temperatures described in paragraph 17.13.2, it is the responsibility of the deputy head, or the designated representative, to take appropriate action to ensure that environmental conditions do not subject employees to undue stress or discomfort. Corrective measures which shall be considered include, among others, increased frequency of rest periods and temporary relocation of employees to work stations outside the affected area.

(b) An unsatisfactory condition is deemed to exist when the Humidex reading exceeds 40 (Appendix A), or when the air temperature (dry bulb) falls below 17°C. In these cases, operations shall be stopped and employees released from the workplace if relocation is not practicable. If instrumentation capable of accurately measuring Humidex is not practically available within one hour of a complaint being made, a temperature of 29°C or above shall be considered unsatisfactory.

17.3.3 For the purposes of paragraph 17.3.2, conditions shall not be intentionally maintained within the marginal zones of 17°C to 20°C and 26°C to 29°C. Such conditions should result only from occurrences over which departments have no direct control, such as weather extremes or equipment failures.

## 17.4 Hot surfaces

17.4.1 Steam and hot water pipes, heaters and any other hot surfaces having surface temperatures which could injure any person through bodily contact, shall be guarded or covered in such a manner as to prevent such direct contact. Where asbestos lagging is used for insulation purposes, the requirements contained in chapter 2-2, Dangerous substances directive and chapter 4-3, Occupational exposure to asbestos procedures, shall be followed.

## 17.5 Doors and windows

17.5.1 Each glass door, and every other transparent part of a building that could be mistaken for a passageway, shall be appropriately marked with conspicuous warning signs or symbols indicating the presence of the glass or transparent material.

17.5.2 Each double action swinging door used for two-way pedestrian traffic shall be designed and installed in a manner that will permit persons on either side to be seen by persons on the other side of the door.

17.5.3 Where a door or gate extends into a pedestrian or vehicle passageway, appropriate warning, guarding or other measures shall be provided. Such measures shall include marking the floor to indicate clearly the area of the hazard.

17.5.4 Where an open door, gate or other obstruction temporarily reduces the effective width of a pedestrian or vehicle passageway to less than that required for safe passage, action shall be taken by the person in charge to ensure that, while its effective width is so reduced:

(a)     a person is posted near the door or gate to warn employees of the danger; or
(b)     barriers are placed across the passageway to prevent persons from passing while the door or gate is open.

17.5.5 Where a window on any level above the ground floor level of a building is cleaned, the standards set out in CSA Standard Z91-M1980, *Safety Code for Window Cleaning Operations*, the English version of which is dated May, 1980 and the French version of which is dated November, 1983 shall be adopted and implemented.

### 17.6 Awnings and canopies

17.6.1 Any window awning or canopy or any part of a building that projects over an exterior walkway shall be installed in a manner that permits a clearance of not less than 2.2 m between the walkway surface and the lowest projection of an awning or canopy or projecting part of the building.

### 17.7 Floor and wall openings

17.7.1 Where an employee has access to a wall opening from which there is a drop of more than 1.2 m, or to a floor opening, guardrails shall be fitted around the wall opening or floor opening or it shall be covered with material capable of supporting all loads that may be imposed on it.

17.7.2 The material referred to in paragraph 17.7.1 shall be securely fastened to and supported on structural members.

17.7.3 Paragraph 17.7.1 does not apply to the loading and unloading areas of truck, railroad and marine docks.

### 17.8 Open top bins, hoppers, vats and pits

17.8.1 Where an employee has access to an open top bin, hopper, vat, pit or other enclosure from a point directly above that enclosure, the person in charge shall ensure that the enclosure is:

(a)     completely covered with a grating, screen or other cover that will prevent the employee from falling into the enclosure; or

(b)     provided with a walkway that is not less than 500 mm wide and is fitted with guardrails.

17.8.2 Any grating, screen covering or walkway referred to in paragraph 17.8.1 shall be so designed, constructed and maintained that it will support a load that is not less than the maximum load that may be imposed on it, or a live load of 6 kPa, whichever is the greater.

17.8.3 Where an employee is working above an open top bin, hopper, vat, pit or other open top enclosure that is not covered with a grating, screen or other covering, the inside wall of the enclosure shall be fitted with a fixed ladder, except where the operations carried on in the enclosure render such a fitting impracticable.

17.8.4 Every open top bin, hopper, vat, pit or other open top enclosure referred to in paragraph 17.8.1 whose walls extend less than 1.1 m above an adjacent floor or platform used by an employee shall be:

(a)     covered with a grating, screen or other covering;

(b)     fitted with a guardrail; or

(c)     guarded by a person to prevent employees from falling into the enclosure.

17.8.5 Where, due to the temporary removal of any cover, an opening is created into which persons may fall, barriers shall be securely placed around such openings to protect and warn persons of the hazard.

## 17.9 Ladders, stairways and ramps

17.9.1 Where an employee is required to move from one level to another level that is more than 450 mm higher or lower than the first level, the department shall install a fixed ladder, stairway or ramp between the levels.

17.9.2 All fixed stairways shall comply with the American National Standards Institute Standard A64.1-1968, *Requirements for Fixed Industrial Stairs*.

17.9.3 Every ramp, walkway, platform or safety landing shall be fitted with railings and guards as recommended in the American National Standards Institute Standard A12.1-1973, *Safety Requirements for Floor and Wall Openings, Railings, and Toe-Boards*.

17.9.4 All fixed ladders shall comply with the American National Standards Institute Standard A14.3-1984, *Safety Requirements for Fixed Ladders*.

17.9.5 Where one end of a stairway is so close to a traffic route used by vehicles, to a machine or to any other hazard as to be hazardous to the safety of an employee using the stairway, the department shall:

(a) post a sign at that end of the stairway to warn employees of the hazard; and

(b) where possible, install a barricade that will protect employees using the stairway from the hazard.

17.9.6  Subject to paragraph 17.9.8, a fixed ladder that is more than 6 m in length shall be fixed with a cage for that position of its length that is more than 2 m above the base level of the ladder in such a manner that it will catch an employee who loses his or her grip and falls backward or sideways off the ladder.

17.9.7  Subject to paragraph 17.9.9, a fixed ladder that is more than 9 m in length shall have, at intervals of not more than 6 m, a landing or platform that:

(a) is not less than 0.36 m in area; and

(b) is fitted at its outer edges with a guardrail.

17.9.8  A fixed ladder, cage, landing or platform referred to in paragraphs 17.9.6 and 17.9.7 shall be designed and constructed to withstand all loads that may be imposed on it.

17.9.9  Paragraphs 17.9.6 and 17.9.7 do not apply to a fixed ladder that is used with a fall protection system referred to in the *Canada Occupational Safety and Health Regulations*, Part XII, Safety Materials, Equipment, Devices and Clothing, issued pursuant to the *Canada Labour Code*, Part II.

17.9.10  A fixed ladder shall be:

(a) vertical;

(b) securely held in place at the top and bottom and at intermediate points not more than 3 m apart; and

(c) fitted with:
    (i) rungs that are at least 150 mm from the wall and spaced at intervals not exceeding 300 mm, and
    (ii) side rails that extend not less than 900 mm above the landing or platform.

17.9.11  Every ramp shall have the minimum slope that is reasonable for the purpose for which it is used. In no case shall the gradient exceed:

(a) the gradient that is recommended as safe by the manufacturer of mobile equipment used on the ramp; or

(b) such lesser gradient that is safe, having regard to the mechanical condition of mobile equipment used on the ramp, the weight of the loads transported, and the condition of the roadway.

### 17.10  Docks, ramps and dock plates

17.10.1  Every loading and unloading dock shall be:

(a)  of sufficient strength to support the maximum load that may be imposed on it;
(b)  free of surface irregularities that may interfere with the safe operation of mobile equipment; and
(c)  fitted around its sides that are not used for loading or unloading with side rails, curbs or rolled edges of sufficient height and strength to prevent mobile equipment from running over the edge.

17.10.2  Every portable ramp and every dock plate shall be:

(a)  clearly marked or tagged to indicate the maximum safe load that it is capable of supporting; and
(b)  installed so that it cannot slide, move or otherwise be displaced under the load that may be imposed on it.

### 17.11  Guardrails

17.11.1  Every guardrail shall consist of:

(a)  a horizontal top rail not less than 900 mm and not more than 1100 mm above the base of the guardrail;
(b)  a horizontal intermediate rail spaced midway between the top rail and the base; and
(c)  supporting posts spaced not more than 3 m apart at their centres.

17.11.2  Every guardrail shall be designed to withstand a static load of 890 N applied in any direction at any point on the top rail.

### 17.12  Toe boards

17.12.1  Where there is a hazard that tools or other objects may fall from a platform or other raised area onto an employee:

(a)  a toe board that extends from the floor of the platform or other raised area to a height of not less than 125 mm shall be installed; or
(b)  where the tools or other objects are piled to such a height that a toe board referred to in sub-paragraph (a) does not prevent the tools or other objects from falling, a solid or mesh panel shall be installed from the floor to a height of not less than 450 mm.

### 17.13  Housekeeping and maintenance

17.13.1  Nothing shall be left or stored in any passageway or travelled area in a manner that may endanger the safety or health of persons or the safe operation of vehicles moving through that passageway or area.

17.13.2 Every exterior stairway, walkway, ramp, passageway, roof and canopy shall be kept free of accumulations of ice and snow. Where necessary, protection shall be provided from dangerous accumulations of ice which may fall from overhead structures.

17.13.3 All dust, dirt, waste and scrap material in every workplace in a building shall be removed as often as is necessary to protect the safety and health of employees and shall be disposed of in such a manner that the safety and health of persons is not endangered.

17.13.4 Every travelled surface in a workplace shall be:

(a) slip resistant; and
(b) maintained free of splinters, holes, loose boards and tiles or similar defects.

17.13.5 Where a floor in a workplace is normally wet and employees in the workplace do not use non-slip waterproof footwear, the floor shall be covered with a dry false floor or platform treated with a non-slip material or substance.

17.13.6 Electrical power vaults, switch and generator rooms or enclosures, and other similarly dangerous areas shall be kept locked or otherwise made inaccessible except to authorized persons who are qualified to safely enter or perform work in such areas.

17.13.7 Every building shall be kept in such a state of repair and maintenance so as not to endanger the safety or health of any employee.

**17.14  Temporary heat**

17.14.1 Subject to paragraph 17.14.2, where a salamander or other high capacity portable open-flame heating device is used in an enclosed workplace, the heating device shall:

(a) be so located, protected and used that there is no hazard of igniting tarpaulins, wood or other combustible materials adjacent to the heating device;
(b) be used only when there is adequate ventilation provided;
(c) be so located as to be protected from damage or overturning; and
(d) not restrict a means of exit.

17.14.2 Where the heating device referred to in paragraph 17.14.1 does not provide complete combustion of the fuel used in connection with it, it shall be equipped with a securely supported sheet metal pipe that discharges the products of combustion outside the enclosed workplace.

**17.15  Illumination**

17.15.1 The levels of illumination in each building, and the provision of emergency lighting systems, shall comply with the requirements contained in the *Canada Occupational Safety and Health Regulations*, Part VI, Levels of Lighting, issued pursuant to the *Canada Labour Code*, Part II.

## CHAPTER 2-17 - APPENDIX A

## OFFICE ACCOMMODATION

## HUMIDEX TABLE FOR TEMPERATURE AND RELATIVE HUMIDITY READINGS

### Relative Humidity (%)

| Temp (0C) | 100 | 95 | 90 | 85 | 80 | 75 | 70 | 65 | 60 | 55 | 50 | 45 | 40 | 35 | 30 | 25 | 20 |
|---|---|---|---|---|---|---|---|---|---|---|---|---|---|---|---|---|---|
| 35 |  | 58 | 57 | 56 | 54 | 52 | 51 | 49 | 48 | 47 | 45 | 43 | 42 | 41 | 38 | 37 |  |
| 34 | 58 | 57 | 55 | 53 | 52 | 51 | 49 | 48 | 47 | 45 | 43 | 42 | 41 | 39 | 37 | 36 |  |
| 33 | 55 | 54 | 52 | 51 | 50 | 48 | 47 | 46 | 44 | 43 | 42 | 40 | 38 | 37 | 36 | 34 |  |
| 32 | 52 | 51 | 50 | 49 | 47 | 46 | 45 | 43 | 42 | 41 | 39 | 38 | 37 | 36 | 34 | 33 |  |
| 31 | 50 | 49 | 48 | 48 | 45 | 44 | 43 | 41 | 40 | 39 | 38 | 36 | 35 | 34 | 33 | 31 |  |
| 30 | 49 | 47 | 46 | 44 | 43 | 42 | 41 | 40 | 38 | 37 | 36 | 35 | 34 | 34 | 31 | 31 |  |
| 29* | 46 | 45 | 44 | 43 | 42 | 41 | 39 | 38 | 37 | 36 | 34 | 33 | 32 | 31 | 30 |  |  |
| 28 | 43 | 42 | 41 | 41 | 39 | 38 | 37 | 36 | 35 | 34 | 33 | 32 | 31 | 29 | 28 |  |  |
| 27 | 41 | 40 | 39 | 38 | 37 | 36 | 35 | 34 | 33 | 32 | 31 | 30 | 29 | 28 | 28 |  |  |
| 26 | 39 | 38 | 37 | 36 | 35 | 34 | 33 | 32 | 31 | 31 | 29 | 28 | 28 | 27 |  |  |  |
| 25 | 37 | 36 | 35 | 34 | 33 | 33 | 32 | 31 | 30 | 29 | 28 | 27 | 27 | 26 |  |  |  |
| 24 | 35 | 34 | 33 | 33 | 32 | 31 | 30 | 29 | 28 | 28 | 27 | 26 | 26 | 25 |  |  |  |
| 23 | 33 | 32 | 32 | 31 | 30 | 29 | 28 | 27 | 27 | 26 | 25 | 24 | 23 |  |  |  |  |
| 22 | 31 | 29 | 29 | 28 | 28 | 27 | 26 | 26 | 24 | 24 | 23 | 23 |  |  |  |  |  |
| 21 | 29 | 29 | 28 | 27 | 27 | 26 | 26 | 24 | 24 | 23 | 23 | 22 |  |  |  |  |  |
| 20 | 27 | 27 | 26 | 25 | 25 | 24 | 24 | 23 | 22 | 22 | 21 |  |  |  |  |  |  |
| 19 | 25 | 25 | 24 | 24 | 23 | 23 | 22 | 22 | 21 | 21 | 20 |  |  |  |  |  |  |
| 18 | 23 | 23 | 22 | 22 | 21 | 21 | 20 | 20 |  |  |  |  |  |  |  |  |  |
| 17 | 21 | 21 | 21 | 20 | 20 | 19 | 19 |  |  |  |  |  |  |  |  |  |  |

*Acceptable Temperature Range* (spanning 29* to 19)
*Ideal Temperature Range* (spanning 29* to 20)

* 29° C  If instrumentation capable of accurately measuring humidex is not practically available within one hour of a complaint being made, a temperature of 29°C or above shall be considered unsatisfactory.

**Humidex**
- Relocate or Release Staff
- Corrective Measures
- Ideal Operational Range

September 1992
Based on Environment Canada Humidex Charts

## CHAPTER 2-18 - SANITATION DIRECTIVE

### Grievance procedure

In cases of alleged misinterpretation or misapplication arising out of this standard, the grievance procedure, for all represented employees, within the meaning of the *Public Service Staff Relations Act* will be in accordance with Section 7.0 of the *National Joint Council By-Laws*. For unrepresented employees the departmental grievance procedure applies.

This standard is deemed to be part of collective agreements between the parties to the National Joint Council and employees are to be afforded ready access to this standard.

### Application

1. This standard incorporates the minimum requirements of the *Canada Labour Code*, Part II, and applicable regulations issued pursuant to that legislation and applies to all departments and other portions of the Public Service, as defined in Part I of Schedule I of the *Public Service Staff Relations Act*.

### Scope

2. Notwithstanding the scope of other federal government codes or directives concerning sanitation, environmental pollution or control, this standard is primarily concerned with occupational health. This standard shall have application in all government-owned buildings occupied by Public Service employees. Where public servants occupy buildings not owned by the federal government, it shall be applied to the maximum extent that is reasonably practicable.

### Definitions

3. In this standard:

(1)  *ARI* means the Air-Conditioning and Refrigeration Institute of the United States (ARI);

(2)  *Canadian Plumbing Code* means the *Canadian Plumbing Code*, 1985 (Code canadien de la plomberie);

(3)  *change room* means a room that is used by employees to change from their street clothes to their work clothes and back to their street clothes after work; (vestiaire)

(4)  *field accommodation* means fixed or mobile accommodation that is living, eating or sleeping quarters provided by a department for the accommodation of employees at a workplace (logement sur place);

(5)  *food preparation area* means any area that is used for the storage, handling, preparation or serving of food; for example, cafeterias and canteens as defined in Chapter 130 of the Administrative Policy Manual (APM) and as amended from time to time (aire de préparation des repas);

(6)  *hazardous substance* means a controlled product or a chemical, biological or physical agent that by reason of a property that the agent

possesses is hazardous to the safety or health of a person exposed to it (substance dangereuse);

(7) **lunchroom** means a room equipped with tables and chairs in which employees may eat food brought into the premises (salle à manger);

(8) **mobile accommodation** means field accommodation that may be easily and quickly moved (logement mobile);

(9) **personal service room** means a change room, toilet room (excluding outdoor toilets), washroom, shower room, lunchroom, living space, sleeping quarters or any combination thereof (local servant aux besoins personnels);

(10) **potable water** means water of a quality which satisfies the standards or requirements of Health Canada for drinking water (eau potable);

(11) **sanitary condition** means that state of any environment, equipment or object that will not render it injurious to health (salubre);

(12) **sanitary facility** means a toilet or personal cleansing facility, and may include a toilet, urinal, wash basin and shower bath (installation sanitaire);

(13) **toilet room** means a room that contains a toilet or a urinal, but does not include an outdoor privy (lieux d'aisances);

(14) **vermin** means any insect or rodent pest (vermine);

(15) **workplace** means any place where an employee is engaged in work for the employee's department (lieu de travail).

## General responsibilities

4. Every department shall ensure that each personal service room and food preparation area and lunchroom used by employees is maintained in a clean and sanitary condition in accordance with Chapter 130 of the *Administrative Policy Manual* under which Health Canada is authorized to inspect the premises at any time.

5. Personal service rooms, lunchrooms and food preparation areas shall be used by employees in such a manner that the rooms or areas will remain clean and in a sanitary condition.

## Care of premises

6. All janitorial or other work that may cause dusty or unsanitary conditions shall, to the extent that is reasonably practicable, be performed after normal working hours.

7. All cleaning, sweeping and other activities shall be carried out in a manner that will minimize contamination by dust or other injurious substances, and in a manner that will not cause slippery or hazardous conditions.

8. Dirt and waste material shall not be allowed to accumulate to such an extent that unsafe or unsanitary conditions result.

9. Each sanitary facility and personal service room shall be cleaned at least once each 24-hour period following its use by employees.

10. Each container that is used for solid or liquid waste in the workplace shall:

(1) be equipped with a tight-fitting cover;
(2) be so constructed that it can easily be cleaned and maintained in a sanitary condition;
(3) be leak proof; and
(4) where there may be internal pressure in the container, be so designed that the pressure is relieved by controlled ventilation.

11. Each container referred to in paragraph 10 shall be emptied at least once every day that it is used.

12. Each enclosed part of a workplace, each personal service room area and each food preparation area shall be located, constructed, equipped, maintained and isolated in such a manner as to prevent the entrance and harbourage of vermin and animals as well as hazardous substances.

13. Where vermin have entered any enclosed part of an area referred to in paragraph 12, immediate action shall be taken for their elimination and control and for restoration of the area to a sanitary condition.

14. No person shall use a personal service room for the purpose of storing any material unless that material is related to the use of that room and is stored in a proper closet fitted with a door and which is provided in that room for that purpose.

15. With respect to each personal service room and food preparation area:

(1) the floors, partitions and walls shall have a durable, water-resistant finish and be so constructed that they can be easily washed and maintained in a sanitary condition; and
(2) the floor, and the lower 150 mm of any walls and partitions that are in contact with the floor, shall, in any food preparation area or room that contains a sanitary facility, be watertight and impervious to moisture, and the joint between the walls and the floor shall be covered.

16. In each personal service room and food preparation area, the temperature, measured one metre above the floor in the centre of the room or area, shall be maintained at a level of not less than 18°C and, where reasonably practicable, not more than 29°C.

17. Where separate personal service rooms are provided for employees of each sex, each room shall be equipped with a door that is self-closing and is clearly marked to indicate the sex of the employees for whom the room is provided.

**Plumbing system**

18. Every plumbing system that supplies potable water and removes water-borne waste:

(1)  shall meet the standards set out in the *Canadian Plumbing Code*; and
(2)  subject to paragraph 21 shall be connected to a municipal sanitation sewer or water main.

19.  Where, in accordance with the requirements of this standard, a sanitary facility is required on departmental premises, the sanitary facility shall be connected to a municipal sanitary sewer or water main or to both, where it is reasonably practicable to do so, in accordance with the applicable provincial environmental standard or code governing such installations, or the Environment Canada guidelines (EPS-1-ES-76-1) where no such provincial standard exists, or is less stringent.

20.  For the eventuality that the supply of potable water and water for the removal of water-borne waste is temporarily interrupted, departments shall establish contingency procedures.  Such procedures shall be established with the advice of Health Canada and in consultation with the appropriate safety and health committee(s).

21.  Where a sanitary facility is required and municipal sewer or water system or both, are not available, a sewer or water system, or both, as the case may be, shall be installed in accordance with the provincial environmental directive or code governing such installations, or the Environment Canada guidelines (EPS-1-EC-76-1) where no such provincial standard exists, or is less stringent.

**Toilet facilities**

22.  The number of toilet facilities shall, as a minimum, be in accordance with the provisions of the *National Building Code of Canada*, 1985.  For examples of office and industrial workplaces see appendix B.

23.  Toilet rooms shall be located not more than 60 m (197 ft) from and not more than one storey above or below each workplace.

24.  Where toilet facilities are provided for employees of each sex, the maximum number of employees of each sex means the number of each sex usually at the workplace on any one shift, but does not include employees who are employees away from the workplace for more than 75 per cent of the time and do not normally use the toilet room at the workplace.

25.  Notwithstanding paragraph 22, where more than one toilet is required for male employees, urinals may be provided in place of not more than one-half the number of toilets required.

26.  Where it is not reasonably practicable to install a water closet-type toilet connected to a sewage disposal system, a chemical recirculating or combustion toilet or an outdoor "privy" may be installed, provided the facility is constructed and maintained in accordance with the American National Standards Institute Standard ANSI Z4.3-1979 *Minimum Requirements for Nonsewered Disposal Systems*, or another standard acceptable to Health Canada.

27. There shall be an adequate supply of toilet paper, with a holder, in each toilet compartment, and a covered receptacle for the disposal of sanitary pads shall be provided in each toilet room used by female employees.

## Washrooms and wash basins

28. Subject to the provisions of paragraph 32, at least one wash basin with a supply of water, including hot water, where practicable, shall be provided in every room containing one or two toilets or urinals, and at least one additional wash basin with a supply of water shall be provided for every two additional toilets or urinals.

29. The following minimum requirements shall be adhered to wherever washing facilities are provided:

(1)  soap or another acceptable skin-cleansing agent shall be provided in a suitable dispenser adjacent to each wash basin or shower;
(2)  sufficient sanitary hand drying facilities to adequately serve the number of employees using the washroom shall be provided; and
(3)  separate non-combustible receptacles shall be provided for the disposal of used cloth towels and paper towels.

30. Where hot water is provided for washing purposes, it shall be maintained at a temperature of not less than 35°C and not more than 43°C and in no case shall water be heated by mixing with steam.

31. Where an outdoor privy, chemical or other toilet is provided, washing facilities shall be located as close to it as is reasonably practicable.

32. Notwithstanding the provisions of paragraph 28, where employees may be exposed to skin contamination from toxic, infectious, irritating or noxious substances which are potential safety or health hazards, a washroom with individual wash basins supplied with both hot and cold water shall be provided near the workplace, as follows:

| Number of employees | Number of wash basins |
|---|---|
| 1 — 5 | 1 |
| 6 — 10 | 2 |
| 11 — 15 | 3 |
| 16 — 20 | 4 |
| More than 20 | 4, plus 1 for every 15 employees, or portion thereof, in excess. |

33. Industrial wash troughs or circular wash basins may be used in place of the required individual wash basins, if they provide equivalent facilities for personal cleansing purposes.

34. A wash basin shall not be installed nearer than 600 mm from any toilet or urinal unless it is separate from the toilet or urinal by a waterproof partition.

## Showers and shower rooms

35. A shower room with at least one shower bath for every 10 employees or proportion of that number on a work shift shall be provided for the use of those employees who, during the course of their duties, regularly perform strenuous physical work in a high temperature or high humidity or are exposed to body contamination by toxic, infectious, irritating or any other substance hazardous to safety or heath.

36. Shower stalls shall meet the following specifications:

(1) Every shower receptor shall be constructed and arranged in such a way that water cannot leak through the walls or floor.
(2) No more than six shower heads shall be served by a single shower drain.
(3) Where two or more shower heads are served by a shower drain, the floor shall be sloped and the drain so located that water from one head cannot flow over the area that serves another head.
(4) Except for column showers, where a battery of shower heads is installed, the horizontal distance between two adjacent shower heads shall be at least 750 mm (30 inches).
(5) Waterproof finish shall be provided to a height of not less that 1.8 m (6 feet) above the floor in shower rooms and shall consist of ceramic, plastic or metal tile, sheet vinyl, tempered hardboard, lamented thermosetting decorative sheets or linoleum.
(6) Finished flooring in shower rooms shall consist of resilient flooring, felted-synthetic fibre floor coverings, concrete terrazzo, ceramic tile, mastic or other types of flooring providing similar degrees of water resistance.

37. Every shower shall be provided with cold water and hot water that meets the requirements of paragraph 30.

38. Soap or another approved cleansing agent shall be made available, and for each occasion that an employee is required to take a shower, a clean towel shall be provided.

39. Where duck-boards are used in showers, they shall not be made of wood.

## Ventilation

40. Each personal service room and each food preparation area shall be ventilated to provide at least two changes of air per hour:

(1) by mechanical means, where the room is used by ten or more employees; or
(2) by mechanical means or natural ventilation through a window or similar opening, where the room is used by fewer than ten employees if
   (a) the window or similar opening is located on an outside wall of the room, and

(b)   not less than 0.2 square metres of unobstructed opening is provided for each employee who normally uses the room at any one time.

41.  Where the ventilation of a personal service room is by mechanical means, pursuant to paragraph 40(1), and that room contains lockers, showers or sanitary facilities, the amount of air provided shall be not less than that specified in appendix A.

42.  Where the ventilation of a food preparation area or a lunchroom is by mechanical means, the rate of ventilation shall be that which is specified in ASHRAE Standard 62-1981, *Ventilation for Acceptable Indoor Air Quality*.

43.  An exhaust system from a personal service room containing a sanitary facility shall be connected in such a manner than an exchange of air from the personal service room to another room cannot occur at any time.

**Potable water**

44.  Water for drinking, personal washing and food preparation shall be potable and meet the standards set out in the *Guidelines for Canadian Drinking Water Quality 1987*, published by authority of the Minister of Health.

45.  Where it is necessary to transport water for drinking or washing, only sanitary containers and sanitary methods of handling the water shall be used.

46.  Wherever a storage container for drinking water is used it shall be:

(1)   securely covered and closed;
(2)   used only for the purpose of storing potable water;
(3)   maintained in a sanitary condition;
(4)   used in such a way that, when water is drawn from the container, the water does not become contaminated; and
(5)   disinfected in a manner approved by Health Canada at least once each 7 days while in use, and before the container is used following storage.

47.  Except where drinking water is provided by a fountain, there shall be provided:

(1)   an adequate supply of single-use drinking cups in a sanitary container located near the water container; and
(2)   a non-combustible covered receptacle for the disposal of used drinking cups.

48.  The use of a common drinking cup is prohibited.

49.  Ice that is added to drinking water or used for the contact refrigeration of foodstuffs shall be made from potable water and shall be stored and handled so as to prevent it from becoming contaminated. Ice handling equipment, as well as the storage area, should be regularly disinfected.

50.  Where drinking water is supplied by a drinking fountain:

(1)    the fountain shall meet the standards set out in ARI Standard 1010-82, *Standard for Drinking-Fountains and Self-Contained, Mechanically-Refrigerated Drinking-Water Coolers*; and

(2)    the fountain shall not be installed in a personal service room containing a toilet.

## Clothing storage

51. Change rooms shall be provided where:

(1)    the nature of the work engaged in by an employee makes it necessary for the employee to change from street clothing to work clothing for safety, health or occupational cleanliness reasons; or

(2)    an employee is normally engaged in work in which the employee's clothing becomes wet or contaminated by a hazardous substance to the extent that it could constitute a health or safety hazard to that employee or other persons.

52. In each change room:

(1)    a floor area of at least 0.4 m² shall be provided for each of the employees who normally use the room at any one time; and

(2)    where it is necessary for the employees to change footwear, seats shall be provided in sufficient numbers to accommodate them.

53. To the extent that this is reasonably practicable, a clothing change room shall be located:

(1)    near the workplace and connected thereto by a completely covered route to the workplace;

(2)    near a shower room provided pursuant to paragraph 35;

(3)    near a toilet room;

(4)    on a direct route to the entrance to the workplace.

54. Where contaminated work clothing referred to in paragraph 51(2) is changed, it shall be stored, handled or disposed of in such a manner that the contaminant does not come in contact with uncontaminated clothing.

55. Departments shall ensure that employees do not wear away from the workplace any clothing that has been contaminated.

56. Facilities shall be provided or arrangements shall be made to clean, to the extent required for decontamination, and to dry such clothes before they are worn again.

57. Clothing storage facilities shall be provided by the employer for the storage of overcoat and outer clothes not worn by employees while they are working.

## Lunchrooms

58. No person shall eat, prepare or store food:

(1)　in a place where a hazardous substance is likely to contaminate food, dishes or utensils;
(2)　in a personal service room that contains a toilet, urinal or shower bath;
(3)　in any place which, according to Health Canada, is unsuitable for this purpose;
(4)　in laboratories.

59. Where a lunchroom is provided for employees:

(1)　it shall be physically separated or isolated from any place where there is a possibility of contamination by hazardous substances;
(2)　it shall not be used for any purpose that is incompatible with its use as a lunchroom;
(3)　it shall be provided with non-combustible, covered receptacles for the disposal of waste food or other waste material;
(4)　dishes or other food utensils shall not be washed in lavatory or sanitary facility wash basins;
(5)　it shall not have any dimension of less than 2.3 m;
(6)　it shall have a minimum floor area of 9 m²;
(7)　it shall have 1.1 m² of floor area for each of the employees who normally use the room at any one time;
(8)　it shall be furnished with a sufficient number of tables and chairs to accommodate adequately the number of employees normally using the lunchroom at any one time.

## Field accommodation

60. A field accommodation shall not be used as such unless:

(1)　it is so constructed that it can be adequately cleaned and disinfected;
(2)　the food preparation area and the lunchroom are separately partitioned from the sleeping quarters;
(3)　where a potable water plumbing system is provided, such system complies with the requirements of Health Canada;
(4)　sewage, refuse and garbage disposal facilities are provided;
(5)　all such accommodation, including sanitary facilities, is located, constructed and maintained in a clean and sanitary condition;
(6)　it is located on well-drained ground;
(7)　toilet rooms are maintained in a sanitary condition;
(8)　vermin prevention, heating, ventilation and sanitary sewage systems are provided.

61. Departments shall ensure that, in any field accommodation used for sleeping quarters:

(1) a separate bed or bunk is provided for each occupant;
(2) the beds or bunks provided are not more than double-tiered and are so constructed that they can be adequately cleaned and disinfected;
(3) all mattresses, sheets, pillows, pillow cases, blankets, sleeping bags and bed covers are maintained in a clean and sanitary condition and are regularly inspected under the responsibility of the person in charge at the location;
(4) clean laundered sheets and pillow cases are, under normal circumstances, supplied to each occupant at least once each week;
(5) blankets are laundered at least once every three months of normal use, or upon a change of bed occupant if sheets are not used; and
(6) a locker capable of being locked and at least one shelf in addition to the locker are provided for each bed or bunk.

62. An environmental health officer of Health Canada may direct that additional or other measures be taken, where required, in order to maintain sanitary and healthful conditions in a field accommodation.

63. To the extent that is practicable, mobile accommodation shall meet the standards set out in CSA Standard Z240.2.1-1979, amended from time to time.

**Food preparation, storage and serving of food**

64. The storage, supply and handling of food shall comply with:

(1) the *Sanitation Code for Canada's Food Service Industry* issued by the Canadian Restaurant Association, Toronto, dated 1988; and
(2) Where reasonably practicable, toilet rooms and wash basins separate from those used by other employees shall be provided for food handlers.

65. Where, in the opinion of an authorized official of Health Canada, a code, procedure or condition referred to in this standard, or utilized by a department or agency, does not provide a sufficient degree of health protection, or may otherwise be inappropriate, he or she may make directions in writing to the department or agency concerning the specific codes or procedures to be applied.

66. Information or advice concerning applicable codes, procedures and good industrial sanitation and health practices with respect to a specific situation may be obtained from the appropriate Regional Office of Medical Services Branch, Health Canada.

**Accommodation standards**

67. Information concerning accommodation data may be obtained through Public Works and Government Services Canada.

## CHAPTER 2-18 - APPENDIX A - VENTILATION FOR LOCKER ROOMS, SHOWER SPACES AND TOILET SPACES

| Column 1<br>Type of location | Column 2<br>Minimum requirements |
|---|---|
| 1. Locker rooms: | |
| (a) change room for employees with clean work clothes; | (a) 5 L/s per m² of floor area |
| (b) change room for employees with wet or sweaty clothes; | (b) 10 L/s per m² of floor area; 3 L/s exhausted from each locker; |
| (c) change room for employees who work or clean where clothes will be wet or where clothes will pick up heavy odours. | (c) 15 L/s per m² of floor area; 4 L/s exhausted from each locker. |
| 2. Shower spaces | 10 L/s per m² of floor area. At least 20 L/s per shower head, 90 L/s minimum for shower spaces. |
| 3. Toilet spaces | 10 L/s per m² of floor area. At least 10 L/s per sanitary facility, 90 L/s minimum for toilet spaces. |

## CHAPTER 2-18 - APPENDIX B - TOILET FACILITIES

### Office workplace

| Number of persons of each sex | Minimum number of water closets for each sex |
|---|---|
| 1-25 | 1 |
| 26-50 | 2 |
| over 50 | 3 plus 1 for each additional increment of 50 persons of each sex |

**Industrial workplace**

| Number of persons of each sex | Minimum number of water closets for each sex |
|---|---|
| 1-10 | 1 |
| 11-25 | 2 |
| 26-50 | 3 |
| 51-75 | 4 |
| 76-100 | 5 |
| Over 100 | 6 plus 1 for each additional increment of 30 persons of each sex |

## CHAPTER 2-19 - REFUSAL TO WORK DIRECTIVE

### Grievance procedure

In cases of alleged misinterpretation or misapplication arising out of this standard, the grievance procedure, for all represented employees, within the meaning of the *Public Service Staff Relations Act*, will be in accordance with Section 7.0 of the *National Joint Council By-Laws*. For unrepresented employees, the departmental grievance procedure applies.

This standard is deemed to be part of collective agreements between the parties to the National Joint Council and employees are to be afforded ready access to this standard.

### Application

1. This standard applies to all departments and other portions of the Public Service, as defined in Part I of Schedule I of the *Public Service Staff Relations Act*.

### Definitions

2. In this standard:

(1) *board* means the Public Service Staff Relations Board (commission);
(2) *danger* means any hazard or condition that could reasonably be expected to cause injury or illness to a person exposed thereto before the hazard or condition can be corrected (danger);
(3) *safety and health committee* means a committee established pursuant to chapter 2-20, Committees and representatives directive (comité de santé et de sécurité);
(4) *safety and health representative* means a person appointed as a safety and health representative pursuant to chapter 2-20, Committees and representatives directive (représentant pour la santé et la sécurité);
(5) *safety officer* means a person designated as a safety officer pursuant to the *Canada Labour Code*, Part II, and includes a regional safety officer (agent de sécurité);

(6)    *workplace* means any place where an employee is engaged in work for the employee's department (lieu de travail).

## Refusal to work

3.  Where an employee, while at work, has reasonable cause to believe that:

(1)    the use or operation of a machine or thing constitutes a danger to the employee or another employee; or
(2)    a condition exists in any place that constitutes a danger to the employee,

the employee may refuse to use or operate the machine or thing or to work in that place.

4.  An employee may not, pursuant to this standard, refuse to use or operate a machine or thing, or to work in a place where:

(1)    the refusal puts the life, health or safety of another person directly in danger; or
(2)    the danger referred to in paragraph 3 is inherent in the employee's work or is a normal condition of employment.

5.  For purposes of paragraph 4(2), "inherent in the employee's work" means employment risks normally related to a specific task, trade or occupation. In such cases, an employee shall be qualified, through knowledge, training and experience, to perform the assigned work and wear or use the prescribed protective devices and equipment.

## Employees on ships and aircraft

6.  Where an employee on a ship or an aircraft that is in operation has reasonable cause to believe that:

(1)    the use or operation of a machine or thing on the ship or aircraft constitutes a danger to the employee or another employee; or
(2)    a condition exists in a place on the ship or aircraft that constitutes a danger to the employee,

the employee shall forthwith notify the person in charge of the ship or aircraft of the circumstances of the danger and the person in charge shall, as soon as practicable thereafter, having regard to the safe operation of the ship or aircraft, decide whether or not the employee may discontinue the use or operation of the machine or thing or to work in that place and shall inform the employee accordingly.

7.  An employee who, pursuant to paragraph 6, is informed that he or she may not discontinue the use or operation of a machine or thing or to work in a place shall not, while the ship or aircraft on which he or she is employed is in operation, refuse, pursuant to this standard, to operate the machine or thing or to work in that place.

8.  For the purposes of paragraphs 6 and 7:

(1)  a ship is in operation from the time it casts off from a wharf in any Canadian or foreign port until it is next secured alongside a wharf in Canada; and
(2)  an aircraft is in operation from the time it first moves under its own power for the purpose of taking off from any Canadian or foreign place of departure until it comes to rest at the end of its flight to its first destination in Canada.

## Reporting

9.  Where an employee refuses to use or operate a machine or thing or to work in a place pursuant to paragraph 3, or is prevented from acting in accordance with paragraph 7, the employee shall forthwith report the circumstances of the matter to his or her supervisor, or to the person in charge, and to:

(1)  a member of the safety and health committee, if any, established for the workplace affected; or
(2)  the safety and health representative, if any, appointed for the workplace affected.

10.  A department shall forthwith, on receipt of a report under paragraph 9, investigate the report in the presence of the employee who made the report and in the presence of:

(1)  at least one member of the safety and health committee, if any, to which the report was made under paragraph 9, who does not exercise managerial functions;
(2)  the safety and health representative, if any; or
(3)  where no safety and health committee or safety and health representative has been established or appointed for the workplace affected, in the presence of at least one person selected by the employee.

## Continuing refusal

11.  Where a department disputes a report made to it by an employee pursuant to paragraph 9, or where the department takes steps to make the machine or thing or the place safe in respect of which such report was made, and the employee has reasonable cause to believe that:

(1)  the use or operation of the machine or thing continues to constitute a danger to the employee or another employee; or
(2)  a condition continues to exist in the place that constitutes a danger to the employee,

the employee may continue to refuse to use or operate the machine or thing or to work in that place.

## Investigation by a safety officer

12. Where an employee continues to refuse to use or operate a machine or thing or to work in a place pursuant to paragraph 11, the department and the employee shall each forthwith notify a safety officer and the safety officer shall forthwith, on receipt of either notification, investigate or cause another safety officer to investigate the matter in the presence of a departmental representative and the employee or the employee's representative.

13. A safety officer shall, on completion of an investigation made pursuant to paragraph 12, decide whether or not:

(1)    the use or operation of the machine or thing in respect of which the investigation was made constitutes a danger to any employee; or
(2)    a condition exists in the place in respect of which the investigation was made that constitutes a danger to the employee referred to in paragraph 12,

and the safety officer shall forthwith notify the department and the employee of the safety officer's decision.

## Continued work

14. Prior to the investigation and decision of a safety officer pursuant to paragraphs 12 and 13:

(1)    the department may require that the employee concerned remain at a safe location near the place in respect of which the investigation is being made or assign the employee reasonable alternate work; and
(2)    the department shall not assign any other employee to use or operate the machine or thing or to work in that place pending resolution of the situation.

## Safety officer decision

15. Where a safety officer decides that the use or operation of a machine or thing constitutes a danger to an employee or that a condition exists in a place that constitutes a danger to an employee:

(1)    the safety officer shall notify the department of the danger and issue directions in writing to the department directing it immediately or within such period of time as the safety officer specifies:
    (a)   to make measures for guarding the source of danger; or
    (b)   to protect any person from the danger; and
(2)    the safety officer may, if he or she considers that the danger cannot otherwise be guarded or protected against immediately, issue a direction in writing to the department directing that the place, machine or thing in respect of which the direction is made shall not be used or operated until the safety officer's directions are complied with, but nothing in this paragraph prevents the doing of anything necessary for the proper compliance with the direction.

16. An employee may continue to refuse to use or operate the machine or thing or to work in that place until the safety officer's direction is complied with or until it is varied or rescinded under this standard.

17. When a safety officer decides that the use or operation of a machine or thing does not constitute a danger to an employee or that a condition does not exist in a place that constitutes a danger to an employee, an employee is not entitled under this standard to continue to refuse to use or operate the machine or thing or to work in the place, but the employee may, by notice in writing given within seven days of receiving notice of the decision of the safety officer, require the safety officer to refer the decision to the Board, and thereupon the safety officer shall refer the decision to the Board.

**Board inquiry**

18. Where a decision of a safety officer is referred to the Board pursuant to paragraph 17, the Board shall, without delay and in a summary way, inquire into the circumstances of the decision and the reasons therefor and may:

(1)   confirm the decision; or
(2)   give any direction that it considers appropriate in respect of the machine, thing or place in respect of which the decision was made that a safety officer is required or entitled to give under paragraph 15.

19. Where the Board gives a direction under paragraph 18, it shall cause to be affixed to or near the machine, thing or place in respect of which the direction is given a notice in the form approved by the Minister of Human Resources Development, and no person shall remove the notice unless authorized by a safety officer or the Board.

20. Where the Board directs, pursuant to paragraph 18, that a machine, thing or place not be used until its directions are complied with, the department shall discontinue the use thereof, and no person shall use such machine, thing or place until the directions are complied with, but nothing in this paragraph prevents the doing of anything necessary for the proper compliance therewith.

**Compensation**

21. The fact that a department or employee has complied with or failed to comply with any of the provisions of this standard shall not be construed to affect any right of an employee to compensation under any statute relating to compensation for employment injury, or to affect any liability or obligation of any department or employee under any such statute.

**General prohibition**

22. No department shall dismiss, suspend, lay off or demote an employee or impose any financial or other penalty on an employee or refuse to pay the employee remuneration in respect of any period of time that the employee, but for the exercise of his or her rights under this standard, would have worked, or take any disciplinary action against, or threaten to take such action against an employee because that employee:

(1) has testified or is about to testify in any proceeding taken or inquiry held under this standard;

(2) has provided information to a person engaged in the performance of duties regarding the conditions of work affecting the safety or health of that employee or any of his or her fellow employees; or

(3) has acted in accordance with this standard or has sought the enforcement of any of the provisions of this standard.

## Right to complain

23. Where an employee alleges that a department has taken action against the employee in contravention of paragraph 22 because the employee has acted in accordance with paragraphs 3 to 17, the employee may, subject to paragraph 25, make a complaint in writing to the Board of the alleged contravention.

24. A complaint made pursuant to paragraph 23 shall be made to the Board not later than ninety days from the date on which the complainant knew, or in the opinion of the Board ought to have known, of the action or circumstances giving rise to the complaint.

25. An employee may not make a complaint if he or she has failed to comply with paragraphs 9 and 12 in relation to the matter that is the subject matter of the complaint.

26. Notwithstanding any law or agreement to the contrary, a complaint referred to in paragraph 23 may not be referred by an employee to adjudication.

## Board action

27. On receipt of a complaint made under paragraph 23, the Board:

(1) may assist the parties to the complaint to settle the complaint; and

(2) where the Board does not act under paragraph 27(1) or the complaint is not settled within such period as the Board considers to be reasonable in the circumstances, shall hear and determine the complaint.

28. A complaint made pursuant to paragraph 23 in respect of an alleged contravention of paragraph 22 by a department is itself evidence that such contravention actually occurred and, if any party to the complaint proceedings alleges that such contravention did not occur, the burden of proof thereof is on that party.

29. Where, under paragraph 27, the Board determines that a department has contravened paragraph 22, the Board may, by order, require the department to cease contravening that provision and may, where applicable, by order, require the department to:

(1) permit to return to the duties of his or her employment any employee who has been affected by the contravention;

(2) reinstate any former employee affected by the contravention;

(3)    pay to any employee or former employee affected by the contravention, compensation not exceeding such sum as, in the opinion of the Board, is equivalent to the remuneration that would, but for the contravention, have been paid by the department to that employee or former employee; and

(4)    rescind any disciplinary action taken in respect of, and pay compensation to, any employee affected by the contravention, not exceeding such sum as, in the opinion of the Board, is equivalent to any financial or other penalty imposed on the employee by the department.

## CHAPTER 2-20 - COMMITTEES AND REPRESENTATIVES DIRECTIVE

### Grievance procedure

In cases of alleged misinterpretation or misapplication arising out of this standard, the grievance procedure, for all represented employees, within the meaning of the *Public Service Staff Relations Act*, will be in accordance with Section 7.0 of the *National Joint Council By-Laws*. For unrepresented employees, the departmental grievance procedure applies.

This standard is deemed to be part of collective agreements between the parties to the National Joint Council and employees are to be afforded ready access to this standard.

### Application

1. This standard incorporates the minimum requirements of the *Canada Labour Code*, Part II, and applicable regulations issued pursuant to that legislation and applies to all departments and other portions of the Public Service, as defined in Part I of Schedule I of the *Public Service Staff Relations Act*.

### Definitions

2. In this standard:

(1)    ***bargaining agent*** means an employee organization: (agent négociateur)
      (a)    that has been certified by the Public Service Staff Relations Board as bargaining agent for a bargaining unit, and
      (b)    the certification of which has not been revoked;

(2)    ***regional office*** means the regional office of Human Resources Development Canada for the administrative region of the Department in which the workplace is situated (bureau régional);

(3)    ***safety and health committee*** means a committee established pursuant to this standard (comité de la santé et de la sécurité);

(4)    ***safety and health representative*** means a representative appointed pursuant to this standard (représantant à la sécurité et à la santé);

(5)    ***safety officer*** means a person designated as a safety officer pursuant to the *Canada Labour Code*, Part II, and includes a regional safety officer (agent de sécurité);

(6)    ***workplace*** means any place where an employee is engaged in work for the employee's department, i.e., the place of assignment as a function of employment (lieu de travail).

## Safety and health committees

### National and regional committees

3. Agreements between departments and bargaining agents for the formation and operation of national and regional safety and health committees remain valid. Chapter 4-9, National and regional safety and health committees, outlines various requirements respecting the establishment and operation of these committees.

### Workplace committees

4. Every department shall, for each departmental workplace at which twenty or more employees are normally employed, establish a safety and health committee consisting of at least two persons, one of whom is an employee or, where the committee consists of more than two persons, at least half of whom are employees who:

(1)   do not exercise managerial functions; and
(2)   have been selected by the bargaining agent representatives.

5. For the purposes of paragraph 4(1), an employee cannot be a step in the grievance procedure or a managerial exclusion.

6. A department is not required under this standard to establish a safety and health committee pursuant to paragraph 4, for a workplace that is on board a ship in respect of employees whose base is the ship.

### Additional committees

7. Where a department controls more than one workplace referred to in this standard or the size or nature of the operations of the department or the workplace precludes the effective functioning of a single safety and health committee for those workplaces, the department shall, in consultation with the bargaining agents and subject to the approval of or in accordance with the direction of a safety officer, establish a safety and health committee for such of those workplaces as are specified in the approval or direction.

### Members

8. A department shall select the member or members of a safety and health committee to represent the department from among persons who have management authority.

9. A safety and health committee shall have two chairpersons selected from among the members of the committee, one being selected by the bargaining agent representatives and the other by the departmental representatives.

10. The chairpersons referred to in paragraph 9 shall act alternately for such period of time as the safety and health committee specifies in its rules of procedure.

11. A person may be selected as a member of a safety and health committee for more than one term.

12. Where a member of a safety and health committee resigns or ceases to be a member for any other reason, the vacancy shall be filled within 30 days after the next regular meeting of the committee.

13. The quorum of a safety and health committee shall consist of the majority of the members of the committee, of which at least half are bargaining agent representatives and at least one is a departmental representative.

## Exemptions

14. Where a department, in consultation with the bargaining agents, is satisfied that the nature of work being done by employees at a workplace is relatively free from risks to safety and health, the department may be exempted from the requirements of paragraph 4, in respect of the workplace. In such cases, a request for exemption shall be referred by the department to the appropriate regional or district office of Human Resources Development Canada.

15. Where, pursuant to a collective agreement or any other agreement between the employer and employees, a committee of persons has been appointed in respect of a workplace controlled by a department and such committee has, in the opinion of a safety officer, a responsibility for matters relating to safety and health in the workplace to such an extent that a safety and health committee established under paragraph 4 for that workplace would not be necessary:

(1) the safety officer may, by order, exempt the department from the requirements of paragraph 4 in respect of that workplace;
(2) the committee of persons that has been appointed for the workplace has, in addition to any rights, functions, powers, privileges and obligations under the agreement, the same rights, functions, powers, privileges and obligations as a safety and health committee under this standard; and
(3) the committee of persons so appointed shall, for the purposes of this standard, be deemed to be a safety and health committee established under paragraph 4 and all rights and obligations of departments and employees under this standard and the provisions of this standard respecting a safety and health committee apply, with such modifications as the circumstances require, in respect of the committee of persons so appointed.

## Posting names

16. A department shall post and keep posted the names and work locations of all the members of the safety and health committee established for the workplace controlled by the department in a conspicuous place or places where they are likely to come to the attention of the department's employees.

## Committee powers

17. A safety and health committee:

(1)  shall receive, consider and expeditiously dispose of complaints relating to the safety and health of the employees represented by the committee;
(2)  shall maintain records pertaining to the disposition of complaints relating to the safety and health of the employees represented by the committee;
(3)  shall cooperate with any occupational health service established to serve the workplace;
(4)  may establish and promote safety and health programs for the education of the employees represented by the committee;
(5)  shall take part in all inquiries and investigations pertaining to occupational safety and health including such consultations as may be necessary with persons who are professionally or technically qualified to advise the committee on such matters. The committee's involvement in these activities shall include the participation of committee members appointed by the chairpersons;
(6)  should develop, establish and maintain programs, measures and procedures for the protection or improvement of the safety and health of employees;
(7)  shall monitor on a regular basis programs, measures and procedures related to the safety and health of employees. Such monitoring includes workplace inspections and accident investigations. Where unsafe conditions or practices requiring immediate attention are found, the committee shall forthwith notify the person in charge;
(8)  shall ensure that adequate records are kept on work accidents, injuries and health hazards and shall monitor data relating to such accidents, injuries and hazards on a regular basis;
(9)  shall cooperate with safety officers;
(10) may request from a department such information as either party of the committee considers necessary to identify existing or potential hazards with respect to materials, processes or equipment in the workplace;
(11) shall have full access to all correspondence and reports relating to the safety and health of employees represented by the committee, but shall not have access to the medical records of any employee, except with the consent of that employee;
(12) shall assist in the determination of requirements for personal protective equipment pursuant to the Personal protective equipment directive, chapter 2-14; and
(13) may provide advice in planning and implementing changes in the workplace where occupational safety and health may be a factor, including work processes and procedures.

## Minutes

18. The chairperson selected by the departmental representatives shall provide a copy of the minutes of each safety and health committee meeting to the department and to each committee member as soon as possible after the meeting.

19. As soon as possible after a copy of committee minutes is received, the department shall post same in the place or places referred to in paragraph 16 and keep it posted there for a period of two months.

20. A copy of the committee minutes shall be kept by the department at the workplace to which it applies or at departmental headquarters for a period of two years from the day on which the committee meeting was held in such a manner that it is readily available for examination by a safety officer.

21. The chairperson selected by the departmental representatives shall:

(1)  not later than March 1 in each year, submit to the regional safety officer at the regional office, a report of the committee's activities during the 12-month period ending on December 31 of the preceding year, signed by both chairpersons on the Human Resources Development Canada Form No. 499 (Rev. 2/86) entitled *Safety and Health Committee Report* and containing the information required by that form; and
(2)  as soon as possible after submitting the report referred to in paragraph 21(1), post a copy of the report in the place or places referred to in paragraph 16 and keep it posted there for a period of two months.

### Records

22. A safety and health committee shall keep accurate records of all matters that come before it pursuant to paragraph 17 and shall keep minutes of its meetings and shall make such minutes and records available to a safety officer on his or her request.

### Meetings

23. A safety and health committee shall meet during regular working hours at least once each month and, where meetings are required on an urgent basis as a result of an emergency or other special circumstance, the committee shall meet as required whether or not during regular working hours.

24. Members of a safety and health committee are entitled to such time from their work as is necessary to attend meetings or to carry out any other functions as members of the committee, including reasonable meeting preparation time, and any time spent by the member while carrying out any of his or her functions as a member of the committee shall, for the purposes of calculating wages owing to him or her, be deemed to have been spent at work.

### Liability

25. No member of a safety and health committee is personally liable for anything done or omitted to be done by him or her in good faith under the authority of this standard.

### Committee rules

26. A safety and health committee may establish its own rules of procedure in respect of the terms of the office, not exceeding two years, of its members, the

time, place and frequency of regular meetings of the committee, and such procedures for its operations as it considers advisable.

## General prohibition

27. No department shall fail or neglect to provide a safety and health committee with any information requested by it pursuant to paragraph 17(10).

## Safety and health representatives

## Representatives

28. Every department shall, for each workplace controlled by it at which five or more employees are normally employed and for which no safety and health committee has been established, appoint the person selected pursuant to paragraph 29 as the safety and health representative for that workplace.

29. The employees at a workplace referred to in paragraph 28 who do not exercise managerial functions, and have bargaining agent representatives, shall select from among those employees a person to be appointed as the safety and health representative of that workplace and shall advise the department in writing of the name of the person so selected.

30. For the purposes of paragraph 29, an employee cannot be a step in the grievance procedure or a managerial exclusion.

31. The term of office of a safety and health representative shall be not more than two years.

32. An employee may be selected as a safety and health representative for more than one term.

33. Where a safety and health representative resigns or ceases to be a representative for any other reason, the vacancy shall be filled within 30 days.

## Additional representatives

34. Where a department controls more than one workplace referred to in this standard or the size or nature of the operations of the department or the work precludes the effective functioning of a single safety and health representative for those workplaces, the department shall, in consultation with the bargaining agents and subject to the approval of or in accordance with the direction of a safety officer, appoint a safety and health representative for such of those workplaces as are specified in the approval or direction.

## Posting of name

35. A department shall post and keep posted, in a conspicuous place or places where it is likely to come to the attention of its employees, the name and work location of the safety and health representative appointed for the workplace controlled by that department.

**Representative powers**

36. A safety and health representative:

(1) shall receive, consider and expeditiously dispose of complaints relating to the safety and health of employees represented by the representative;
(2) shall take part in all inquiries and investigations pertaining to occupational safety and health including such consultations as may be necessary with persons who are professionally or technically qualified to advise the representative on such matters;
(3) shall monitor on a regular basis, programs, measures and procedures related to the safety and health of employees;
(4) shall ensure that adequate records are kept on work accidents, injuries and health hazards and shall monitor data relating to such accidents, injuries and hazards on a regular basis;
(5) may request from a department such information as the representative considers necessary to identify existing or potential hazards with respect to materials, processes or equipment in the workplace; and
(6) shall have full access to all correspondence and reports relating to safety and health of the employees represented by the representative, but shall not have access to the medical records of any employee, except with the consent of that employee.

**Wages**

37. A safety and health representative is entitled to such time from his or her work as is necessary to carry out his or her functions as a representative and any time spent while carrying out any of those functions shall, for the purpose of calculating wages owing to him or her, be deemed to have been spent at work.

**Liability**

38. No safety and health representative is personally liable for anything done or omitted to be done by him or her in good faith under the authority of this standard.

**General prohibition**

39. No department shall fail or neglect to provide a safety and health representative with any information requested by him or her pursuant to paragraph 36(5).

## CHAPTER 3 - FIRE PROTECTION SERVICES

### Introduction

Fire protection in the Public Service is subject to the requirements of the OSH directives and standards which, in turn, reflect the *Canada Labour Code*, Part II and pursuant regulations. Additionally, it is subject to the Fire Protection Standards developed by the Fire Commissioner of Canada (FCC) and issued by the Treasury Board for Public Service establishments.

The office of the FCC, located within Human Resources Development Canada (HRDC), provides services such as fire safety inspections and fire investigations, and monitors the design and construction of buildings for conformity with the standards.

Fire protection standards are published in the following chapter. In addition, the standards listed below, which are also currently in force, will be published as they are revised in cooperation with HRDC. These standards may be obtained from the office of the FCC:

- Construction operations
- Welding and cutting
- Records storage
- Piers and wharves
- General storage
- Fire extinguishers (motor vehicles only)
- Sprinkler systems

An additional policy - Fire Protection, Investigation and Reporting - can be found in the "Materiel Services and Risk Management" volume of the *Treasury Board Manual*.

## CHAPTER 3-1 - STANDARD FOR FIRE SAFETY PLANNING AND FIRE EMERGENCY ORGANIZATION

### 1. General

#### 1.1 Purpose

This standard establishes the minimum requirements for fire safety plans including the organization of designated staff for fire emergency purposes.

#### 1.2 Application

This standard applies to all:

(a) departments and agencies listed in schedules A and B of the *Financial Administration Act* (FAA) with the exception of the Department of National Defence;
(b) branches designated as departments for the purposes of the FAA; and
(c) those departments and other portions of the Public Service as defined in part I of schedule I of the *Public Service Staff Relations Act*.

#### 1.3 Scope

This standard describes the procedures to be followed by departments and agencies in planning and organizing for fire emergencies in Government of Canada property. It is to be read and implemented in conjunction with Part II of the *Canada Labour Code*, Part XVII of the *Canada Occupational Safety and Health Regulations*, and the requirements of the *National Fire Code of Canada* regarding emergency planning. (See appendix A.)

### 1.4 **Administration**

(a) The Fire Commissioner of Canada or his authorized representative is responsible for the administration and enforcement of this standard.
(b) This standard is not to be interpreted as permitting practices specifically prohibited by provincial, municipal, or other federal legislation.

### 1.5 **Definitions**

Certain terms used in this standard are defined to ensure understanding of their meaning and intent.

(1) *Designated staff* means those occupants of a building who have some delegated responsibility for the fire safety of other occupants under the fire safety plan (personnel désigné);
(2) *fire emergency organization* means a formal organization of designated staff to perform specified duties in the event of a fire emergency (see appendix A) (organisation des secours en cas d'incendie);
(3) *Government of Canada property* means real or personal property under the administration and control of a federal government department or agency, including property leased to the government (bien du gouvernement du Canada);
(4) *senior officer* means the highest ranking official of any department or agency occupying space in a building or facility and being responsible for the preparation and administration of the fire safety plan. The person so identified is to be located in that building (fonctionnaire supérieur).

### 1.6 **Abbreviations**

In this standard:

(a) **"FC"** means Fire Commissioner of Canada or the authorized representative of the Fire Commissioner of Canada;
(b) **"NBC"** means the latest edition of the *National Building Code of Canada*, as amended periodically;
(c) **"NFC"** means the latest edition of the *National Fire Code of Canada*, as amended periodically.

### 1.7 **Standards**

(a) Unless otherwise noted, a reference to another standard means to the latest edition of that standard, as amended periodically.
(b) In the event of an inconsistency between any standard incorporated by reference in this standard and any other provision of this standard, that other provision shall prevail to the extent of the inconsistency.

### 2. **The fire safety plan**

### 2.1 **Preparation**

(a) A fire safety plan shall be prepared for all Government of Canada properties by the senior officer.
(b) The fire safety plan shall include:

(i)    emergency procedures to be used in case of fire, including
- sounding the fire alarm;
- notifying the fire department;
- instructing occupants on procedures to be followed when the fire alarm sounds;
- evacuating endangered occupants, including special provisions for the disabled;
- confining, controlling and extinguishing a fire; and
- the time required to complete evacuation.

(ii)   the appointment and organization of designated staff to carry out fire safety duties;

(iii)  the instruction of designated staff and employees so that they are aware of their responsibilities for fire safety;

(iv)  the preparation of diagrams showing the name, if any, and address of the building and the type, location and operation of the building fire emergency systems;

(v)   the holding of fire drills;

(vi)  the control of fire hazards in the building;

(vii) the inspection and maintenance of building facilities provided for the safety of occupants;

(viii) a record of the fire emergency systems installed in the building and instructions to the designated staff and fire department on the operation of the systems;

(ix)  the procedures for reporting fires and false alarms; and

(x)   the procedures to facilitate fire department access to the building and to the fire location within the building.

(c)  In buildings as designated in Subsection 3.1, in addition to the requirements of (b), the fire safety plan shall include:

(i)    the establishment of a fire emergency organization;

(ii)   the duties and responsibilities of members of the fire emergency organization;

(iii)  the organization chart of the emergency organization;

(iv)  the appointment and instruction of designated staff in the use of the voice communication system;

(v)   a plan of the building, showing:
- the name, if any, and the address of the building,
- the name and address of the owner of the building,
- the names and locations of the tenants of the building,
- the date of preparation of the plan,
- the scale of the plan,
- the location of the building in relation to nearby streets and in relation to all buildings and other structures located within 30 m of the building,
- the maximum number of persons normally occupying the building at any time,
- a horizontal projection of the building, showing thereon its principal dimensions, and
- the number of floors above and below ground level;

(vi)  a plan of each floor of the building, showing:
- the name, if any, and the address of the building,
- the date of preparation of the plan,

- the scale of the plan,
- a horizontal projection of the floor, showing thereon its principal dimensions,
- the number of the floor to which the plan applies,
- the maximum number of persons normally occupying the floor at any time,
- the location of all fire escapes, fire exits, stairways, elevating devices, main corridors and other means of exit,
- the location of all fire protection equipment, and
- the location of the main electric power switches for the lighting system, elevating devices, principal heating, ventilation and air-conditioning equipment and other electrical equipment,

(vii) the names, room numbers and telephone numbers of the chief fire emergency warden and the deputy chief fire emergency warden of the building appointed by the senior officer.

(d) In institutional occupancies, as defined in the NBC, a sufficient number of designated staff shall be on duty to perform the tasks outlined in the fire safety plan.

(e) In group A, division 1 occupancies, as defined in the NBC, and containing more than 60 occupants, there shall be at least 1 designated staff member on duty in the building to perform the tasks outlined in the fire safety plan whenever the building is open to the public.

(f) In high buildings within the scope of Subsection 3.2.6 of the NBC, the fire safety plan shall include the following:

(i) the instruction of designated staff on the use of the voice communication system,

(ii) the procedures for the use of elevators and for the evacuation of the disabled,

(iii) the action to be taken by designated staff in initiating any smoke control or other fire emergency systems installed in the building until the fire department arrives, and

(iv) the procedures established to facilitate fire department access to the building and fire location within the building.

(g) In buildings or parts of buildings used for the short or long term storage of raw materials, goods in process, or finished goods, the fire safety plan shall identify:

(i) the product classifications, as per the NFC, for each part of the building where products of different classification are stored,

(ii) the method of storage, including aisle widths for rack storage,

(iii) the maximum permitted height of storage for the building or part of the building, if different,

(iv) the maximum permitted size of individual storage areas,

(v) in sprinklered buildings, the sprinkler system design criteria, inside and outside hose allowances, and results of the benchmark sprinkler system main drain and water flow tests,

(vi) the storage method and maximum height of storage as described in (ii) and (iii) shall be prominently posted in the storage area.

(h) In buildings where radioactive materials are stored or handled, the fire safety plan shall include:

(i) methods to control a fire emergency and to recover radioactive materials and equipment containing radioactive materials safely and efficiently,

(ii) the names, addresses, and telephone numbers of persons to be contacted in case of fire during non-operating hours,

(iii) the names, addresses and telephone numbers of primary and alternative sources of expert radiation safety advice and assistance,

(iv) the location of primary and alternative sources of radiation survey instruments, and

(v) the location for decontamination.

## 2.2 Administration

(a) The fire safety plan shall be prepared and administered by the senior officer.

(b) In buildings occupied by a number of departments, the senior officer of the department having the largest number of employees shall prepare and administer the fire safety plan.

(c) The senior officers of the remaining departments referred to in (b) shall cooperate in the formation and operation of the fire safety plan and shall be responsible for providing the necessary fire emergency wardens for the areas occupied by their respective departments.

(d) The senior officer preparing and administering the fire safety plan is encouraged to seek the cooperation of private sector and public sector employers and employees working in the building to whom this standard does not apply.

## 2.3 Review and acceptance

(a) The fire safety plan is to be prepared in cooperation with the fire department.

(b) The fire safety plan shall be reviewed and signed by the senior officer.

(c) The signed fire safety plan shall be forwarded to the District Office of Human Resources Development Canada for review and acceptance prior to implementation.

## 2.4 Distribution and posting

(a) A copy of the fire safety plan shall be distributed in manual form, to each member of the fire emergency organization, and shall be surrendered on leaving the organization.

(b) The fire emergency procedures shall be prominently posted on each floor and all Government employees are expected to be familiar with the instructions contained therein.

## 2.5 Availability

A copy of the fire safety plan shall be kept in the building in a specified location for inspection by Human Resources Development Canada officials or by an officer from the local fire department and for reference by designated staff in the building. Where there is a central alarm and control facility, it shall be kept at this location.

### 3. The fire emergency organization

#### 3.1 Requirement

(a) A fire emergency organization is required in all buildings that are more than three storeys, including storeys below grade, or when the occupant load usually in a building exceeds the number given in appendix A for the classification of the major occupancy of the building, except as required in (b).

(b) The FC may require a fire emergency organization in buildings other than as stipulated in (a) to accommodate unique situations.

(c) In buildings not requiring a fire emergency organization, one person plus an assistant shall be designated responsible for fire safety matters in the building. These two persons shall be known as the "fire emergency warden" and "deputy fire emergency warden".

#### 3.2 Administration

(a) The fire emergency organization shall be established and administered by the senior officer.

(b) In buildings occupied by a number of departments, the senior officer of the department having the largest number of employees shall establish and administer the fire emergency organization.

(c) The senior officers of the remaining departments referred to in (b) shall cooperate in the formation and operation of the fire emergency organization and shall be responsible for providing the necessary fire emergency wardens for the areas occupied by their respective departments.

(d) The senior officer establishing and administering the fire emergency organization is encouraged to seek the cooperation of private sector and public sector employers and employees working in the building to whom this standard does not apply in the formation and operation of the fire emergency organization.

#### 3.3 Personnel appointment and training

(a) Personnel for the fire emergency organization shall be employees who are normally employed in the building and shall be recruited on an equitable basis from staff of all occupying departments (see appendix B).

(b) Every fire emergency warden shall be instructed and trained in the responsibilities under the fire safety plan and in the use of fire protection equipment. A record of all instruction and training provided shall be kept by the employer for a period of two years from the date on which instruction or training is provided.

(c) Every employee shall be instructed and trained in the procedures to be followed in the event of an emergency and the location, use, and operation of fire protection equipment and emergency equipment provided by the employer.

### 3.4 **Chief fire emergency warden**

(a) In buildings occupied by a single department, the chief fire emergency warden shall be appointed by the senior officer of that department.

(b) In buildings occupied by more than one department, the chief fire emergency warden shall be appointed by the senior officer of the department having the largest number of employees in the building.

(c) During a fire emergency, the chief fire emergency warden shall assume full authority for and control of the fire emergency organization and for the evacuation of building occupants until such time as the emergency is terminated or the fire department arrives at the scene and assumes responsibility.

### 3.5 **Deputy chief fire emergency warden**

(a) In buildings occupied by a single department, a deputy chief fire emergency warden shall be appointed by the senior officer of that department upon agreement and/or recommendation of the chief fire emergency warden.

(b) In buildings occupied by more than one department, the deputy chief fire emergency warden may be appointed by the senior officer of the second largest occupying department; but by arrangement with the senior officer of the major occupying department in consultation with the chief fire emergency warden.

### 3.6 **Qualifications**

Chief fire emergency wardens and deputy chief fire emergency wardens should be appointed from supervisory staff, and should possess or acquire the skill and knowledge necessary to fulfill the duties of their positions.

### 3.7 **Floor fire emergency warden**

(a) A floor fire emergency warden shall be appointed for each floor area occupied by Government of Canada departments.

(b) The floor fire emergency warden shall be appointed by the chief fire emergency warden subject to the approval of the senior officer of the appointee's department, and the appointee's supervisor shall be so advised and shall also be informed of the responsibilities entailed.

### 3.8 **Deputy floor fire emergency warden**

A deputy floor fire emergency warden and additional deputy floor fire emergency wardens, as may be required, shall be appointed by the floor fire emergency warden, subject to the approval of the senior officer of the appointee's department. The appointee's supervisor shall be so advised.

### 3.9 **Monitors for disabled**

Monitors shall be appointed by the floor fire emergency warden to assist with the evacuation procedures for the disabled, subject to the agreement of the appointee's department and the disabled person. The disabled and his or her supervisor shall be so advised.

### 3.10 Other fire emergency wardens

Other fire emergency wardens may be required for special duties because of the configuration, location or occupancy of the property, and shall be appointed by the chief fire emergency warden subject to the approval of the senior officer of the appointee's department. The appointee's supervisor shall be so advised.

### 3.11 Temporary absence

When the chief fire emergency warden and the deputy chief fire emergency warden are to be absent from the building at the same time, they shall make arrangements to appoint acting fire emergency wardens during their absence and the senior officers of their respective department and shall be so advised. It is important to consider such factors as shift operations, meal and break periods when making provisions for temporary absences.

### 3.12 Resignations

When a chief fire emergency warden or a deputy fire chief emergency warden resigns, a replacement shall be appointed at the earliest possible time.

### 3.13 Fire emergency organization meetings

(a)  At least once every year and after any change is made in the fire safety plan for the building, all fire emergency wardens shall meet for the purpose of ensuring that they are familiar with the fire safety plan and their responsibilities thereunder.

(b)  The senior officer shall keep a record of each meeting for a period of two years from the date of the last meeting. The record shall contain the date of the meeting, the names of those present, and a summary of the matters discussed.

### 3.14 Visual inspections

Daily visual inspections by fire emergency wardens shall be made of floor areas and obvious faulty conditions reported to the appropriate officials for corrective action. Faulty conditions include: fire doors wedged or blocked open; exits, stairways and corridors obstructed; exit lights out; fire fighting equipment inoperative or obstructed; and obvious hazards such as the improper disposal of smokers' material, the unnecessary accumulation of combustibles, the improper use of flammable liquids, temporary or unsafe electric wiring and other unsafe conditions and practices.

### 3.15 Other duties

(a)  Where buildings are required to have voice communication systems, the fire emergency organization shall be responsible for the operation of the emergency voice communication facility pending the arrival of the fire department. The senior officer is responsible to arrange for instruction in the use of such equipment for certain members of the fire emergency organization designated by the chief fire emergency warden.

(b)  Personnel of the fire emergency organization may be requested to assist in other emergencies.

### 3.16 Fire department responsibility

(a) All fire emergencies shall be under the control of the officer in charge of the fire department upon his arrival at the scene. The fire emergency organization shall be placed under the direction of such officer.

### 3.17 Visual identifiers

(a) All wardens of the fire emergency organization shall be provided with a cap or other means of identification.
(b) The caps or other means used for identification shall be distinctive in colour for the chief, deputy chief, wardens and monitors for the disabled.
(c) The caps shall be provided by the department appointing the fire emergency wardens.

### 3.18 Loudspeakers

(a) In buildings occupied by 100 or more occupants, or in buildings where, due to building configuration it is deemed necessary by the FC, a battery-operated portable loudspeaker shall be provided to assist in crowd control.
(b) The loudspeaker shall be under the custody and control of the chief fire emergency warden.
(c) The loudspeaker shall be provided by the department responsible for the appointment of the chief fire emergency warden.

### 3.19 Flashlights

(a) In high-rise buildings, and other buildings as designated by the FC, flashlights shall be provided to each fire emergency warden.
(b) The flashlights shall be provided and maintained in good operating condition by the department appointing the fire emergency wardens.

## 4. Evacuation drills

### 4.1 Requirements

(a) At least one evacuation drill involving all occupants shall be conducted annually in all Government of Canada occupied areas of buildings except as required in (b).
(b) In certain occupancies, evacuation drills shall be held as follows:
   (i) **Non-residential schools** (attended by children)
      – a minimum of 6 drills each year, held at irregular intervals: 3 drills in the fall term starting shortly after the school opening, and 3 drills between January and June;
   (ii) **Non-residential schools** (adult)
      – at least 2 drills each year: 1 drill in the fall and 1 drill in the spring;
   (iii) **Residential schools** (attended by children)
      – at least 1 drill per month;
   (iv) **Residential schools** (adult)
      – 1 drill per course session, resulting in at least 1 drill every 3 months;

(v) **Daycare centres**
- 1 drill per month;

(vi) **Hospitals**
- 1 drill per month involving permanent staff on all shifts, and patients to the maximum extent possible taking into consideration their medical condition;

(vii) **Passenger terminals**
- 1 drill routine to be held at least every 3 months which may be conducted on a zone basis and may include passengers;

(viii) **Correctional institutions**
- 1 drill every 3 months involving permanent staff on all shifts, and inmates where possible;

(ix) **High buildings within the scope of subsection 3.2.6 of the NBC**
- designated staff drill every two months;
- a total evacuation drill once yearly;
- 1 drill every 3 months involving groups of adjoining floors.

Private sector occupants shall be encouraged to participate to the maximum extent possible.

## 4.2 Participation

All occupants shall participate in evacuation drills unless specifically exempted by the chief fire emergency warden. The disabled shall participate as much as reasonably practical.

## 4.3 Notification

The local fire chief shall be notified at least one week in advance if fire department participation is desired during a drill. Also, the local fire department has to be notified immediately prior to any fire alarm system activation for the purpose of carrying out a drill and afterwards as soon as the fire alarm system has been restored to normal operating condition.

## 4.4 Record of evacuations and drills

All evacuations and drills shall be recorded and the results maintained on file for a period of two years.

## 5. Evacuation procedures for disabled

## 5.1 Responsibility

(a) Specialized procedures shall be implemented as a part of the fire safety plan in order to provide for the safe evacuation of persons whose mobility is impaired (including both employees and visitors) in the event of a fire emergency, and shall include the following:

   (i) A register shall be maintained in every building containing the location and number of disabled employees in the building, with a brief description of the impairment. In high buildings, the register shall be located in the control room, and in other buildings, in a central location. The names of the individuals should be registered, subject to the individual's agreement;

(ii) The disabled persons are to identify themselves to their immediate supervisor. The supervisor is responsible to ensure that the disabled persons are registered; this includes temporary disabled individuals;

(iii) At least two monitors shall be assigned to each disabled person;

(iv) The procedures to be taken for the evacuation of the disabled shall be discussed with the disabled individual.

5.2 The procedures for the evacuation of the disabled should be practised with the disabled person(s) to the extent practicable.

## CHAPTER 3-1 - APPENDIX A - TABLE OF MAXIMUM OCCUPANT LOAD

| Group* | No. of Division | Occupancy Classification | Occupant Load** |
|--------|-----------------|--------------------------|------------------|
| A | 1,2,3,4 | Assembly: e.g. theatres, passenger terminals, museums, libraries, and schools. | All |
| B | 1,2 | Institutional: e.g. hospitals, correctional institutions (occupants under restraint). | All |
| C | — | Residential: e.g. residential schools, barracks (excludes 1 and 2 family dwellings). | 10 |
| D | — | Offices. | 50 |
| E | — | Mercantile: e.g. retail stores. | 50 |
| F | 1 | Industrial, high hazard: e.g. grain elevators, spray painting shops. | 25 |
| F | 2,3 | Industrial, light and moderate hazard: e.g. mail processing plants, laboratories, aircraft hangars, warehouses. | 50 |

\* Group occupancy classification as per NBC.
\*\* Includes employees and patrons/visitors.

## CHAPTER 3-1 - APPENDIX B - EXPLANATORY MATERIAL

### A – 1.3

The *Canada Occupational Safety and Health Regulations*, Part XVII, contains requirements for emergency procedures, emergency warden appointment and training, and emergency warden meetings and drills.

The emergency procedures to be prepared by the employer are for a variety of emergencies of which fire emergency is one. Therefore, the chief emergency warden, deputy chief emergency warden and other emergency wardens appointed as members of the emergency organization are usually the same individuals identified in this standard as the chief fire emergency warden, deputy chief fire emergency warden, and fire emergency wardens as members of the fire emergency organization.

This standard includes all fire safety requirements related to emergency planning and procedures that are contained in the *Canada Labour Code,* Part II, as well as the requirements of the NFC and the Office of the Fire Commissioner of Canada.

During the preparation and implementation of emergency procedures, the fire safety plan is to be prepared as an independent document which can be part of the entire *Emergency Procedures Manual*.

**A – 3.3**

The officers of the fire emergency organization should be appointed from supervisory staff to the extent possible.

## CHAPTER 3-1 - APPENDIX C - REFERENCE PUBLICATIONS

This standard makes reference to the following codes and standards which are not found in this volume:

**Fire Commissioner of Canada standards**

Construction operations
Welding and cutting
Computer systems
Record storage
Piers and wharves
General storage
Sprinkler systems

These standards may be obtained from the Fire Commissioner of Canada, Operations Program, Human Resources Development Canada, Ottawa, Ontario, K1A 0J2 or from regional offices of Human Resources Development Canada.

**Treasury Board publication**

*Treasury Board Manual*
"Materiel, Services and Risk Management" volume
Fire Protection, Investigation and Reporting (Part III, chapter 5)

This volume may be purchased from the Canada Communication Group — Publishing, 45 Sacré-Coeur Blvd., Hull, Québec, K1A 0S9.

**National Research Council of Canada publications**

*National Building Code of Canada*
*National Fire Code of Canada*

These codes may be purchased from the Secretary, Associate Committee on the *National Building Code*, National Research Council, Ottawa, K1A 0R6.

**Human Resources Development Canada publications**

*Canada Labour Code*, Part II
*Canada Occupational Safety and Health Regulations*

C.L.C. Part II and Regulations may be purchased from Public Works and Government Services Canada, Queen's Printer for Canada.

## CHAPTER 3-2 - FIRE PROTECTION STANDARD FOR DESIGN AND CONSTRUCTION

### 1. General

#### 1.1 Purpose

This standard establishes a satisfactory level of fire safety to be incorporated into the design and construction of Government of Canada property to minimize risks to life and property and to protect and conserve the Government's financial position.

#### 1.2 Application

This standard applies to all:

(a) departments and agencies listed in Schedules A and B of the *Financial Administration Act* (FAA) with the exception of the Department of National Defence;
(b) branches designated as departments for the purposes of the FAA; and
(c) those departments and other portions of the Public Service as defined in Part I of Schedule I of the *Public Service Staff Relations Act*.

#### 1.3 Scope

This standard describes the procedures to be followed by departments and agencies in incorporating fire protection engineering requirements into:

(a) the design, construction and occupancy of new buildings or structures;
(b) the alteration, extension, reconstruction, demolition, removal, relocation, and occupancy of existing buildings or structures; and
(c) the design and installation of fixed fire protection systems such as automatic sprinkler systems, standpipe and hose systems, fire hydrants, and fire alarm systems.

(See appendix B)

#### 1.4 Administration

(a) The Fire Commissioner of Canada or his authorized representative is responsible for the administration and enforcement of this standard.
(b) This standard is not to be interpreted as permitting practices specifically prohibited by provincial, municipal, or other federal legislation.

## 1.5 Definitions

In this standard, *Government of Canada property* means real or personal property under the administration and control of a federal government department or agency, including property leased to the government.

## 1.6 Abbreviations

In this standard:

(a)  **FC** means Fire Commissioner of Canada or the authorized representative of the Fire Commissioner of Canada;
(b)  **NBC** means the latest edition of the *National Building Code of Canada*; and
(c)  **NFC** means the latest edition of the *National Fire Code of Canada*.

## 2. Standards and requirements

### 2.1 New construction

(a)  New buildings and structures shall conform, at a minimum, to the following requirements:
    (i)   the *National Building Code of Canada*; and
    (ii)  FC standards.
(b)  In the case of situations not covered by the codes and standards stipulated in (a), the FC may refer to other nationally-recognized codes and standards to determine the fire protection engineering requirements.

### 2.2 Additions, alterations, and changes of occupancy

(a)  The requirements of subsection 2.1 apply to additions and alterations to existing buildings and structures, and to changes of occupancy, except as provided in (b).
(b)  When strict application of the requirements of subsection 2.1 would involve a costly retrofit, those requirements that are not considered essential to life safety may be determined on the basis of a risk and cost benefit analysis.

**Note**: The FC should be consulted to determine the extent to which the NBC applies with respect to upgrading buildings to remove unacceptable fire hazards.

### 2.3 Existing buildings

(a)  Existing buildings and structures shall conform to:
    (i)   the *National Fire Code (NFC) of Canada*; and
    (ii)  FC standards, as far as is reasonably practicable.

**Note**: The FC should be consulted to determine the extent to which FC standards apply to existing buildings.

### 2.4 **Construction and demolition operations**

(a) Fire safety at construction and demolition sites shall conform to:
  (i) Part 8 of the *National Building Code of Canada*; and
  (ii) FC *Standard for Construction Operations*.
(b) Before construction or demolition begins, an acceptable fire safety plan shall be prepared for the site in accordance with the appropriate Sections of Part 2 of the NFC.

## 3. **Procedures**

### 3.1 **General**

(a) A department or agency proposing to acquire or lease a building or to design, build, or alter a building, structure, or fire protection system shall have the plans and specifications reviewed by the FC if the project falls within the following categories:
  (i) new construction;
  (ii) additions and alterations to existing property when such additions and alterations affect any fire protective features of building design (see appendix B);
  (iii) the installation or alteration of any fire protection system (see appendix B); or
  (iv) proposals to acquire or lease accommodation exceeding a total of 600 m or intended for an occupant load in excess of 50 persons.
(b) Projects other than those described in (a) shall conform to the standards prescribed in Section 2 but need not be submitted to the FC for review unless specifically requested by the FC.

### 3.2 **Submissions to FC**

(a) Preliminary plans for projects described in Article 3.1(a) shall be submitted to the FC for review and comment on fire protection requirements.
  **Note**: The definition of "preliminary" may vary with the cost and complexity of a project. Ordinarily, it should mean when working drawings are begun, and not later than when they are 33% complete. However, on major projects the FC should be consulted during the development of the project brief and/or conceptual design stage.
(b) Final plans and specifications for projects described in Article 3.1(a) shall be submitted to the FC for review before tenders are called.
(c) Shop drawings and calculations for fire detection and suppression systems shall be submitted to the FC for review before the systems are installed or altered.

### 3.3 **Plans, specifications and calculations**

(a) The FC shall be given enough information to demonstrate that the proposed work will conform to this standard and to determine whether or not it may affect adjacent property.
(b) Plans shall be drawn to scale and shall indicate the nature and extent of the work or proposed occupancy in enough detail to establish that, when

completed, the work and the proposed occupancy will conform to this standard.

(c) Site plans shall show:
  (i) by distances from property lines, the location of the proposed building;
  (ii) similar distances that locate every other adjacent existing building on the property;
  (iii) existing and finished ground levels to an established datum at, or adjacent to, the site; and
  (iv) the access routes for fire fighting.

(d) The information submitted to the FC for review shall include the following:
  (i) name of the department or agency;
  (ii) building location, including street address;
  (iii) point of compass;
  (iv) floor plans;
  (v) construction and occupancy of building;
  (vi) building height (in storeys) and building area;
  (vii) size and location of street water mains and water supply connection(s);
  (viii) capacity of water supply (volume, flow rate and pressures);
  (ix) fire walls, fire doors, unprotected vertical openings, and blind spaces;
  (x) the degree of fire separation of storeys, shafts and special rooms or areas, including the location and rating of closures in fire separations;
  (xi) the source for the information on fire-resistance ratings of the elements of construction;
  (xii) location of exits;
  (xiii) fire detection, suppression and alarm systems; and
  (xiv) emergency lighting systems.

## 3.4 Inspection, testing and commissioning

(a) Departments and agencies (or their authorized agents) shall conduct site inspections during construction to ensure compliance with the design documents and the standards prescribed in Section 2.

(b) Departments and agencies (or their authorized agents) shall ensure that all fire protection systems are operationally effective, reliable, and installed in accordance with the applicable codes and standards by:
  (i) conducting a complete inspection of each system;
  (ii) identifying and correcting any deficiencies; and
  (iii) conducting commissioning tests as specified in the design documents and the appropriate reference standards prescribed in Section 2.

(c) The FC may conduct site inspections and witness tests of systems to verify compliance with the requirements of this standard. Departments and agencies shall ensure that the FC is given reasonable notice of appropriate times for the inspection and testing of fire protection features and systems.

(d) Prior to use and occupancy, written notice shall be given to the FC of:
  (i) the completion of all required exits, fire separations, and closures;

(ii) the completion of all fire protection systems including standpipe, sprinkler, fire alarm, and emergency lighting systems; and

(iii) the satisfactory completion of commissioning tests for all fire protection systems.

**Note**: See FC Standard *Construction Operations* for conditions of occupancy of unfinished buildings.

## CHAPTER 3-2 - APPENDIX A - REFERENCE PUBLICATIONS

This standard refers to the following codes and standards which are not found in this volume:

### Fire Commissioner of Canada standards

Construction operations
Welding and cutting
Computer systems
Record storage
Piers and wharves
General storage
Sprinkler systems

These standards can presently be obtained from the Fire Commissioner of Canada, Operations Program, Human Resources Development Canada, Ottawa, Ontario, K1A 0J2 or from regional offices of Human Resources Development Canada. As they are revised, they will be issued in this volume.

### Treasury Board publication

*Treasury Board Manual*
"Materiel, Services and Risk Management" volume
Fire Protection, Investigation and Reporting (Part III, chapter 5)

This volume may be purchased from the Canada Communication Group — Publishing, 45 Sacré-Coeur Blvd., Hull, Québec, K1A 0S9.

### National Research Council of Canada publications

*National Building Code of Canada*
*National Fire Code of Canada*

These codes can be purchased from the Secretary, Associate Committee on the *National Building Code*, National Research Council, Ottawa, K1A 0R6.

## CHAPTER 3-2 - APPENDIX B - FIRE PROTECTIVE FEATURES OF DESIGN

In the context of this standard, fire protection systems and fire protective features of design include, but are not necessarily limited to, the following:

(a) means of egress for all occupants in case of fire;

(b) appropriate fire resistance necessary for structures of a designated occupancy and fire test ratings for the structural assemblies;

(c)     restrictions on fire spread by appropriate design, construction, arrangement and use of building or structure, and materials;

(d)     installation requirements for heating, ventilating, air conditioning, lighting and power equipment;

(e)     safeguards to minimize and control the risk of fire or explosion involving hazardous materials, processes, or operations;

(f)     requirements for the storage and use of flammable and combustible liquids, gases and solids;

(g)     the design of water supplies for fire protection, the pressure and flow at which they are to be delivered and the proper sizing, arrangement, installation, testing and maintenance of piping, both aboveground and underground including valves, hydrants, reservoirs, tanks, pumps and their related controls;

(h)     the application, design, installation, testing and maintenance of manual and automatic fire suppression systems, equipment and devices, including sprinkler systems, standpipe and hose systems, dry chemical systems, foam systems, $CO_2$ systems, Halon systems, and fire extinguishers;

(i)     the application, design, installation, testing and maintenance of fire alarm systems, including those with voice communication capability and systems for the detection of fire;

(j)     the application of lightning protection systems;

(k)     the design, installation, testing and maintenance of emergency electric power systems; and

(l)     measures to minimize the loss or damages subsequent to fire.

## CHAPTER 3-3 - FIRE PROTECTION STANDARD FOR ELECTRONIC DATA PROCESSING EQUIPMENT

### 1. General

#### 1.1 Purpose

This standard establishes minimum requirements for fire protection of electronic data processing equipment (henceforth referred to as EDP equipment) in order to minimize risks to Government of Canada property and operations, and to protect and conserve the Government's financial position.

#### 1.2 Application

This standard applies to:

(a)     departments and agencies listed in Schedules A and B of the *Financial Administration Act (FAA)* with the exception of the Department of National Defence;

(b)     branches designated as departments for the purposes of the FAA; and

(c)     those departments and other portions of the Public Service as defined in Part I of Schedule I of the *Public Service Staff Relations Act.*

#### 1.3 Scope

This standard describes the fire protection engineering requirements for EDP equipment that is:

(a)  essential to the operation of Government of Canada departments and
     agencies; or
(b)  not essential but of a value that warrants protection from the effect of fire.
(See appendix A)

## 1.4 Requirements

This standard applies to:

(a)  new computer room construction and EDP equipment installations;
(b)  additions and alterations to existing computer room and EDP equipment
     installations; and
(c)  existing computer room and EDP equipment installations where conditions
     present a risk to property or operational continuity.  (See appendix A)

## 1.5 Administration

(a)  The Fire Commissioner of Canada or his authorized representative is
     responsible for the administration and enforcement of this standard.
(b)  This standard is not to be interpreted as permitting practices specifically
     prohibited by provincial or other legislation.
(c)  Where reference is made to other codes and standards, unless otherwise
     stipulated, the reference shall be to the latest edition including
     amendments to that code or standard.
(d)  In the event of an inconsistency between this standard and any other
     standard, the requirements of this standard shall prevail.

## 1.6 Definitions

Certain terms used in this standard are defined.

(1)  *Administrative official* means the official designated by the department
     or agency responsible for the administration of the property (agent
     administratif);

(2)  *alarm signal* means an audible signal transmitted throughout a zone or
     zones or throughout a building to advise occupants that a fire emergency
     exists (signal d'alarme);

(3)  *alert signal* means an audible signal to advise designated persons of a
     fire emergency (signal d'alerte);

(4)  *annunciator* means a device to indicate visually a signal received from
     the **fire alarm system** or the fire detection system (annonciateur);

(5)  *combustible* means that a material fails to meet the acceptance criteria
     of CAN4-S114 *Standard Method of Test for Determination of
     Non-Combustibility in Building Materials* (combustible);

(6)  *computer room* means a room in which EDP equipment is located (salle
     des ordinateurs);

(7) **electronic data processing (EDP) equipment** means computer equipment along with all peripherals, supports, memories, programs or other associated equipment electronically interconnected with it (matériel de traitement électronique de l'information (TEI));

(8) **essential**, as determined by the Administrative Official, means vital to the operations of the department and agency. (See appendix A-1.3) (essentiel);

(9) **fire alarm system** means a combination of devices designed to warn the building occupants of an emergency condition (réseau avertisseur d'incendie);

(10) **fire compartment** means an enclosed space in a building that is separated from all other parts of the building by enclosing construction providing a fire separation having a required fire-resistance rating (compartiment étanche au feu);

(11) **fire detector** means a device which detects a fire condition and automatically initiates a signal and includes heat detectors and smoke detectors (détecteur d'incendie);

(12) **fire-resistance rating** means the time that a material or assembly of materials will resist the effects of fire as determined by the appropriate standard fire test prescribed in the NBC (degré de résistance au feu);

(13) **fire separation** means a construction assembly that acts as a barrier against the spread of fire (séparation coupe-feu);

(14) **flame-resistant** means that such material conforms to CAN/ULC — S109, *Standard for Flame Tests of Flame-Resistant Fabrics and Films* (difficilement inflammable);

(15) **flame-spread rating** means an index or classification indicating the extent of spread-of-flame on the surface of a material or an assembly of materials as determined by the appropriate standard fire test prescribed in the NBC (indice de propagation de la flamme);

(16) **Government of Canada property** means real or personal property under the administration and control of a federal government department or agency, including property leased to the government (propriétés du gouvernement du Canada);

(17) **Halon 1301 system** means a fire suppression system consisting of containers of Halon 1301 gas as the extinguishing medium and shall include total flood systems and local application systems (réseau d'extincteurs au Halon 1301);

(18) **local application system** means a fixed fire suppression system consisting of containers of Halon 1301 gas as the extinguishing medium

so arranged to discharge the Halon gas directly on the burning material (réseau à projection locale);

(19) **non-combustible** means that a material meeting the acceptance criteria of CAN4-S114, *Standard Method of Test for Determination of Non-Combustibility in Building Material* (incombustible);

(20) **records** means media for the storage of information and includes paper records, punch cards, plastic or metal-base tapes, microfilm or other photographic media, magnetic discs, optical discs, memory drums and cores or other means of maintaining or storing information (documents);

(21) **smoke detector** means a fire detector designed to operate when the concentration of airborne products of combustion exceeds a predetermined level (détecteur de fumée);

(22) **smoke developed classification** means an index or classification indicating the level of smoke developed by a material or assembly of materials as determined by the appropriate standard for test prescribed in the NBC (indice de dégagement des fumées);

(23) **total flood system** means a fixed fire suppression system consisting of containers of Halon 1301 gas as the extinguishing medium so arranged to discharge the Halon gas into an enclosed space or room (réseau à saturation).

## 1.7 Abbreviations

(a) The abbreviations in this standard for the names of associations shall have the meanings assigned to them in this clause.

| | | |
|---|---|---|
| CSA | – | Canadian Standards Association |
| FC | – | Fire Commissioner of Canada or the authorized representative of the Fire Commissioner of Canada |
| NBC | – | *National Building Code of Canada* |
| NFC | – | *National Fire Code of Canada* |
| NFPA | – | National Fire Protection Association |
| ULC | – | Underwriters' Laboratories of Canada |

(b) The abbreviations of words and phrases in this standard shall have the meanings assigned to them in this clause.

| | | |
|---|---|---|
| °C | – | degree(s) Celsius |
| HVAC | – | Heating, Ventilating and Air Conditioning |
| h | – | hour(s) |
| m | – | meter(s) |
| lx | – | lux |
| min | – | minute(s) |
| No. | – | number |
| s | – | second(s) |
| $ | – | dollar(s) |

## 2. Classifications and requirements

### 2.1 Classification

EDP equipment shall be protected by one of the following levels of protection according to the value of each unit or units which are in very close proximity to each other:

(a)  Level 3 –  For essential EDP equipment.
(b)  Level 2 –  For non-essential EDP equipment of high value exceeding $1,000,000.
(c)  Level 1 –  For non-essential EDP equipment of medium value, equal to or greater than $250,000 but does not exceed $1,000,000.
(d)  Level 0 –  For non-essential EDP equipment of a value less than $250,000.

### 2.2 Requirements for levels of protection

(a)  EDP equipment requiring Level 3 protection shall conform to this standard.
(b)  EDP equipment requiring Level 2 protection shall conform to this standard with the exception of subsection 6.4.
(c)  EDP equipment requiring Level 1 protection shall conform to this standard with the exception of subsection 3.2, articles 4.1(d), subsection 4.2, articles 4.3(b) to (e), subsection 4.4, article 5.1(e), articles 5.2(b) to (d), subsections 6.2, 6.3, and 6.4.
(d)  EDP equipment requiring Level 0 protection shall be considered as normal office equipment and shall be protected by fire extinguishers in conformance with 6.1.  No additional protection is required other than that described in the NBC and NFC.

(See appendix C – Summary of requirements)

## 3. Location and construction

### 3.1 Location

(a)  EDP equipment shall be located to minimize possible exposure to fire, water, corrosive fumes and smoke from adjacent areas.
(b)  EDP equipment shall not be located above, below or adjacent to areas or structures where hazardous processes are present, unless protective features are provided.
(c)  Where practicable, EDP equipment shall be located in sprinklered buildings.

### 3.2 Building construction

**Note**: This Subsection applies only to EDP equipment requiring Level 2 and Level 3 protection.

(a)  EDP equipment shall be located in buildings of non-combustible construction (as defined by the NBC) except as permitted in (b).

(b) EDP equipment may be located in buildings of combustible construction (as defined by the NBC) if the entire building, including the space occupied by the EDP equipment, is protected by a sprinkler system.

## 4. Computer room

### 4.1 Construction

(a) Except as provided in (b), EDP equipment shall be located in a room separated from the remainder of the building by a fire separation having a fire-resistance rating of:
   (i)   3/4 h where the floor assembly is required to have a fire-resistance rating of not greater than 3/4 h;
   (ii)  1 h where the floor assembly is required to have a fire-resistance rating of 1 h or more; and
   (iii) not less than 2 h, if the computer room is exposed to industrial occupancies as defined by the NBC.
(b) For EDP equipment requiring Level 1 protection, the fire separation in (a) is not required if the equipment is located in a building protected by a sprinkler system or equipped with a fire alarm system.
(c) Openings in the fire separation of the computer room shall be protected by closures as described in the NBC.
(d) The interior finish of walls, partitions and ceilings shall be constructed of non-combustible materials or of materials having a flame-spread rating of 25 or less, and a smoke developed classification of 50 or less.
(e) All concealed spaces above suspended ceilings shall be provided with access panels or hatches designed and located so that all portions of the void space are accessible.
(f) Any floor immediately above the computer room shall be made watertight to minimize possible water damage to the EDP equipment. Any openings in the floor including those for columns, beams, pipes or conduits shall be sealed.
(g) Where there is a possibility of water damage to the EDP equipment, protection shall be provided by means of
   (i)   adequate drainage to remove excess water; or
   (ii)  water detection equipment to sense the presence of water.
(h) Building services, such as fuel lines and water pipes, other than those serving the computer room shall not be located within the computer room.

### 4.2 Raised floors

(a) Supporting members for raised floors shall be constructed of non-combustible material.
(b) Decking for raised floors shall be constructed of:
   (i)   steel, aluminum or other non-combustible material; or
   (ii)  a core material of wood or wood products covered on the top and bottom with metal, with all openings or cut edges covered with metal or plastic clips or grommets so that none of the core is exposed, and the assembly has a flame-spread rating of 25 or less.
(c) Access panels, suitably identified, shall be provided in all raised floors to give ready access to the space beneath.

(d) Except as permitted in (e), high pressure laminates, or carpeting on raised floors shall have a flame-spread rating of 75 or less and a smoke developed classification of 150 or less.

(e) When the computer room is protected by a sprinkler system, commercial grade carpeting is permitted on raised floors.

(f) Carpeting on raised floors as permitted in (d) and (e) shall be installed so as not to obstruct or interfere with openings required for access to the space beneath.

(g) Openings into the space under raised floors shall be covered or screened to prevent the entry of debris.

(h) The floor beneath raised floors shall be curbed by impermeable materials to prevent the entry of water from the adjacent floor areas.

(i) Raised floor of metal construction shall be electrically grounded.

(j) Before the installation of a raised floor is made on combustible flooring, the flooring shall be covered with non-combustible materials.

## 4.3 Computer equipment

(a) All electrical equipment used in EDP equipment shall be certified by CSA or the provincial electrical inspection authority.

(b) All non-electrical parts such as housing, frames and supporting members shall be constructed, as far as practicable, of non-combustible materials.

(c) Air filters for use in individual units of EDP equipment shall be of non-combustible materials or conform to the requirements for Class 1 filters in accordance with CAN4-S111, *Standard Method of Fire Tests for Air Filter Units*.

(d) Fluids required for lubrication, cooling or hydraulic purposes shall have a flash point of 150°C or higher and their containers shall be of sealed construction, equipped with automatic pressure relief devices venting to a safe location.

(e) All sound deadening material used in EDP equipment shall have a flame-spread rating of 25 or less and a smoke developed classification of 50 or less.

## 4.4 Occupancy and furnishing

(a) Only an activity or occupancy directly associated with the EDP equipment shall be permitted to be located in the computer room.

(b) Furniture in the computer room shall be of materials that do not contribute significantly to the combustible contents.

(c) Draperies shall be flame-resistant.

## 4.5 Records

(a) Records other than as described in (b) shall be stored in record storage facilities in accordance with FC No. 311, *Standard for Record Storage*.

(b) Records kept within the computer room shall be limited to the minimum needed for daily requirements and be stored in closed metal containers or cabinets unless the records are essential to operations, in which case they shall be stored in containers having a fire-resistance rating of at least 1 h.

(c) Master records from which operating or current records may readily be reproduced, shall be stored in a different fire compartment or stored off-site in containers conforming to (b).

(d) After each periodic updating of recorded data, the previous record or generation of data shall be removed to the record storage facility referred to in (a).

## 5. Services

### 5.1 Electrical systems

(a) Except as provided in this Subsection, electrical wiring and installations shall conform to the CSA C22.1, *Canadian Electrical Code*, Part I.

(b) Power supply cables may be installed under a raised floor provided the branch circuit power supply conductors to receptacles are in metallic raceways or metal-sheathed cables.

(c) All exposed wiring and cables in plenum and underfloor spaces shall have a flame test classification of FT4 when tested in conformance with CSA C22.2 No. 0.3, *Test Methods for Electrical Wires and Cables*.

(d) Service transformers are not permitted in the computer room unless they are of the dry type or a type filled with dielectric medium having a flash point of 300°C or higher.

(e) Except for lighting and smoke detection systems, a disconnecting means shall be provided to disconnect power to all EDP equipment in the computer room. The disconnecting means shall be controlled from locations readily accessible to the operator at the main egress door of the computer room, and provided with cover plate to prevent accidental operation.

(f) Uninterruptible power supply system (UPS) dedicated to the EDP equipment shall be located away from other building hazards and in a dedicated enclosure having a fire-resistance rating equal to the fire separation provided for the computer room, except that UPS systems with sealed type batteries are permitted to be located within the computer room.

(g) An enclosure for unsealed type battery bank associated with a UPS system shall be used for no other purpose. It shall be ventilated as required by CSA C22.1 for the type of battery used, and the floor and walls shall be water tight and resistant to the battery electrolyte used.

(h) Where the EDP equipment and the UPS system are in different fire compartments, the electrical conductors connecting the EDP equipment and the UPS system shall be protected against fire exposure to ensure continued operation for a duration of at least 1 h.

(i) Emergency lighting to an average level of 10 lx measured at the floor level shall be provided for the computer room for a duration in conformance with the NBC.

### 5.2 HVAC systems

(a) HVAC systems shall conform to the requirements of the NBC, Part 6.

(b) Independent air conditioning systems shall be provided for the computer room, separate from service to other areas, where practicable.

(c) Air ducts serving other areas should not pass through the computer room. Where it is not practicable to re-route such ducts, they shall be enclosed in a construction having a fire-resistance rating equal to that of the computer room.

(d) Air filters for use in an air conditioning system shall be of non-combustible materials or conform to the requirements of Class 1 Filters in accordance with CAN4-S111, *Standard Method of Fire Tests for Air Filter Units*.

## 6. Fire protection

### 6.1 Fire extinguishers

Carbon dioxide extinguishers with a minimum rating of 5-B,C shall be provided in the computer room such that all EDP equipment is within 9 m travel distance of the extinguisher.

### 6.2 Fire alarm systems and smoke detection systems

**Note**: This subsection applies only to EDP equipment requiring Level 2 and Level 3 protection.

(a) A fire alarm system, designed, installed and maintained in accordance with chapter 3-4, Standard for fire alarm systems, shall be provided for the building.

(b) Smoke detectors, connected to a local annunciator, shall be provided in the computer room.

(c) Smoke detectors shall be located throughout the computer room, including rooms housing the uninterruptible power supply system, underfloor spaces and ceiling spaces. The location of detectors in all concealed spaces shall be identified by the local annunciator.

(d) The operation of any smoke detector or fire detector in the computer room shall:
  (i) cause a visual and audible signal at the local annunciator in the computer room; and
  (ii) automatically transmit a signal to the building fire alarm control panel in accordance with chapter 3-4, Standard for fire alarm systems.

(e) The local annunciator in (b) shall be located at or near the main egress door of the computer room.

(f) Where a local control panel is provided for the control of the smoke detectors within the computer room, the power to the control panel shall be supplied from separate branch circuit(s) independent from the other branch circuits supplying power to the computer room.

### 6.3 Sprinkler systems

**Note**: This subsection applies only to EDP equipment requiring Level 2 and Level 3 protection.

(a) Except as otherwise indicated in this subsection, sprinkler systems protecting computer rooms shall be designed, installed and maintained in accordance with FC No. 403, *Standard for Sprinkler Systems*.

(b)  For EDP equipment located in buildings of non-combustible +M construction, a sprinkler system shall be provided for the computer room. (See appendix A)

   **Note**: For EDP equipment located in buildings of combustible construction, the entire building shall be sprinklered as per subsection 3.2.

(c)  Sprinklers may be omitted in an underfloor space constructed in accordance with this standard.

(d)  The sprinkler system for the computer room shall be interconnected with the building fire alarm system and so arranged that the operation of any sprinkler within the computer room shall:

   (i)    cause a visual and audible signal at the local annunciator in the computer room;
   (ii)   automatically transmit a signal to the building fire alarm control panel in accordance with chapter 3-4, Standard for fire alarm systems;
   (iii)  shut down the power to the EDP equipment; and
   (iv)   shut down the air handling system for the computer room.

(e)  The sprinkler system for the computer room shall be provided with a flow switch, a test pipe and an indicating control valve located outside of the room. The locations of the test pipe and the control valves shall be marked and identified.

## 6.4 Halon 1301 systems

(See appendix A)

**Note**: This subsection applies only to EDP equipment requiring Level 3 protection.

(a)  A Halon 1301 system or other acceptable gaseous fire suppression system shall be installed in the computer room where a fire risk analysis concludes that, in addition to sprinkler protection, a gaseous fire suppression system is necessary to minimize potential fire damage to the EDP equipment.

(b)  Plans and specifications shall be submitted to the FC for review in accordance with chapter 3-2, Fire protection standard for design and construction.

(c)  Upon completion of the installation of a Halon 1301 system, the contractor shall:

   (i)    advise the FC in writing that installation has been completed, the piping and associated equipment have been tested for mechanical tightness and freedom from dangerous pipe movements during discharge, and the system is in operating condition and conforms to the requirements of the specifications;
   (ii)   report and identify all unclosable openings and sources of agent loss; and
   (iii)  provide a copy of the operating manual and schematic layout of the system in the computer room.

(d)  Before acceptance of the installation is granted, the completed system shall be tested by the contractor in accordance with the procedures described in NFPA Standard No. 12A, *Halon 1301 Fire Extinguishing*

*Systems*, including procedures described in the appendix A of that standard.

(e) Except as provided in this subsection, Halon 1301 systems shall be designed, installed and maintained in accordance with NFPA Standard No. 12A, *Halon 1301 Fire Extinguishing Systems*.

(f) Halon 1301 systems shall be designed and installed to provide a 5% concentration of Halon 1301 agent within 10 s of actuation and held for:
- (i) 10 min. for total flood systems;
- (ii) 1 min. for individual EDP equipment.

(g) Smoke detectors shall be installed within the spaces protected by total flood systems. The location of detectors concealed in underfloor and ceiling spaces or within computer units shall be identified by the local annunciator.

(h) Except as permitted in (j) for manual activation, total flood systems shall be actuated by the smoke detectors in the computer room.

(i) The operating sequence for the activation of the total flood system shall be as follows:
- (i) The operation of one smoke detector shall activate the visual and audible signals as described in 6.2(d); and
- (ii) The operation of two or more smoke detectors in a protected area or the operation of a manual station shall:
  - (a) cause a visual and audible signal at the local annunciator in the computer room,
  - (b) cause a distinctive signal throughout the computer room,
  - (c) automatically transmit a signal to the building fire alarm control panel in accordance with chapter 3-4, Standard for fire alarm systems,
  - (d) shut down the power to the computer room except as permitted by (j),
  - (e) shut down the air handling system in the computer room, and
  - (f) discharge the Halon gas after a period of not more than 30 s from actuation of the second smoke detector and not more than 5 s from the operation of a manual station.

(j) EDP equipment linked with special operational equipment such as that for air traffic control or an EDP equipment under continuous surveillance of a qualified operator may be permitted to continue to operate until manually shut down. Such an arrangement shall have the approval of the administrative official, and may include the manual operation of a Halon 1301 system.

(k)
- (i) Means to manually actuate the total flood system shall be provided at a readily accessible location at or near each egress door of the computer room.
- (ii) All manual stations that are used to release Halon gas shall be properly identified. Where manual stations for more than one system are provided in close proximity, they shall be clearly marked as to which area they affect.
- (iii) Manual stations shall be provided with tamper covers or shall be of a type requiring double action for activation.

(l)
- (i) When an abort switch is provided for the system, the switch shall be clearly marked and identified as to its use, and shall be installed so

as not to override any manual station. The abort switch shall have a different colour from that of the manual station.

    (ii)    The operation of any abort switch shall cause a visual and audible signal at the local annunciator.

(m)    Local application systems for the protection of individual computer units or similar applications shall be designed and installed to be actuated manually and by fire detectors located within the EDP equipment or within the enclosed space in which the equipment is located.

(n)    (i)    Computer rooms that are protected by Halon 1301 systems shall be provided with a discharge indicator and signage outside the room indicating that it is so protected.

    (ii)    Containers of Halon gas shall be either located within the computer room or in an area protected by a fire suppression system and accessible only to authorized personnel.

    (iii)    Rooms outside of the computer room used for the storage of Halon containers shall not be used for other purposes.

    (iv)    Unless the Halon gas containers are clearly visible within the computer room, the location of the containers shall be marked and identified.

## 6.5 Other fire safety measures

(a)    Where the computer room has a raised floor, a "floor panel lifter" shall be provided for emergency use near each egress door in an accessible location within the computer room.

(b)    A sufficient number of waterproof and flame-resistant treated covers or tarpaulins to protect the EDP equipment and a sufficient number of mops or squeegees to reduce water damage shall be provided and kept outside of, but immediately accessible to, the computer room.

## 7. Fire safety planning

### 7.1 General

The general emergency procedures to be followed shall be in accordance with the requirements of chapter 3-1, Standard for fire safety planning and fire emergency organization.

### 7.2 Pre-planning for continued operations

The effect upon continuity of operations for each EDP equipment should be determined, and when necessary, arrangements made for the use of alternative facilities, including the transportation of personnel, data and supplies.

### 7.3 Fire safety plan

(a)    A fire safety plan for the computer room shall be prepared and posted.

(b)    The following specific duties, in addition to chapter 3-1, Standard for fire safety planning and fire emergency organization, shall be assigned to designated personnel:

    (i)    assure the safe egress of personnel from any area of danger from fire or other emergency;

(ii)    ensure that power to the computer room has been shut off either automatically or manually subject to operational requirements;

(iii)   ensure that the air handling system, when a separate air conditioning system is provided for the equipment area, has been shut off either automatically or manually;

(iv)   conduct fire fighting operations with special equipment if safe to do so;

(v)    direct the fire department to the scene of the fire and stand by to aid and provide information;

(vi)   direct the removal and preservation of equipment and records endangered by fire;

(vii)  undertake salvage operations, including the use of waterproof covers or tarpaulins, to minimize damage due to fire, smoke and water; and

(viii) notify the FC of the incident in accordance with FC No. 11, *Standard For Investigating and Reporting of Fires*.

### 7.4 Personnel training

To ensure that the fire safety plan described in 7.3 is effective, the designated personnel shall receive regular training on the necessary actions to be taken in the event of a fire emergency.

### CHAPTER 3-3 - APPENDIX A - EXPLANATORY MATERIALS

### A – 1.3

The requirements of this standard are intended for certain types of EDP equipment. The word "essential" as defined in this standard is intended to assist the administrative official to determine the strategic importance of the EDP equipment whose loss would paralyse the operation.

The strategic importance of a particular EDP equipment can be evaluated by assessing the function of the equipment and the degree of operational dependence upon such equipment.

There are several major areas where judgement is required in the application of this standard:

(a)   Function: Is the equipment vital/critical to the continuous operation?

(b)   Dependency: Is the processing highly centralized and is this the only facility performing this function? Is an external service provided in case of breakdown?

(c)   Recovery: Is a replacement readily obtainable?

EDP equipment which is vital to life safety such as air traffic control systems or vital to uninterrupted operational requirements is considered to be essential. EDP equipment is not considered to be essential when loss due to fire would result in only temporary inconvenience and minor expense.

The administrative official should also consider the monetary value of the EDP equipment versus the cost of protection in applying the standard (the recommended cost ratio is 100:4). Cost benefit analysis should be carried out to determine the significance of a particular protection system prior to installation.

**A – 1.4(c)**

The extent of application of this standard to existing **computer rooms** to remove an unacceptable risk should be based on judgment and the merits of each case.

**A – 6.3(b)**

Consideration should be given to extending sprinkler protection to the suite in which the computer room is located in order to minimize exposure hazards.

**A – 6.4**

Assessments conducted by the World Meteorological Organization (WMO) and the National Aeronautics and Space Administration (NASA) of the U.S. suggest that Halon gas emissions could contribute to possible depletion of the stratospheric ozone. The stratospheric ozone layer protects humans from exposure to damaging ultraviolet radiation. The depletion of this ozone layer would lead to adverse health and environmental effects.

In March 1985, Canada signed the Vienna Convention for the protection of the ozone layer. In September 1987, Canada signed the Montreal Protocol on Substances that Deplete the Ozone Layer. The Protocol establishes an international framework for reducing emissions of specified ozone-depleting substances, namely certain types of chlorofluorocarbons (CFCs) and Halons.

As a result, consideration should be given to evaluating all existing Halon systems as to their need and viability. Alternative means of fire suppression systems should be used unless a gaseous system such as Halon is absolutely necessary.

**A – 6.4(d)**

Non-recoverable total flooding tests of Halon gas should be avoided where possible. Alternate test agent/method may be used in lieu of Halon gas.

**CHAPTER 3-3 - APPENDIX B - REFERENCE PUBLICATIONS**

This standard refers to the following codes and standards which are not found in this volume:

**Fire Commissioner of Canada standards**

Records Storage
Sprinkler Systems

These standards may be obtained from the Fire Commissioner of Canada, Human Resources Development Canada, Ottawa, Ontario, K1A 0J2, or from regional offices of Human Resources Development Canada. As they are revised, they will be published in this volume.

## Treasury Board publication

*Treasury Board Manual*
"Materiel, Services and Risk Management" volume
Fire Protection, Investigation and Reporting (Part III, chapter 5)

This volume may be purchased from the Canada Communication Group — Publishing, 45 Sacré-Coeur Blvd., Hull, Québec, K1A 0S9.

## CSA Standard C22.1

CSA C22.1      –      *Canadian Electrical Code*, Part I

CSA C22.2 0.3  –      *Test Methods for Electrical Wires and Cables*

Canadian Standards Association publications may be purchased from Canadian Standards Association, 178 Rexdale Blvd., Rexdale, Ontario, M9W 1R3.

## National Research Council of Canada publications

*National Building Code of Canada*
*National Fire Code of Canada*

These publications may be purchased from the Secretary, Associate Committee on the *National Building Code*, National Research Council, Ottawa, Ontario, K1A 0R6.

## NFPA standard

NFPA No. 12A — *Halon 1301 Fire Extinguishing Systems*

National Fire Protection Association publications may be purchased from the National Fire Protection Association, Batterymarch Park, Quincy, Mass. 02269 or from FIPRECAN, 7-1590 Liverpool Court, Ottawa, Ontario, K1B 4L2.

## ULC standards

CAN4-S109   –   *Standard for Flame Tests of Flame Resistant Fabrics and Films*
CAN4-S111   –   *Standard Method of Fire Tests for Air Filter Units*
CAN4-S114   –   *Standard Method of Test for Noncombustibility of Building Materials*

Underwriters' Laboratories of Canada publications may be purchased from the Underwriters' Laboratories of Canada, 7 Crouse Road, Scarborough, Ontario, M1R 3A9.

## CHAPTER 3-3 - APPENDIX C - SUMMARY OF REQUIREMENTS

| Protection requirements | Level of protection | | | |
|---|---|---|---|---|
| | Level 0 | Level 1 | Level 2 | Level 3 |
| NBC and NFC requirements | Yes | Yes | Yes | Yes |
| 1.   General | Yes | Yes | Yes | Yes |
| 2.   Classifications | Yes | Yes | Yes | Yes |
| 3.1  Location | No | Yes | Yes | Yes |
| 3.2  Building construction | No | No | Yes* | Yes* |
| 4.1  Room construction | No | In part* | Yes | Yes |
| 4.2  Raised floor | No | No | Yes | Yes |
| 4.3  Computer equipment | No | No | Yes | Yes |
| 4.4  Occupancy | No | No | Yes | Yes |
| 4.5  Records | No | Yes | Yes | Yes |
| 5.1  Electrical system | No | In part* | Yes | Yes |
| 5.2  HVAC system | No | No | Yes | Yes |
| 6.1  Extinguishers | Yes | Yes | Yes | Yes |
| 6.2  Alarm systems | No | No | Yes | Yes |
| 6.3  Sprinkler system | No | No | Yes* | Yes* |
| 6.4  Halon system | No | No | No | Yes* |
| 6.5  Other safety measures | No | Yes | Yes | Yes |
| 7.   Fire safety planning | Yes | Yes | Yes | Yes |
| Appendices | Yes | Yes | Yes | Yes |

| | | |
|---|---|---|
| * | : | See text for details and exceptions |
| Yes | : | Requirements applicable |
| No | : | Requirements not applicable, see NBC or NFC for details |

Note:    This is **ONLY** a summary of the requirements and should **NOT** be used to substitute the requirements in the standard.

## CHAPTER 3-4 - STANDARD FOR FIRE ALARM SYSTEMS

### 1. General

#### 1.1 Purpose

This standard describes the requirements for the design, construction, installation, inspection, testing and maintenance of fire alarm systems in Government of Canada property to minimize risks to life and property and to protect and conserve the Government's financial position.

#### 1.2 Application

This standard applies to all:

(a) departments and agencies listed in Schedules A and B of the *Financial Administration Act* (FAA) with the exception of the Department of National Defence;

(b)    branches designated as departments for the purposes of the FAA; and

(c)    those departments and other portions of the Public Service as defined in Part I of Schedule I of the *Public Service Staff Relations Act*.

## 1.3 Scope

(a)    This standard describes the requirements for the
    (i)    design, construction and installation of fire alarm systems in new buildings;
    (ii)    alteration and reconstruction of existing buildings; and
    (iii)    upgrading of a fire alarm system to remove an unacceptable fire hazard (see appendix A).

(b)    This standard describes the requirements for the inspection, testing and maintenance of fire alarm systems in all buildings.

## 1.4 Administration

(a)    The Fire Commissioner of Canada or the authorized representative of the Fire Commissioner (henceforth referred to as FC) is responsible for the administration and enforcement of this standard.

(b)    This standard is not to be interpreted as permitting practices specifically prohibited by provincial or other legislation.

(c)    Where reference is made to other codes and standards, unless otherwise stipulated, the reference shall be to the latest edition including amendments to that code or standard.

(d)    In the event of an inconsistency between this standard and any other standard, the requirements in this standard shall prevail.

## 1.5 Review procedures

(a)    Plans and specifications for all fire alarm systems and modifications to existing fire alarm systems shall be submitted to the FC for review in accordance with chapter 3-2, Fire protection standard for design and construction.

(b)    All new fire alarm systems and modifications to existing fire alarm systems shall be tested and verified by the manufacturer, an authorized agent of the equipment manufacturer, or a ULC listed fire alarm service company in accordance with the requirements of CAN/ULC-S537, *Standard for the Verification of Fire Alarm Systems* and a copy of the verification certificate and report shall be submitted to the FC.

(c)    After verification of a fire alarm system installation or modification, the system shall be subject to an inspection and test performed by the installer and witnessed by the FC for final acceptance, except as permitted in (d).

(d)    Subject to the concurrence of the FC, the acceptance tests may be witnessed by a representative of the department or agency concerned with a written report or certificate forwarded to the FC.

## 1.6 Definitions

Certain terms used in this standard are defined to ensure understanding of their meaning and intent.

*Administrative official* means the official designated by the department or agency responsible for the administration of the property (agent administratif);

*alarm signal* means an audible signal transmitted throughout a zone or zones or throughout a building to advise occupants that a fire emergency exists (signal d'alarme);

*alert signal* means an audible signal to advise designated persons of a fire emergency (signal d'alerte);

*ancillary system* means a system actuated by the fire alarm system but which is not a required part of the fire alarm system (système auxiliaire);

*annunciator* means a device to indicate visually a signal received from the fire alarm system (annonciateur);

*audible signal appliance* means a device to indicate, by means of sound output, the actuation of the fire alarm system (audible signal appliances include air horns, bells, buzzers, sirens, gongs, chimes and loudspeakers) (appareil à signal sonore);

*central station system* means an independent facility to receive, verify and transmit alarms to the appropriate fire fighting service, and which conforms to NFPA Standard No. 71, *Installation, Maintenance, and Use of Central Station Signalling Systems* (poste central);

*control unit* means a unit which provides the central control and logic processing for the fire alarm system (tableau de commande);

*exit* means that part of a means of egress that leads from the floor area it serves, including any doorway leading directly from the floor area, to an open public thoroughfare or to an exterior open space protected from fire exposure from the building and having access to an open public thoroughfare (issue);

*fire alarm system* means a combination of devices designed to warn the building occupants of an emergency condition (réseau avertisseur d'incendie);

*fire detector* means a device which detects a fire condition and automatically initiates an electrical signal to actuate an alert signal or an alarm signal and includes heat detectors and smoke detectors (see also heat detector, and smoke detector) (détecteur d'incendie);

*fire suppression system* means a system intended to automatically detect and extinguish a fire (réseau d'extinction d'incendie);

*floor area* means the space on any storey of a building between exterior walls and required fire walls, including the space occupied by interior walls and partitions, but not including exits and vertical service spaces that pierce the storey (aire de plancher);

**Government of Canada property** means real or personal property under the administration and control of a federal government department or agency, including property leased to the government (propriété du gouvernement du Canada);

**heat detector** means a fire detector designed to operate at a predetermined temperature or rate of temperature rise (détecteur thermique);

**heritage building** means any federally owned building that has been designated as either "Classified" or "Recognized" for its heritage significance upon recommendation by the Federal Heritage Buildings Review Office (immeuble patrimonial);

**high building** means a building falling within the scope of high building in the NBC (immeuble de grande hauteur);

**historic building** means any federally owned building that has been declared of national significance by the Minister of Environment Canada on the advice of the Historic Sites and Monuments Board of Canada (immeuble historique);

**listed** means equipment or materials included in a list published by a nationally recognized organization concerned with product evaluation, that maintains periodic inspection of production of listed equipment or materials and whose listing states either that the equipment or material meets appropriate standards or has been tested and found suitable for use in a specified manner (énuméré);

**manual pull station** means a device designed to initiate a signal when operated manually (poste manuel);

**proprietary control centre** means a facility located on the premises to receive, verify and transmit alarms to the appropriate fire fighting service and which conforms to NFPA No. 72D, *Installation, Maintenance, and Use of Proprietary Protective Signalling Systems* (central d'alarme privé);

**smoke alarm** means a combination smoke detector and audible signal appliance designed to sound an alarm within the room or space in which it is located upon the detection of smoke within that room or suite (avertisseur de fumée);

**smoke detector** means a fire detector designed to operate when the concentration of airborne combustion products exceeds a predetermined level (détecteur de fumée);

**trouble signal** means a signal warning of a fault condition in the fire alarm system (signal de défaillance électrique);

**visual signal appliance** means a device which utilizes light to alert occupants of a building to an emergency situation in a direct or indirect manner (appareil à signal);

*zone* means a subdivision of a building intended to identify the origin of either an alarm signal or an alert signal, the location of which is readily identifiable (zone).

### 1.7 Abbreviations

(a) The abbreviations in this standard for the names of associations shall have the meanings assigned to them in this Article.

| | | |
|---|---|---|
| CSA | – | Canadian Standards Association |
| FC | – | Fire Commissioner of Canada or the authorized representative of the Fire Commissioner |
| NBC | – | National Building Code of Canada |
| NFC | – | National Fire Code of Canada |
| NFPA | – | National Fire Protection Association |
| ULC | – | Underwriters' Laboratories of Canada |

(b) The abbreviations of words and phrases in this standard shall have the meanings assigned to them in this article.

| | | |
|---|---|---|
| h | – | hour(s) |
| m | – | meter(s) |
| min. | – | minute(s) |
| No. | – | number |
| % | – | percent |

## 2. Requirements

### 2.1 General

(a) Except as otherwise stated, the design, construction and installation of fire alarm systems shall conform to the National Building Code of Canada.

(b) A fire alarm system shall be provided for a building
  (i) where it is required to have a fire alarm system by the NBC;
  (ii) where it is required to have a fire alarm system by any of the standards listed in appendix C; or
  (iii) where conditions present a serious life hazard as determined by the FC (see appendix A).

(c) In addition to (b), and except as permitted in (d), a fire alarm system shall be provided for a building
  (i) in which there is a large property loss potential or a potential for a serious interruption of essential services as determined by the Administrative Official, in consultation with the FC;
  (ii) that has been designated as a historic building or a heritage building; or
  (iii) that contains works of art, historical artifacts or scientific specimens which are considered to form part of the nation's cultural heritage and are irreplaceable.

(d) When it can be determined on the basis of a cost/benefit analysis, that a complete fire alarm system as required in (c) is not economically feasible or for technical reasons would not be practical, alternative fire protection measures may be provided as recommended by the FC in consultation with the Administrative Official.

## 2.2 Locations of heat detectors

(a) Where a fire alarm system is required, and except as permitted in (b), heat detectors shall be installed in
   (i) areas specified by the NBC;
   (ii) workshops, such as machine shops, electrical shops, carpenter shops, paint shops, and maintenance shops; and
   (iii) laboratories in which flammable or combustible materials are stored or handled.
(b) Heat detectors in (a) need not be provided where the area is protected by sprinkler systems conforming to FC No. 403, *Standard for Sprinkler Systems*.

## 2.3 Locations of smoke detectors

(a) Where a fire alarm system is required, smoke detectors, including duct-type smoke detectors, shall be installed in
   (i) areas specified by the NBC;
   (ii) computer rooms in accordance with Chapter 3-3, Fire protection standard for electronic data processing equipment;
   (iii) record storage facilities in accordance with FC No. 311, *Standard for Record Storage*; and
   (iv) every room containing valuable or irreplaceable materials as designated by the Administrative Official.

## 2.4 Locations of smoke alarms

(a) Except as permitted in (b), smoke alarms shall be installed
   (i) in areas as specified in the NBC; and
   (ii) on each storey of a multi-level dwelling unit, such that there is at least one smoke alarm on each storey.
   **Note**: In split-level dwelling units, one smoke alarm can serve an adjacent lower level if there is no intervening door between the two levels.
(b) Smoke alarms are not required in rooms or suites in which smoke detectors are required and provided in accordance with 2.3 (a).

## 3. Types of fire alarm systems

### 3.1 General

(a) Every fire alarm system shall be a single stage system conforming to subsection 3.2, except as required in (b) or permitted in (c).
(b) A two-stage fire alarm system conforming to subsection 3.3 shall be installed in a building that
   (i) is a high building; or
   (ii) contains a Group B occupancy having accommodation for more than 10 persons detained or receiving care or treatment (see appendix A).
(c) Subject to the concurrence of the FC, a two-stage fire alarm system may be used in lieu of a single stage system where
   (i) it is requested by the Administrative Official for operational reasons;

(ii) the building is sprinklered or incorporates such other features of design and construction that would make immediate total evacuation unnecessary; and

(iii) during all times that the building is occupied, there are sufficient staff on duty trained to respond to an alarm and direct or conduct an orderly evacuation of all occupants in the event of an emergency.

## 3.2 Single-stage fire alarm systems

(a) A single-stage fire alarm system shall be designed to provide an alarm signal throughout the building when actuated as stipulated in (b).

(b) The operation of any manual pull station, fire detector, or fire suppression system, shall
   (i) cause an alarm signal on all audible signal appliances throughout the building and at the control unit;
   (ii) indicate the floor or zone from which the fire alarm system was actuated by means of a visual signal at the control unit and at the annunciator(s) in accordance with subsections 4.9 and 4.11;
   (iii) transmit a signal to the fire department as required in subsection 4.12;
   (iv) cause any recirculating air handling system to shut down or function so as to provide the required control of smoke movement, when such system serves an area as described in the NBC;
   (v) cause all required fire doors and smoke control doors, if normally held open, to close automatically; and
   (vi) cause all locking devices on exit doors, if in the locked position, to release.

(c) Audible signal appliances as required in (b) shall remain in operation until the system has been restored to normal or until silenced from the control unit as stipulated in subsection 4.9 except as permitted in (d).

(d) In the case of buildings where at certain times there may be no persons on duty to respond to an alarm, the audible signal appliances may be silenced automatically after a period of not less than 20 min.

## 3.3 Two-stage fire alarm systems

(a) A two-stage fire alarm system shall be designed to cause an alert signal at the 1st stage as stipulated in (b), and an alarm signal at the 2nd stage as stipulated in (c).

(b) The operation of any manual pull station, fire detector, or fire suppression system shall at the 1st stage
   (i) cause an alert signal on all audible signal appliances throughout the building and at the central alarm and control facility, except as stipulated in (iii) and (iv);
   (ii) indicate the floor or zone from which the fire alarm system was actuated by means of a visual signal at the central alarm and control facility and at the annunciator(s), where provided, in accordance with subsections 4.10 and 4.11 respectively;
   (iii) in buildings where the occupants are under close supervision and control of staff, such as hospitals and penitentiary living units, the alert signal may be arranged to be heard only at approved supervisory locations;

(iv) in the case of a high building, cause an alarm signal in lieu of an alert signal on all audible signal appliances throughout the zone and adjacent zones on the same floor level from which the fire alarm system was actuated and throughout the corresponding zone or zones on the floor level immediately above and the floor level immediately below;

(v) transmit a signal to the fire department as stipulated in subsection 4.12;

(vi) cause any recirculating air handling system to shut down or function so as to provide the required control of smoke movement, when such system serves an area as described in the NBC;

(vii) cause all required fire doors and smoke control doors, if normally held open, to close automatically;

(viii) cause all locking devices on exit doors, if in the locked position, to release (see appendix A);

(ix) cause any background music system to be automatically silenced; and

(x) initiate the operation of the recording equipment as stipulated in subsection 4.16 where such recording equipment is provided.

(c) Except as required in (d) or permitted in (e), the operation of any manual pull station by means of keys accessible to authorized persons only or the actuation of the means provided at the central alarm and control facility shall cause an alarm signal on all audible signal appliances throughout the building in lieu of the alert signal.

(d) In high buildings the central alarm and control facility shall be provided with means to initiate alarm signals selectively by floors to facilitate phased evacuation procedures.

(e) In large complexes such as penitentiaries or air terminal buildings in international airports, the fire alarm system may provide for selective activation of alarm signals by zones or groups of zones if the building incorporates such features of design and construction that partial or phased evacuation procedures may be safely used.

(f) The alert signal or alarm signal shall be capable of being silenced from the central alarm and control facility, but only after a minimum period of operation of 1 min. from the initial actuation of the alert signal.

(g) The alert signal as required in (b)(i) shall automatically be changed to an alarm signal as required in (c) after a period of not more than 5 min., unless

(i) the alert signal has been manually acknowledged at the central alarm and control facility, cancelling the automatic alarm signal, but allowing the alert signal to continue to operate;

(ii) the alert signal has been manually silenced at the central alarm and control facility;

(iii) the fire alarm system has been manually switched from the 1st stage to the 2nd stage by zones as described in 3.3(e); or

(iv) the fire alarm system has been restored to normal.

(h) Audible signal appliances shall continue to operate until the system has been restored to normal or until silenced from the central alarm and control facility, as stipulated in subsection 4.10, except as permitted in (i).

(i) In the case of buildings where at certain times there may not be any persons on duty to respond to an alarm, the audible signal appliances may be silenced automatically after a period of not less than 20 min.

## 3.4 Emergency voice communication systems

(a) A fire alarm system in a high building shall include provision for voice communication capability consisting of
  (i) a central alarm and control facility conforming to subsection 4.10;
  (ii) a loudspeaker system conforming to subsection 4.14; and
  (iii) an emergency telephone system conforming to subsection 4.15.
(b) For two-stage fire alarm systems other than in high buildings, a partial or complete voice communication system shall be provided where it is necessary, as determined by the Administrative Official in consultation with the FC, for effective emergency response and evacuation procedures. In such cases the voice communication system may consist of one or more of the elements described in (a) or such other communication systems as may be recommended by the FC.

# 4. Installation

## 4.1 General

(a) All fire alarm systems shall be designed and installed in accordance with CSA C22.1, *Canadian Electrical Code*, Part I, and CAN/ULC-S524, *Standard for the Installation of Fire Alarm Systems*.
(b) In the case of devices or site conditions that are not covered by the standards in (a), the design and installation shall comply with NFPA No. 72E, *Automatic Fire Detectors*, or other standards of good fire protection engineering practice.
(c) All equipment and devices shall be of a listed type.
(d) All components of a fire alarm system shall be compatible. Devices connected to the control unit shall be either those specified by the manufacturer's approved wiring diagram for the control unit; or substitute devices that are functionally compatible with the control unit and provide equivalent circuit loading and equivalent actuation of the control unit.
(e) Fire alarm systems may be combined with building security and/or environmental control systems provided the complete system is of a listed type (see appendix A).

## 4.2 Power supply

(a) Every fire alarm system including those with voice communication capability shall be provided with a main power and an emergency power supply.
(b) The main power supply shall be provided in accordance with CSA C22.1, *Canadian Electrical Code*, Part I.
(c) The emergency power supply shall be provided in accordance with
  (i) CAN/ULC-S524, *Standard for the Installation of Fire Alarm Systems*,
  (ii) CAN/ULC-S527, *Standard for Control Units for Fire Alarm Systems*, and
  (iii) the NBC.

## 4.3 Wiring methods

(a) Wiring for fire alarm systems and emergency voice communication equipment shall be installed in accordance with CSA C22.1, *Canadian Electrical Code*, Part I.

(b) In high buildings, electrical conductors used in connection with fire alarm systems and emergency equipment, including emergency voice communication equipment, shall meet the requirements of (a) and be protected in conformance with the NBC.

## 4.4 Electrical supervision

(a) Every fire alarm system and emergency voice communication system shall be electrically supervised in accordance with CAN/ULC-S524, *Standard for the Installation of Fire Alarm Systems*.

(b) When a fire alarm system consists of more than 1 zone as required in 4.11(a) the zone in which the fault condition occurs shall be indicated.

(c) Audible trouble signals shall remain in operation until the system has been restored to normal or until silenced from the control unit or from the central alarm and control facility.

## 4.5 Manual pull stations

(a) Manual pull stations shall conform to ULC-S528, *Manually Actuated Signalling Boxes For Fire Alarm Systems*.

(b) In the case of a single stage fire alarm system, manual pull stations shall be designed to activate the system as required in 3.2(b).

(c) In the case of a two-stage fire alarm system, manual pull stations shall be designed to activate all devices as described in subsection 3.3 and be equipped such that the use of a key or similar device will activate the fire alarm system as required in 3.3 (c), (d) and (e).

(d) Manual pull stations shall be installed in each floor area, adjacent to each required exit door and in the path of egress from the building. They shall be so located as to be readily visible and accessible and not likely to be obstructed, except as permitted in (e).

(e) In a building where the occupants are under supervision or restraint manual pull stations may be installed at designated locations where they are under continuous surveillance.

(f) All manual pull stations in a building shall be of the same general type with similar operational features.

## 4.6 Audible signal appliances (See also subsection 4.14 "Loudspeakers")

(a) Audible signal appliances shall conform to the requirements of ULC-S525, *Audible Signal Appliances for Fire Alarm Systems*.

(b) The audible signal shall consist of a bell sound unless there are environmental conditions requiring the use of a different device or tone.

(c) Audible signal appliances shall be installed throughout the building except within exits, so as to be effectively heard in all parts of the floor area above all other sounds, except as permitted in (f).

(d) In the case of a 2 stage fire alarm system, the same audible signal appliances may be used to provide both the alert signal and the alarm signal.

(e) Audible signal appliances shall not be concealed above ceilings or behind walls or partitions. They may be recessed in walls or ceilings if located in a metal box with the appliances readily visible and accessible for inspection and maintenance.

(f) In a building where the occupants are under supervision or restraint, audible signal appliances may be located only in designated areas which are under continuous surveillance.

(g) Audible signal appliances shall only be sounded for fire or other emergency purposes.

(h) Visual signal appliances conforming to CAN/ULC-S526, *Standard for Visual Signal Appliances for Fire Alarm Systems* shall be installed in addition to audible signal appliances where there is an abnormally high ambient noise level, or in a building or portion thereof intended for use primarily by persons with hearing impairments.

(i) In the case of a fire alarm system with voice communication capability, the alert signal and the alarm signal may be transmitted by means of the loudspeakers in lieu of bells.

(j) When the alert signal and alarm signal are transmitted by means of loudspeakers, such signals shall not be transmitted through the loudspeakers located in exit stairways.

(k) All audible signal appliances in a building shall be of the same general type.

## 4.7 Fire detectors

(a) Fire detectors shall conform to the following standards
   (i) heat detectors — ULC-S530, *Heat Actuated Fire Detectors for Fire Alarm Systems*;
   (ii) smoke detectors — CAN/ULC-S529, *Smoke detectors for Fire Alarm Systems*.

(b) Fire detectors shall be installed in accordance with CAN/ULC-S524, *Standard for the Installation of Fire Alarm Systems* and shall be located as specified according to their respective laboratory certifications.

(c) Fire detectors shall be of the appropriate temperature or sensitivity rating.

(d) A heat detector or smoke detector in a suite or dwelling unit shall be installed in the central portion of such suite or dwelling unit.

(e) Duct-type smoke detectors in air recirculating systems shall be
   (i) listed for air duct installation; and
   (ii) installed at a location in the main supply air duct on the downstream side of the filter units; and at a location in the return air duct prior to exhausting from the building or prior to being diluted by outside fresh air.

## 4.8 Smoke alarms

(a) Smoke alarms shall conform to the requirements of CAN/ULC-S531, *Smoke Alarms*.

(b) Smoke alarms shall be installed in accordance with the requirements of the NBC, except as permitted in (c).

(c)    In the case of a retrofit installation in an existing building, smoke alarms
       may be of the battery operated type, and in such case need not be
       interconnected.

## 4.9 Control units (See appendix A)

(a)    Control units shall conform to CAN/ULC-S527, *Standard for Control Units
       for Fire Alarm Systems*.
(b)    The control unit shall be located inside the building at or near the main
       entrance so as to be accessible at all times to authorized personnel and
       the fire department. When a remote control unit is provided near the
       main entrance, the main control unit is permitted to be located in other
       parts of the building.
(c)    The control unit shall be provided with manual means
       (i)    to silence the audible signal appliances throughout the building as
              required in 3.2(b)(i) and 3.3(b)(i); and
       (ii)   to silence the trouble signal as required in 4.4(b).
(d)    The actuation of the means as stipulated in (c) shall silence the audible
       signal but shall retain the visual signal.
(e)    The control unit shall be provided with means for the transmission of a
       signal to the fire department in accordance with subsection 4.12.
(f)    Spare fuses corresponding to those used in the control unit shall be
       provided at the control unit and shall be accessible only to authorized
       personnel.
(g)    When a building is part of a complex of buildings located in the vicinity of
       each other and under single management or administration, the control
       units in these buildings may be interconnected to a central control unit
       located at or near a main entrance to the complex so as to be accessible
       at all times to authorized personnel and the fire department.

## 4.10 Central alarm and control facilities (See appendix A)

(a)    Central alarm and control facilities shall conform to the requirements of
       CAN/ULC-S527, *Standard for Control Units for Fire Alarm Systems*.
(b)    The central alarm and control facility shall be located
       (i)    on the street entrance floor of the building to which the fire
              department would normally respond; and
       (ii)   within a sound resistant room or enclosure of adequate size to
              accommodate equipment and operating personnel.
(c)    In a building complex consisting of 2 or more communicating buildings or
       towers and under common ownership and management, integrated fire
       alarm and voice communication systems throughout the complex shall be
       controlled from a single central alarm and control facility.
(d)    The central alarm and control facility shall be provided with
       (i)    means to indicate visibly and audibly signals from the fire alarm
              system and fire suppression systems;
       (ii)   means to indicate visibly and audibly the floor or zone from which a
              signal is initiated;
       (iii)  means to silence the audible signal at the central alarm and control
              facility;
       (iv)   means to silence all audible signal appliances throughout the
              building, but only after a minimum operating period of 1 min.;

392

(v)   means to cancel the automatic changeover of the alert signal to an alarm signal, thus allowing the alert signal to continue to operate;

(vi)   distinctive visual signal devices to indicate when the means as required in (iii), (iv) and (v) have been actuated;

(vii)   means to activate all audible signal appliances throughout the building either selectively or collectively;

(viii)   means to activate loudspeakers as required in 4.14, either selectively or collectively;

(ix)   means to control telephone communication as required in subsection 4.15;

(x)   means to record all voice traffic, alert signals, and alarm signals as detailed in subsection 4.16 when a recording means is required;

(xi)   means for the transmission of a signal to the fire department in accordance with subsection 4.12;

(xii)   means to control air handling systems as required in subclause 3.3(b)(vi);

(xiii)   means to cause the top vent to open when there is a smoke shaft;

(xiv)   means to cause all required fire doors, and smoke control doors, if normally held open, to close automatically; and

(xv)   means to silence any background music systems.

## 4.11 Zoning

(a)   Fire alarm systems shall be zoned in accordance with the NBC (see appendix A).

(b)   An annunciator shall be provided for every fire alarm system comprising more than 1 zone as required in (a).

(c)   In lieu of a separate annunciator, the annunciator may be integrated with the control unit or the central alarm and control facility.

(d)   Additional annunciators may be provided at specified locations such as security offices and mechanical equipment rooms.

## 4.12 Transmission of alarms

(a)   Fire alarm systems, if required, shall provide for the transmission of an alarm to the fire department, except as permitted in (b), (c) and (d), by means of a connection
(i)   directly to the fire department,
(ii)   to a central station system, or
(iii)   to a proprietary control centre.

(b)   When facilities as stipulated in (a) are not available, the alarm may be transmitted to the fire department by means of any other independent agency capable of providing this service.

(c)   When the automatic transmission of an alarm to the fire department as stipulated in (a) and (b) cannot be provided, permanently mounted signs shall be posted at each manual pull station giving instructions that the fire department or the fire authority should be notified and including the telephone number of the fire department or the fire authority.

(d)   In areas where the fire department is located beyond the range of effective response, provision shall be made for the transmission of the alarm to a continuously manned facility. If the continuously manned facility is not available, other means of alerting the general public such as bells or horns on the exterior of the building shall be considered.

### 4.13 Connections to fire suppression systems

(a) A sprinkler system or other fire suppression system, when provided, shall be connected to the fire alarm system.

(b) The discharge of a fire suppression system shall activate the fire alarm system as specified in subsections 3.2 and 3.3.

(c) The operation of each of the supervisory devices on a fire suppression system (e.g. valve tamper switches, loss of power to fire pumps) shall cause an audible signal and a separate visual trouble signal at the control unit (see appendix A-4.11(a)).

### 4.14 Loudspeakers

(a) Loudspeakers shall conform to the requirements of CAN/ULC-S541, *Standard for Speakers for Fire Alarm Systems*.

(b) Loudspeakers shall be so located within floor areas as to provide effective voice communication throughout the building, including exit stairways, and shall be controlled from the central alarm and control facility, except as permitted in (c).

(c) In a building where the occupants are under supervision or restraint, subject to the concurrence of the FC, the loudspeakers may be located in designated locations only and which are under continuous surveillance.

(d) Loudspeakers shall be arranged so that they can be controlled by floors or zones including stairways, both selectively and collectively from the central alarm and control facility.

(e) A fire alarm system incorporating voice communication capability shall be designed so that when the loudspeakers are actuated selectively in any zone or zones the alert signal or alarm signal can be maintained in the other zones of the building.

(f) Each loudspeaker circuit shall be provided with means to disconnect the circuit automatically in the event of a short circuit and cause a signal at the central alarm and control facility to indicate the defective circuit.

(g) Loudspeakers shall not be concealed above ceilings or behind walls or partitions but may be recessed in walls and ceilings provided the devices can be readily identified and are visible and accessible for inspection and maintenance.

### 4.15 Emergency telephones

(a) Emergency telephone systems shall be installed in accordance with CAN/ULC-S524, *Standard for the Installation of Fire Alarm Systems* and the requirements of this subsection.

(b) Emergency telephones shall be provided and located within every floor area and at the central alarm and control facility.

(c) Emergency telephones shall be located within 6 m of each exit, except as permitted in (d).

(d) In the case of a building where the occupants are under supervision or restraint, the emergency telephones may be installed in designated locations where they will be under continuous surveillance.

(e)  Emergency telephones shall be identified as to their location and use and coloured red.

(f)  The emergency telephone system shall be arranged to provide 2-way telephone communication between the central alarm and control facility and any floor emergency telephone as follows

   (i)   the raising of any emergency telephone handset from its bracket on any floor shall cause a distinctive audible and visible flashing signal at the central alarm and control facility, indicating the floor or zone where the telephone is located;

   (ii)  the raising of the telephone handset at the central alarm and control facility shall cancel the audible signal and replace the visible flashing signal by a continuous signal; and

   (iii) the raising of another telephone handset on any other floor or zone after the initial call has been answered at the central alarm and control facility shall cause a busy signal at the telephone on the floor or zone and a visible flashing signal at the central alarm and control facility as described in (i).

(g)  Means shall be provided at the central alarm and control facility to provide for either private communication between the telephone at the central alarm and control facility and any other telephone on the system or open line communication among all telephones on the system.

(h)  Where vandalism or misuse of equipment may be a problem, a telephone jack may be provided in a floor area in lieu of a telephone subject to the concurrence of the FC. In such a case at least 2 portable telephones equipped with jack plugs shall be located at the central alarm and control facility for use by building emergency personnel or the fire department.

### 4.16  Recording systems

(a)  In the case of a two-stage fire alarm system, a two-track tape recording system using a cassette of 90 min. duration may be provided at the central alarm and control facility to record simultaneously alert signals, alarm signals and voice traffic over the loudspeakers on one track and voice traffic over the telephones on the other track.

(b)  The tape recording device shall be automatically actuated upon the operation of

   (i)   a manual pull station, fire detector or fire suppression system; or

   (ii)  the voice communication system.

(c)  The tape recording device shall be arranged so that once started, it shall continue to operate for the full duration of the tape.

(d)  The time lapse of the cassette shall be indicated by a running time meter that shall be actuated whenever the cassette is operating.

(e)  When the tape recording device is operating, a continuous visual signal shall be actuated. At the conclusion of the operating period, the signal shall become a flashing signal.

(f)  The tape recording device shall be locked in the console and accessible to authorized persons only.

(g)  The removal of the cassette shall be indicated by a visual flashing signal which shall be cancelled when the cassette is replaced.

(h)  At least 2 spare cassettes shall be kept at the central alarm and control facility for use by authorized persons only.

## 5. Inspection, testing and maintenance

### 5.1 General

(a)  Except as otherwise stated, the inspection, testing and maintenance of fire alarm systems shall conform to the *National Fire Code of Canada*.

(b)  The inspection, testing and maintenance of fire alarm systems shall be the responsibility of the department or agency responsible for the administration of the property, except as stipulated in (c).

(c)  In the case of leased property, the inspection, testing and maintenance of fire alarm systems shall be the responsibility of the owner of the property unless otherwise stipulated in the leasing agreement.

(d)  The inspection, testing and maintenance of fire alarm systems shall be carried out either by
   (i)   qualified staff,
   (ii)  the manufacturer of the equipment,
   (iii) an authorized agent of the equipment manufacturer, or
   (iv)  a ULC listed fire alarm service company.

(e)  All persons who would be required to take action in the event of an alarm shall be notified before any fire alarm system is tested to avoid an unnecessary response.

(f)  Immediate action shall be taken to have all defects and deficiencies observed during testing procedures repaired with the minimum of delay.

(g)  Faults or deficiencies which cannot be resolved within 24 h shall be reported to the FC.

(h)  A record of all inspections and tests shall be kept on file for at least 2 years for review by the FC when requested.

### 5.2 Impairment and interruption

(a)  Routine maintenance or an alteration to the fire alarm system which could possibly inadvertently actuate the fire alarm system shall not be carried out during normal working hours, except for emergency service or repairs as permitted in (b).

(b)  When emergency service or repairs are required, the Chief Fire Emergency Warden or the Deputy shall be notified, and the fire alarm system shall be shut off temporarily to prevent a possible false alarm.

(c)  Should the shut-off period exceed 24 h, the FC shall be so advised and arrangements shall be made to provide temporary fire alarm service by means of mechanical or electrical gongs, horns, sirens, or otherwise as required by the FC.

### 5.3 Procedures

(a)  Fire alarm systems shall be inspected, tested and maintained in accordance with CAN/ULC-S536, *Standard for the Inspection and Testing of Fire Alarm Systems*.

(b)  Periodic inspections and testing of fire alarm systems shall conform to the schedules described in CAN/ULC-S536, *Standard for the Inspection and Testing of Fire Alarm Systems*.

(c) In the case of fire alarm systems incorporating voice communication capability, in addition to the requirements of (b) the monthly tests shall include

    (i) the operation of the loudspeakers from the central alarm and control facility to ensure that they are functioning properly and can be heard throughout all parts of the building,

    (ii) the transmission of telephone communication between the central alarm and control facility and the telephones throughout the building to ensure that the telephones are operative, and

    (iii) the operation of the recording means where provided in accordance with subsection 4.16 and if requested to do so, the submission of the used cassettes to the FC for checking.

(d) The yearly tests described in CAN/ULC-S536, *Standard for Inspection and Testing of Fire Alarm Systems*, including fire alarm systems incorporating voice communication capability, shall be conducted under emergency power.

(e) When the emergency power supply consists of an engine driven generator, it shall be regularly tested in accordance with the requirements of CSA C282, *Emergency Electrical Power Supply for Buildings*.

(f) Ancillary systems shall be tested in conjunction with tests of the fire alarm system.

## CHAPTER 3-4 - APPENDIX A - EXPLANATORY MATERIALS

### A – 1.3(c)

The extent of application of this standard to the upgrading of fire alarm systems to remove an unacceptable hazard should be based on the judgement and the merits of each case.

### A – 2.1(b)(iii) and A – 3.1(b)(ii)

The intent of the requirement is to provide a fire alarm early warning system in buildings such as RCMP detachment buildings with detention facility for 10 persons or less and nursing stations offering care or treatment with overnight accommodation for 10 persons or less. Fire alarm systems in these buildings/occupancies are permitted to be a single-stage fire alarm system.

### A – 3.3(b)(viii)

In large buildings such as air terminal buildings, the locking devices on exit doors may be released within the zone in which the alert or alarm signal is originated. The floor area of each zone in a storey should not be less than 2000 $m^2$.

### A – 4.1(e)

#### Integrated systems

It is possible to combine a fire alarm system with building security and/or environmental control systems with approved equipment. However there are many potential operational problems with such combined systems. A decision to use an integrated system should only be made on the basis of a thorough value engineering study of a specific project. Such an evaluation should take into consideration the following points:

1. Conventional design practice for security and fire alarm systems has long called for completely independent systems, even including separate conduit for wiring, to reduce the likelihood of simultaneous interruption or failure, and to reduce unauthorized interference with these systems. With an integrated system, loss of one system would result in loss of building security, fire alarm, and environmental control systems simultaneously. The designer should include an analysis of this risk in his evaluation.

2. The major fire alarm manufacturers all have multiplex fire alarm systems available now, so it is no longer necessary to purchase a combined system in order to obtain the benefits of multiplex signalling technology.

3. At one time there was a substantial saving in equipment cost with combined systems. However, the equipment cost of multiplex signalling systems has declined dramatically in the past decade along with other computer-based technology. Today there is often little or no equipment cost advantage to combined systems. The designer should ensure that his cost analysis is based on current data.

4. The useful life expectancies of the three subsystems often differ substantially. The designer should therefore determine if making major changes to one subsystem would necessitate changes to the entire integrated system, and how this will affect the total life-cycle cost of the systems.

5. The designer should consider the effect of any increased system complexity on maintenance cost; training of system operators; and overall system reliability.

6. A combined system generally increases the number of people who will have to have access to the system, as it serves several functions. This may present security-of-access problems for both building security and fire alarm. The designer should consider if the degree of access control that the client can exercise will satisfy the security needs of the installation.

7. In a combined system it may be possible for persons authorized to work on one subsystem to inadvertently or deliberately tamper with the other two subsystems. The designer should determine this possibility and weigh it against the degree of security and reliability required for each sub-system.

8. The designer should consider the complexity of the software that may be required to combine three functions that operate on different priorities and have different information-processing needs. The designer should be satisfied that adequate software support will be available throughout the life expectancy of the system.

9. The designer should be aware that with separate systems one has the option of either separate locations for the control units or co-location. With an integrated system, of course, only one location is possible, and this location will usually be dictated by the fire protection authority. The designer should also be

aware that information exchange between systems can be accomplished with either separate or integrated systems.

## A – 4.9 and A – 4.10

Where it is considered necessary for operational and maintenance requirements and with the concurrence of the FC, supervised bypass switches may be provided for the purpose of inhibiting the activation of the following functions:

- release of fire extinguishing agent in a fire suppression system;
- shut down of the power supply to electronic data processing (EDP) equipment;
- shut down of air circulation systems; or
- other ancillary systems.

## A – 4.10

A central alarm and control facility is a facility which houses the controls as described in the NBC and 4.10 of this standard. The NBC and 3.4 of this standard describe where a central alarm and control facility is required and the location of the facility.

## A – 4.11(a)

Zoning of supervisory devices (off-normal condition) of fixed fire suppression systems is prescribed by the applicable standards on fire suppression systems.

## CHAPTER 3-4 - APPENDIX B - REFERENCE PUBLICATIONS

This standard refers to the following codes and standards which are not found in this volume:

### Fire Commissioner of Canada standards

Records storage
Sprinkler systems
General storage

These standards may be obtained from the Fire Commissioner of Canada, Human Resources Development Canada, Ottawa, Ontario, K1A 0J2 or from Offices of Human Resources Development Canada. As they are revised, they will be published in this volume.

### Treasury Board publication

*Treasury Board Manual*
"Materiel, Services and Risk Management" volume
Fire Protection, Investigation and Reporting (Part III, chapter 5)

This volume may be purchased from the Canada Communication Group — Publishing, 45 Sacré-Coeur Blvd., Hull, Québec, K1A 0S9.

**CSA Standards**

CSA Standard C22.1 – *Canadian Electrical Code* Part I

CSA Standard C282 – *Emergency Electrical Power Supply for Buildings*

These standards may be purchased from the Canadian Standards Association, 178 Rexdale Blvd., Rexdale, Ontario, M9W 1R3.

**National Research Council of Canada publications**

*National Building Code of Canada*
*National Fire Code of Canada*

These publications may be purchased from the Secretary, Associate Committee on the National Building Code, National Research Council, Ottawa, Ontario, K1A 0R6.

**NFPA standards**

NFPA 71 – *Installation, Maintenance, and Use of Central Station Signalling Systems*

NFPA 72D – *Installation, Maintenance, and Use of Proprietary Protective Signalling Systems*

NFPA 72E – *Automatic Fire Detectors*

These standards may be purchased from FIPRECAN, 7-1590 Liverpool Court, Ottawa, K1B 4L2.

**ULC standards**

CAN/ULC-S524 – *Standard for the Installation of Fire Alarm Systems*

ULC-S525 – *Standard for Audible Signal Appliances for Fire Alarm Systems*

CAN/ULC-S526 – *Standard for Visual Signal Appliances for Fire Alarm Systems*

CAN/ULC-S527 – *Standard for Control Units for Fire Alarm Systems*

ULC-S528 – *Standard for Manually Actuated Signalling Boxes for Fire Alarm Systems*

CAN/ULC-S529 – *Standard for Smoke Detectors for Fire Alarm Systems*

ULC-S530 – *Standard for Heat Actuated Detectors for Fire Alarm Systems*

CAN/ULC-S531 – *Standard for Smoke Alarms*

| CAN/ULC-S536 | – | *Standard for the Inspection and Testing of Fire Alarm Systems* |
| CAN/ULC-S537 | – | *Standard for the Verification of Fire Alarm Systems* |
| CAN/ULC-S541 | – | *Standard for Speakers for Fire Alarm Systems.* |

These standards may be purchased from the Underwriters' Laboratories of Canada, 7 Crouse Road, Scarborough, Ontario, M1R 3A9.

## CHAPTER 3-5 - STANDARD FOR FIRE INSPECTIONS

### 1. General

#### 1.1 Purpose

This standard establishes minimum requirements for fire inspections of Government of Canada property to minimize fire risks to life and property and to protect and conserve the Government's financial position.

#### 1.2 Application

This standard applies to all:

(a) departments and agencies listed in Schedule I and II of the *Financial Administration Act* (FAA) with the exception of the Department of National Defence;
(b) branches designated as departments for the purposes of the FAA; and
(c) those departments and other portions of the Public Service as defined in Part I of Schedule I of the *Public Service Staff Relations Act*.

#### 1.3 Scope

This standard identifies the requirements of a fire inspection of Government of Canada property and describes the roles and responsibilities of Treasury Board, custodian departments, tenant departments, Human Resources Development Canada, and the Office of the Fire Commissioner of Canada in ensuring that periodic fire inspections are carried out.

#### 1.4 Administration

(a) The Fire Commissioner of Canada or his authorized representative is responsible for the administration and enforcement of this standard.
(b) The requirements of this standard are not to be interpreted as permitting practices specifically prohibited by provincial, municipal, or other federal legislation.

#### 1.5 Definitions

In this standard:

(1) ***custodian department*** means a department or agency with responsibility to administer real property (ministère administrateur);

(2)    ***Government of Canada property*** means real or personal property under the administration and control of a federal government department or agency, including property leased to the government (bien du gouvernement du Canada);

(3)    ***heritage building*** means any federally owned building that has been designated as either "Classified" or "Recognized" for its heritage significance upon recommendation by the Federal Heritage Building Review Office (bâtiment du patrimoine);

(4)    ***historic building*** means any federally owned building that has been declared of national significance by the Minister of Environment Canada on the advice of the Historic Sites and Monuments Board of Canada (bâtiment historique);

(5)    ***tenant department*** means a department which does not have custodial responsibility over real property, but rather is responsible to administer and control personnel of that department, and such machinery and equipment over which its personnel has control (ministère locataire).

## 1.6  Abbreviations

In this standard:

(a)    "FC" means Fire Commissioner of Canada or the authorized representative of the Fire Commissioner of Canada;

(b)    "NBC" means the latest edition of the *National Building Code of Canada*;

(c)    "NFC" means the latest edition of the *National Fire Code of Canada*.

## 2.  Background

In accordance with the Memorandum of Understanding between Treasury Board and Human Resources Development Canada respecting fire protection services, Human Resources Development Canada acting on behalf of the Treasury Board, is responsible to conduct fire inspections.  Inspections are essential to maintain a satisfactory standard for fire safety in buildings and to ensure the protection, conservation and consequent minimization of risks to life, property and the Government's financial position.  Inspections limit the risk of life and property losses from fire by identifying and causing the correction of those conditions which contribute to the occurrence and spread of fire.

## 3.  Roles and responsibilities

### 3.1  Custodian department

(a)    Custodian departments are accountable for the protection of property. Custodian departments are to comply with Treasury Board fire protection policy and to cooperate with Human Resources Development Canada and municipal fire authorities, on matters pertaining to fire protection.

(b)    Custodian departments are responsible to arrange for the inspection, testing and maintenance of all fire protection equipment and systems in properties under their administration and control in accordance with the requirements of the NBC, the NFC, and Treasury Board fire protection standards.

## 3.2 Tenant department

(a) Tenant departments are accountable for the protection of contents, processes and operations. Tenant departments are to comply with Treasury Board fire protection policy and to cooperate with Human Resources Development Canada on matters pertaining to fire protection.

(b) Tenant departments are responsible to arrange for the inspection, testing and maintenance of additional or specialized fire protection equipment and systems protecting contents, processes and operations under their administration and control in accordance with the requirements of the NBC, the NFC, and Treasury Board fire protection policy. Where fire protection systems are installed to meet tenant department operational requirements, the tenant department is responsible for their effective interface with the buildings' existing fire protection systems.

## 3.3 Treasury Board

(a) Treasury Board approves and communicates to departments and agencies fire protection standards which may be required in excess of those prescribed pursuant to the *Canada Labour Code*, Part II.

(b) Treasury Board reviews instances where Human Resources Development Canada reports that a custodian or tenant department is unable or unwilling to comply with Treasury Board fire protection policy and determines the appropriate action to be taken.

## 3.4 Human Resources Development Canada

(a) Human Resources Development Canada provides fire inspection services (as detailed in sections 4 and 5), analyses fire risks, advises Treasury Board on fire protection policies and standards, and reports to Treasury Board situations where custodian or tenant departments are unable or unwilling to comply with standards.

(b) The FC, within Human Resources Development Canada, is responsible for the administration and enforcement of Treasury Board fire protection standards and those portions of the NBC and the NFC that cover fire protection.

(c) The FC may make arrangements with municipal fire authorities for the provision of fire inspections on Government of Canada properties.

## 4. Fire inspection requirements

Fire inspections are to be carried out by Human Resources Development Canada on behalf of Treasury Board to monitor compliance to codes and standards, to evaluate the danger to life from fire and to determine ways for minimizing fire danger to properties, contents and Government operations. Fire protection items and fire protection systems to be inspected by Human Resources Development Canada include, but are not necessarily limited to, the following:

(a) Building design and construction;
(b) building exposures;
(c) building services;

(d)    hazardous materials, processes and operations;
(e)    materials storage and handling;
(f)    fire detection and alarm systems;
(g)    water supply and fire suppression systems;
(h)    portable fire extinguishers;
(i)    special extinguishing systems;
(j)    fire emergency systems;
(k)    fire emergency procedures, organizations and evacuation plans;
(l)    fire department response and equipment; and
(m)    measures to minimize loss or damage subsequent to fire.

## 5. Priorities and frequencies for fire inspection services

5.1    In order to establish priority and frequency of inspection, the FC will take the following items into consideration for each property/building and associated occupancy:

(a)    The fire inspection is requested by a federal government department or agency and the nature of the request is considered urgent and important for the safety of the building occupants, protection of the property, or protection of operations;
(b)    there is a high number of fire significant protection deficiencies identified in previous inspections not rectified, or re-occurring, which are resulting in an undue hazard to building occupants, property, or operations;
(c)    the type of building and occupancy has a fire loss frequency and severity above the normal or acceptable level; or
(d)    a fire risk analysis of the property identifies:
    (i)    potential large life risk;
    (ii)    potential for large property fire loss, where a fire could result in a loss of more than $1,000,000 to buildings and/or contents;
    (iii)    potential for loss of an historic or heritage building; or
    (iv)    potential for an unacceptable interruption of essential government operations or services.

5.2    Properties falling into any one of the four items identified in subsection 5.1 are to be given a high priority in conducting fire inspections on a periodic basis.

## CHAPTER 3-6 - FIRE PROTECTION STANDARD FOR CORRECTIONAL INSTITUTIONS

### 1. General

### 1.1 Purpose

This standard prescribes fire protection requirements for correctional institutions to minimize risks to life and property and to protect and conserve the Government's financial position.

### 1.2 Application

(a)    This standard applies to correctional institutions under the administration of the Correctional Service of Canada.

(b)   A building or structure in an institution that is used for farming purposes and not as a residence is exempt from the requirements of this standard. (See appendix A)

(c)   Community residential facilities are exempt from the requirements of this standard.  (See appendix A)

## 1.3  Scope

(a)   The requirements of this standard apply to all new construction, including additions and alterations.

(b)   The requirements of this standard apply to existing buildings to the extent practicable.  (See appendix A)

## 1.4  Administration

(a)   The Fire Commissioner of Canada or the authorized representative of the Fire Commissioner (henceforth referred to as FC) is responsible for the administration and enforcement of this standard.

(b)   This standard is not to be interpreted as permitting practices specifically prohibited by provincial, municipal, or other federal legislation.

(c)   Where reference is made to other codes and standards, unless otherwise stipulated, the reference shall be to the latest edition including amendments to that code or standard.

## 1.5  Definitions

(a)   Certain terms used in this standard are defined to ensure understanding of their meaning and intent.

**Contained use area** means a supervised area containing one or more rooms in which occupant movement is restricted to a single room by security measures not under the control of the occupant (see appendix A) (zone de détention cellulaire);

**free access** means access without barriers that require keys, special devices, remote releasing devices, or specialized knowledge of the opening mechanism, to an approved safe area (see appendix A) (accès libre);

**impeded egress zone** means a supervised area in which occupants have free movement but require the release, by security personnel, of security doors at the boundary before they are able to leave the area, but does not include a contained use area (see appendix A) (zone à sortie contrôlée);

**listed** means equipment or materials included in a list published by a nationally recognized organization concerned with product evaluation, that maintains periodic inspection of production of listed equipment or materials and whose listing states either that the equipment or material meets appropriate standards or has been tested and found suitable for use in a specified manner (répertorié);

*living unit* means a building or portion thereof containing sleeping accommodation for inmates, and may include ancillary areas such as lounges, kitchenettes, showers, and janitor's closets. Family visiting units are not considered to be living units (unité résidentielle).

(b) Unless otherwise stated, the definitions of words and phrases given in the NBC and NFC also apply to this standard.

## 1.6 Abbreviations

(a) The abbreviations of proper names in this standard shall have the meanings assigned to them in this clause.

| | | |
|-----|---|---|
| CAN | – | National Standard of Canada |
| CER | – | Communication Equipment Room |
| CGSB | – | Canadian General Standards Board |
| CSA | – | Canadian Standards Association |
| CSC | – | The Correctional Service of Canada |
| FC | – | Fire Commissioner of Canada or the authorized representative of the Fire Commissioner |
| MCCP | – | Main Communication and Control Post |
| NBC | – | National Building Code of Canada |
| NFC | – | National Fire Code of Canada |
| NFPA | – | National Fire Protection Association |
| TB | – | Treasury Board |
| ULC | – | Underwriters' Laboratories of Canada |

(b) The symbols and other abbreviations in this standard shall have the meanings assigned to them in this clause.

| | | |
|-----|---|---|
| $^{\circ}$C | – | degree(s) Celsius |
| h | – | hour(s) |
| kg | – | kilogram(s) |
| kPa | – | kilopascal(s) |
| L | – | litre(s) |
| m | – | metre(s) |
| mm | – | millimetre(s) |
| No. | – | number(s) |
| o.c. | – | on centre |
| s | – | second(s) |

## 2. General requirements

## 2.1 Codes and standards

Except as otherwise provided in this standard, fire protection in all correctional institutions shall conform to the applicable requirements of:

    (i) the National Building Code of Canada (NBC);
    (ii) the National Fire Code of Canada (NFC);
    (iii) Treasury Board fire protection standards; and
    (iv) FC standards.

## 2.2 Review procedures

Plans and specifications for all new construction, including additions and alterations, shall be submitted to the FC for review in accordance with the Treasury Board Fire protection standard for design and construction (chapter 3-2).

## 3. Construction

### 3.1 Buildings without free access

(a) Buildings that do not have free access shall conform to the requirements of the NBC for buildings of Group B Division 1 occupancy.
(b) Buildings that do not have free access shall be of noncombustible construction.

### 3.2 Buildings with free access

(a) Buildings that have free access shall conform to the requirements of the NBC for the major occupancies appropriate to their use. (See appendix A)
(b) A building that has free access shall be of noncombustible construction if the building area exceeds the limits permitted in appendix C. (See also appendix A)

### 3.3 Means of egress

(a) Means of egress shall conform to the requirements of the NBC.
(b) A door in a means of egress, which is required to be locked for reasons of security, shall be such as to be openable by a key from both sides of the door. (See appendix A)
(c) If in the opinion of the FC in consultation with CSC for reasons of security it may be difficult to gain access to a locked egress door in the event of an emergency, remote releasing devices shall be provided for such doors in addition to the key release prescribed in 3.3(b). (See appendix A)

### 3.4 Separation of major occupancies

Major occupancies shall be separated from each other by fire separations in accordance with the requirements of the NBC.

### 3.5 Separation within floor areas (see appendix A)

(a) Except as provided in clause 3.5(b), the following areas shall be separated from the remainder of the building by a fire separation having a fire-resistance rating of at least 2 h:
  (i) living units in either contained use areas or impeded egress zones; and,
  (ii) any other contained use areas.
(b) The fire separation in (a) may be reduced to 1 h in a sprinklered building provided the adjacent area is not an industrial use. (See appendix A)

(c) The following rooms or areas shall be separated from the remainder of the building by a fire separation having a fire-resistance rating of at least 1 h:
   (i) maintenance shops;
   (ii) storage rooms;
   (iii) shipping and receiving areas; and
   (iv) kitchen, cafeteria and dining areas.

### 3.6 Interior finish materials

(a) Interior finish materials shall conform to the requirements of the NBC and this subsection.
(b) Except as provided in (c), interior wall and ceiling finishes in a contained use area shall have a flame spread rating of not more than 25 and a smoke developed classification of not more than 50.
(c) Interior padding used for walls or ceilings shall be of fire-retardant materials, suitable for institutional use. (See appendix A)
(d) Interior floor finish materials shall have a flame spread rating of not more than 300 and a smoke developed classification of not more than 500.

### 4. Occupancy hazards

### 4.1 General

Unless otherwise specified in this standard, hazardous materials, processes and operations shall conform to the requirements of the NFC. (See appendix A)

### 4.2 Explosives and ammunition

The storage, handling and use of explosives and ammunition shall be in accordance with the *Explosives Act*, R.S., c.102, S.1 and the *Explosive Regulations* published by Natural Resources Canada.

### 4.3 Mattresses, pillows and furniture (see appendix A)

(a) Mattresses and pillows shall be of a fire retardant type, suitable for institutional use. (See appendix A)
(b) Upholstered or plastic furniture shall not be used in a contained use area unless the floor area is sprinklered.
(c) Except as permitted in (d), upholstered or plastic furniture in contained use areas or impeded egress zones shall be of a fire retardant type suitable for institutional use. (See appendix A)
(d) Furniture in an impeded egress zone is not required to be fire retardant provided:
   (i) the floor area is sprinklered; and
   (ii) the floor area is not occupied by inmates.

### 4.4 Kitchen cooking equipment

(a) Commercial kitchen cooking equipment such as a range, broiler, frying unit or other equipment capable of producing grease-laden vapours shall be provided with a listed grease removal system.

(b)   Systems for the ventilation of commercial kitchen cooking equipment shall
      be designed, constructed and installed to conform to NFPA Standard
      No. 96, *Installation of Equipment for the Removal of Smoke and
      Grease-Laden Vapors from Commercial Cooking Equipment*.
(c)   An automatic fire suppression system (see appendix A) shall be provided
      for the protection of:
      (i)    commercial kitchen cooking equipment; and,
      (ii)   residential ranges in living units.

## 5. Fire protection systems

### 5.1 Fire alarm systems

(a)   A fire alarm system shall be installed in every building except as permitted
      in (b).
(b)   For a building having free access and a total occupant load of 10 or less,
      a fire alarm system is not required if the building is 2 storeys in height and
      less than 50 m² in building area, or 1 storey in height and less than
      100 m² in building area.
(c)   Except as provided in this Section, fire alarm systems shall conform to the
      Treasury Board Standard for fire alarm systems (chapter 3-4).
(d)   In a building having either a contained use area or an impeded egress
      zone, the fire alarm system shall be a 2 stage-system. (See appendix A)
(e)   The fire alarm system shall be zoned in accordance with the NBC. Zones
      shall also correspond to functional divisions of the institution as
      determined by CSC in consultation with the FC.
(f)   Manual fire alarm stations in inmate occupied areas may be located in
      secure areas, except in buildings intended for use by occupants having
      free access.
(g)   Smoke alarms shall be installed in sleeping rooms and corridors serving
      such rooms in buildings not requiring a fire alarm system.
(i)   Where a fire alarm system is required, smoke detectors shall be installed
      in every sleeping room and every corridor serving as part of a means of
      egress from sleeping rooms.

### 5.2 Sprinkler systems

(a)   Sprinkler systems shall be installed throughout all living units. (See
      appendix A)
(b)   Except as permitted in (d) and (e), sprinkler systems shall be installed
      throughout all other buildings.
(c)   Except as otherwise provided in this Subsection, sprinkler systems shall
      be designed and installed in accordance with the requirements of FC
      Standard No. 403, *Sprinkler Systems*.
(d)   Sprinklers may be omitted from small detached buildings in which in the
      opinion of the FC, there is no life hazard and which do not constitute an
      exposure hazard to the principal functional areas of the institution.
(e)   Unless otherwise required by the NBC or NFC, sprinklers may be omitted
      from detached buildings intended for use by occupants having free
      access, where the building is not more than:
      (i)    2 storeys in height and 500 m² in building area, or
      (ii)   1 storey in height and 1000 m² in building area.

(f) Main sprinkler control valves shall be located in a secure area accessible only to authorized personnel. (See appendix A)

(g) Secondary sprinkler control valves such as zone control valves, test connections, and auxiliary drains shall either:
  (i) be located in secure areas accessible only to authorized personnel; or
  (ii) be locked. (See appendix A)

(h) The outside water motor gong may be omitted from buildings when all parts of the perimeter of the building are within the secured area, and the transmission of an alarm to a continuously manned facility is provided.

## 5.3 Standpipe and hose systems

(a) (See appendix A) Except as required in (b), standpipe and hose systems shall be installed in all buildings of more than 3 storeys or 14 m in height and of 3 storeys or 14 m or less in height when the building area exceeds the following:

| Height (Storeys) | Building area (m²) |
|---|---|
| 1 | 2000 |
| 2 | 1500 |
| 3 | 1000 |

(b) Standpipe and hose systems shall be installed in all living units not having free access.

(c) Except as otherwise provided in this Section, standpipe and hose systems shall be designed and installed in accordance with the NBC.

(d) Except as permitted in (e), every hose station shall be equipped with a 65 mm hose connection for fire department use and a 38 mm connection. In lieu of a 38 mm fire hose, a suitable easily removable adapter connected to a 25 mm rubber hose may be provided. The rubber hose shall not exceed 30 m in length and shall be mounted on a continuous flow hose reel.

(e) The 65 mm hose connections may be omitted in a detached, single-storey building not exceeding 4000 m² in building area. (See appendix A)

(f) Subject to consultation between the FC and the Administrative Official, 38 mm hose stations may be supplied from sprinkler branch lines in any fully sprinklered building. (See appendix A)

(g) An electrically controlled shut-off valve may be installed in the branch water supply line leading to a hose station, provided the valve is
  (i) of a type listed for fire protection service,
  (ii) located in a secure area, accessible only to authorized personnel,
  (iii) capable of being manually overridden,
  (iv) electrically supervised,
  (v) connected to an emergency power supply,
  (vi) arranged to control not more than one hose station in any one fire zone and not more than two hose stations in a building,
  (vii) normally open, and

(viii) inspected on a regular basis, with the frequency prescribed by the NFC for valves controlling sprinkler water supplies.

## 5.4 Portable fire extinguishers

(a) Except as otherwise provided in this subsection, fire extinguishers shall be selected and installed in accordance with the requirements of the NBC and NFC.

(b) Alkali base dry chemical extinguishers shall be installed in areas with commercial cooking equipment.

(c) Carbon dioxide extinguishers shall be provided for the protection of sensitive electrical and electronic equipment.

## 5.5 Hydrants

(a) Hydrant protection shall be provided for all buildings.

(b) Municipal hydrants may be considered as meeting all or part of the requirements of this Section, subject to the approval of the FC.

(c) Hydrants shall be equipped with two 65 mm hose outlets and a fire department pumper connection, sized and threaded to local fire department requirements.

(d) Hydrants shall be located such that they are not less than 1.5 m nor more than 3 m from access roads and readily accessible to fire department apparatus.

(e) Hydrants shall be so located that all parts of the perimeter of the building can be reached by hose streams with not more than 75 m of hose attached to a hydrant.

(f) Hydrants shall be located so that they are no closer than 15 m and no farther than 75 m from any building they are intended to protect.

## 5.6 Fire department connections

(a) Fire department connections shall be located in supervised areas and accessible to fire department apparatus at all times.

(b) Fire department connections shall be located so that the distance from a fire department connection to a hydrant does not exceed 45 m and is unobstructed.

(c) Fire department siamese connections shall be provided for all sprinkler systems and standpipe and hose systems.

## 5.7 Fire department access routes

(a) At least two separate access routes for fire department vehicles shall be provided through the main security fence of a correctional institution.

(b) Where practicable, fire department access routes conforming to the requirements of the NBC shall be provided to the principal entrance of every building. (See appendix A)

## 6. Water supplies

### 6.1 Capacity

(a) The water supply for correctional institutions shall be not less than the fire flow required for the building having the largest demand, as calculated in clause (b).

(b) Except as otherwise permitted by the FC, the minimum water supply for each building shall be the greater of:
   (i) the demand determined in accordance with *Water Supply for Public Fire Protection — A Guide to Recommended Practice* published by Fire Underwriters' Survey; or,
   (ii) the demand for sprinkler systems, including inside and outside hose allowance.

(c) The water supply stipulated in (a) shall be available for a period of not less than 2 h.

(d) The water supply system shall be designed so that the available flow rate at any one hydrant is not less than 30 L/s of water at a residual pressure of not less than 450 kPa (gauge). (See appendix A)

### 6.2 Water supply systems

(a) Where available, the water supply shall consist of 2 separate connections from a municipal water works system.

(b) Where a municipal water works system is not adequate to meet the requirements stipulated in this Section, it shall be augmented by an on-site water supply from tanks or reservoirs conforming to FC Standard No. 403, *Sprinkler Systems*.

(c) The installation of fire pumps and booster pumps shall conform to the appropriate requirements of FC Standard No. 403, *Sprinkler Systems*.

## 7. Electrical

### 7.1 Exit signs

Exit signs shall be provided and placed over every exit door other than the main entrance to a building, and installed in accordance with the NBC.

### 7.2 Emergency lighting

(a) Emergency lighting shall conform to the requirements of the NBC.

(b) Emergency lighting shall be provided for:
   (i) means of egress and other areas prescribed by the NBC,
   (ii) the MCCP and other control posts, and
   (iii) other security areas as determined by CSC.

### 7.3 Emergency power

(a) Emergency power shall be provided for
   (i) electric fire pumps where required by reference in clause 6.2(c),
   (ii) fire alarm systems, and
   (iii) emergency lighting systems.

(b)　Emergency generator systems shall conform to CSA Standard C282, *Emergency Electrical Power Supply for Buildings* and be of sufficient capacity to operate for a period of not less than 2 h.

## 8. Emergency procedures

### 8.1 General

(a)　A fire emergency organization shall be established with an official and deputies appointed by the administration of CSC.

(b)　The general responsibilities and formation of the organization shall be in accordance with the Treasury Board Standard for Fire safety planning and fire emergency organization.

(c)　A fire safety plan as described in subsection 8.2 shall be prepared to be put into immediate action in the event of a fire emergency.

### 8.2 Fire safety plan

(a)　A fire safety plan shall be prepared and administered by the senior departmental officer or a designated official.

(b)　The fire safety plan shall comply with the requirements of the Treasury Board Standard for Fire safety planning and emergency organization (chapter 3-1).

(c)　Duplicate keys for locked doors shall be provided and kept in a secure location that is readily accessible at all times to authorized personnel.

### 8.3 Investigation and reporting of fires and false alarms

(a)　An investigation shall be made by CSC of the cause, origin, and circumstances of every fire or false alarm occurring in a correctional institution.　(See appendix A)

(b)　Records shall be kept by CSC of all fires and false alarms for a period of not less than 2 years.

(c)　Fires shall be reported to the FC as prescribed in appendix A of the Treasury Board Policy on Fire protection, investigation and reporting.

## 9. Main communication and control post/communication equipment room (MCCP/CER)

### 9.1 Scope

This section applies to the Main Communication and Control Post/ Communication Equipment Room (MCCP/CER) containing the main communication and security equipment, and facilities contained therein, such as washrooms and entrance lobby.　(See appendix A)

### 9.2 General requirements

(a)　Except as otherwise provided in this section, the design, construction, and operation of the MCCP/CER shall conform to the requirements prescribed for the protection of Level 2 EDP equipment in the Treasury Board Fire protection standard for electronic data processing equipment.

(b)　The MCCP/CER shall be of noncombustible construction.

(c) The MCCP shall be separated from the CER by a fire separation having a fire-resistance rating of at least 1 h.

(d) Sprinkler systems in the MCCP/CER may be of the pre-action type.

### 9.3 Operations and housekeeping

(a) The MCCP/CER shall have at least 1 attendant on duty at all times.

(b) Facilities such as kitchenettes, hot plates, sinks except lavatory in washroom, refrigerators, stoves, toasters, and microwave ovens shall not be provided within the MCCP/CER.

(c) Housekeeping shall be carried out on a regular basis to ensure no hazards or unnecessary combustibles accumulate in the area.

## CHAPTER 3-6 - APPENDIX A - EXPLANATORY MATERIAL

A-1.2(b)  The FC should be consulted for direction on fire protection requirements for such buildings. As a minimum, farm buildings must comply with the applicable requirements of the NBC and Canadian Farm Building Code.

A-1.2(c)  Community residential facilities include both community residential centres and community correctional centres. Although sometimes referred to as institutions for administrative purposes by CSC, they are not institutional occupancies as defined by the NBC. Usually these facilities are considered to be residential occupancies as defined by the NBC.

A-1.3(b)  The extent of application of this standard to the upgrading of existing buildings should be based on judgement and the merits of each case. The FC should be consulted for advice in assessing risk and determining priorities.

A-1.5(a)  Any building with detention rooms is a contained use area. The most common examples are medium and maximum security living units.

In order to qualify as having free access, **all** required means of egress from a building or area must be without barriers. Otherwise, egress is considered to be impeded.

Impeded egress zones are buildings or portions thereof in which the occupants have some freedom of movement, but from which they cannot exit freely. Typical examples are operational buildings such as dining halls, recreation buildings, and training shops. Note that if **any** of the required means of egress are locked, egress is considered to be impeded.

A-3.2 (a)  The following are examples of occupancy classifications that would apply to buildings having free access:

| | | |
|---|---|---|
| Kitchen and dining facilities | Group A | Division 2 |
| Recreation (Gymnasia) | Group A | Division 2 |
| Socialization (Chapels, Libraries, Meeting rooms) | Group A | Division 2 |

| Medical services | Group B | Division 2 |
| Living units | Group C | |
| Administration | Group D | |
| Maintenance and servicing | Group F | Division 2 |
| Shops – Industrial | Group F | Division 2 |
| Shops – Stores – Garages | Group F | Division 2 |

A-3.2(b) These limits are placed on the size of combustible buildings for fire risk management reasons. The presence or absence of sprinkler protection has no effect on the application of the area limits in this table.

A-3.3(b) Individual cell doors may be keyed from one side only.

A-3.3(c) CSC has issued directives giving guidance and priorities for installing remote release devices on existing buildings.

A-3.5 Fire separations should be provided between all major functional areas. The fire resistance ratings of these separations should be consistent with the ratings required by the NBC for the separation of major occupancies. For example: an industrial workshop having impeded egress is by definition a Group B Division 1 occupancy. However, for the purposes of determining an appropriate fire separation, Table 3.1.3.A of the NBC should be applied as if the workshop was an F2 occupancy.

A-3.5(b) The term *industrial use* is used in this clause in lieu of *industrial occupancy* to avoid confusion with the occupancy classification as defined by the NBC. As noted above, an industrial use such as a workshop is still by definition a B1 occupancy if the means of egress are locked.

A-3.6(c) Padded materials should be evaluated on the basis of full-scale room fire tests that approximate the room dimensions and severe ignition scenarios to be expected in an institutional setting. One such test is California Standard No. 12-42-100, *Room Fire Test For Wall and Ceiling Materials*.

A-4.1 CSC has issued directives giving guidance on fire hazard control in inmate-occupied areas. For further details see CSC Commissioner's Directive No. 345 — Fire Safety.

A-4.3 CSC Commissioner's Directive No. 345 — Fire Safety — contains further guidance on combustible furnishings permitted in inmate-occupied areas.

A-4.3(a) In the past, the following materials have been found to be acceptable:

  (i) core materials of neoprene foam, flame retardant treated cotton, or Cordelan staple fibre, and

(ii)    covers of flame retardant treated cotton ticking or a flame retardant synthetic fabric.

Other materials should be evaluated on the basis of tests representative of the severe ignition scenarios to be expected in an institutional setting. One such test is California Technical Bulletin No. 121, *Flammability Test Procedure for Mattresses for Use in High Risk Occupancies.*

A-4.3(c)  The fire retardancy of furniture should be evaluated on the basis of full-scale fire tests representative of the severe ignition scenarios to be expected in an institutional setting. One such test is California Technical Bulletin No. 133, *Flammability Test Procedure for Seating Furniture for Use in High Risk and Public Occupancies.*

A-4.4(c)  In a sprinklered floor area, the preferred method of fire suppression for cooking equipment is automatic sprinklers.

A-5.1(d)  In areas with 2-stage fire alarm systems, a listed voice communication system is not required if there are other communication systems available that are adequate and reliable. In living units these systems should include all of the following: portable radios; public address system; and telephones at control posts.

A-5.2(a)  For the purposes of this clause, family visiting units are not considered to be living units.

A-5.2(f)  Where practicable, main valves should be located in a secure room or area having access directly from the exterior.

A-5.2(g)  These security requirements are intended to be in addition to normal valve supervision. Supervision alone is not sufficient to prevent tampering In this type of occupancy.

A-5.3(a)  The standpipe requirements exceed those of the NBC because: firefighting access to buildings is generally limited by security constraints; the available public fire department resources at most institution locations are limited; and fire hose systems are used by institutional staff to provide an initial firefighting response.

A-5.3(e)  It is assumed that such buildings would have adequate fire department access and coverage from fire hydrants. If this is not the case, then the 65 mm connections should be installed.

A-5.3(f)  Combined systems are a cost-efficient measure, particularly in view of the requirements of clauses (a) and (b). However, approval must be site-specific to ensure that the combined system can meet both fire protection and security criteria.

A-5.7(b)  It is recognized that strict compliance may not be practical with some institution designs. In such cases the FC, the Fire Department Official, and the Administrative Official should consult on alternative firefighting provisions.

A-6.1(d)  Due to the limitations on fire department response at most institutions, a high residual pressure is specified so that one or two hose streams can be supplied directly from a hydrant.

A-8.3(a)  CSC Commissioner's Directive No. 345 — Fire Safety — includes requirements for the investigation and reporting of fires and false alarms by CSC staff.

A-9.1  There is usually only one MCCP in an institution.  The requirements of this Section are not intended to apply to local control posts.

## CHAPTER 3-6 - APPENDIX B - REFERENCE PUBLICATIONS

This standard refers to the following codes and standards which are not found in this volume:

### Fire Commissioner of Canada standards

| | | |
|---|---|---|
| 301 | – | Construction operations |
| 302 | – | Welding and cutting |
| 403 | – | Sprinkler systems |

These standards may be obtained from the Fire Commissioner of Canada, Human Resources Development Canada, Ottawa, Ontario, K1A OJ2, or from regional offices of Human Resources Development Canada.

### Treasury Board publication

*Treasury Board Manual*
"Materiel, Services and Risk Management" volume
Fire Protection, Investigation and Reporting (Part III, chapter 5)

This volume may be purchased from the Canada Communication Group — Publishing, 45 Sacré-Coeur Blvd., Hull, Québec, K1A 0S9.

### Correctional Service of Canada publication

Commissioner's Directive No. 345 — Fire Safety

This directive may be obtained from the Correctional Service of Canada, Ottawa, Ontario, K1A 0P9.

### CSA standards

| | | |
|---|---|---|
| C22.1 | – | *Canadian Electrical Code* |
| C282 | – | *Emergency Electrical Power Supply for Buildings* |

Canadian Standards Association Standards may be purchased from the Canadian Standards Association, 178 Rexdale Blvd., Rexdale, Ontario, M9W 1R3.

**NFPA standards**

96   –   *Installation of Equipment for Removal of Smoke and Grease-Laden Vapors from Commercial Cooking Equipment*

National Fire Protection Association standards may be purchased from FIPRECAN, 2425-1 Don Reid, Ottawa, Ontario K1H 1A4.

**National Research Council of Canada publications**

*National Building Code of Canada*
*National Fire Code of Canada*

These codes may be purchased from the Institute for Research in Construction, National Research Council, Ottawa, Ontario K1A OR6.

## CHAPTER 3-6 - APPENDIX C - MAXIMUM BUILDING AREA PERMITTED FOR COMBUSTIBLE CONSTRUCTION

|            | BUILDING HEIGHT (STOREYS) | | | |
|------------|---------|---------|---------|--------|
| OCCUPANCY  | 1       | 2       | 3       | 4      |
| A-2        | 1600 m² * | 800 m²  | NP      | NP     |
| B-2        | 1000 m²  | 500 m²  | NP      | NP     |
| C          | 2400 m²  | 1200 m² | 800 m²  | NP     |
| D          | 4800 m²  | 2400 m² | 1600 m² | NP     |
| F-1        | 800 m²   | 400 m²  | NP      | NP     |
| F-2        | 3200 m²  | 1600 m² | 1070 m² | 800 m² |

NP   –   Not permitted.
\*   –   May be up to 3200 m² with heavy timber roof assembly

## CHAPTER 4 - PROCEDURES

### Introduction

Occupational safety and health (OSH) procedures in the Public Service specify uniform practices to follow in the operation of OSH programs. Depending on its mandate, a department may establish additional requirements for specific operations, provided they conform to Treasury Board requirements and are generally accepted as good OSH practices.

**Application**

1. These procedures incorporate the minimum requirements of the *Canada Occupational Safety and Health Regulations*, Part XV, Hazardous Occurrence Investigation, Recording and Reporting, and apply to all departments and other portions of the Public Service, as defined in Part I of Schedule I of the *Public Service Staff Relations Act*.

**Definitions**

2. In these procedures:

***accident*** means an event which results in a fatality, work injury, property damage or material loss arising out of, linked with or occurring in the course of employment (accident);

***disabling injury*** means any work injury or occupational disease that

(a)   prevents an employee from reporting for work or effectively performing all the duties connected with his or her regular work on any day subsequent to the day on which the injury occurred, whether or not that subsequent day is a working day, or

(b)   results in the loss by an employee of a body member or part thereof or in a complete loss of its usefulness or in the permanent impairment of a body function (blessure entraînant l'invalidité).

***first-aid*** means the emergency care that is rendered to an injured or ill employee pursuant to chapter 2-5, First-aid directive (premiers soins);

***minor injury*** means any work injury for which medical treatment is provided, and that excludes a disabling injury (blessure légère);

***person in charge*** means a qualified person appointed by management to ensure the safe and proper conduct of an operation or of the work of employees (personne responsable);

***qualified person*** means, in respect of a specified duty, a person who, because of knowledge, training and experience, is qualified to perform that duty safely and properly (personne qualifiée);

***regional safety officer*** means a person designated as a regional safety officer pursuant to the *Canada Labour Code*, Part II (agent régional de sécurité);

***safety and health committee*** means a committee established pursuant to chapter 2-20, Committees and representatives directive (comité de la sécurité et de la santé);

***safety and health representative*** means a representative appointed pursuant to chapter 2-20, Committees and representatives directive (représentant de la sécurité et de la santé);

***safety officer*** means a person designated as a safety officer pursuant to the *Canada Labour Code*, Part II, and includes a regional safety officer (agent de sécurité);

***work injury*** means any injury, disease or illness incurred by an employee in the course of employment (accident de travail);

***workplace*** means any place where an employee is engaged in work for the employee's department (lieu de travail).

## Reports by employee

3. Where an employee becomes aware of an accident or other occurrence arising in the course of or in connection with the employee's work that has caused or is likely to cause injury to that employee or to any other person, the employee shall, without delay, report the accident or other occurrence to the department, orally or in writing.

## Investigations

4. Subject to paragraph 5, where a department or a manager becomes aware of an accident, occupational disease or other hazardous occurrence affecting any of that department's employees in the course of employment, the department shall, without delay,

(1) appoint a qualified person or persons to carry out an investigation of the hazardous occurrence;

(2) notify the safety and health committee or the safety and health representative, if any, of the hazardous occurrence and of the name of the person or persons appointed to investigate it. The committee or representative, if any, shall take part in the accident investigation pursuant to chapter 2-20, Committees and representatives directive;

(3) take necessary measures to prevent a recurrence of the hazardous occurrence; and

(4) ensure that the requirements of paragraphs 23 and 24 are met.

5. Where the hazardous occurrence referred to in paragraph 4 is an accident involving a motor vehicle on a public road that is investigated by a police authority, the investigation referred to in paragraph 4(1) shall be carried out by obtaining from the appropriate police authority a copy of its report respecting the accident.

6. As soon as possible after receipt of the report referred to in paragraph 5, the department shall provide a copy thereof to the safety and health committee or the safety and health representative, if any.

## Telephone or telex reports

7. The department shall report to a safety officer, by telephone or telex, the date, time, location and nature of any accident, occupational disease or other hazardous occurrence referred to in paragraph 4 that had one of the following results as soon as possible but not later than 24 hours after becoming aware of that result, namely:

(1)  the death of an employee;
(2)  a disabling injury to two or more employees;
(3)  the loss by an employee of a body member or part thereof or in the complete loss of the usefulness of the body member;
(4)  the permanent impairment of a body function of an employee;
(5)  an explosion;
(6)  damage to a boiler or pressure vessel that results in fire or the rupture of the boiler or pressure vessel; or
(7)  a free fall of, or damage to, an elevating device that renders it unserviceable.

## Records

8. The department shall, within 72 hours after an accident referred to in paragraph 7(6) or (7), ascertain the causes of the accident and record in writing

(1)  a description of the accident and the date, time and location of the accident;
(2)  the causes of the accident; and
(3)  the corrective action that was taken or that will be taken, or the reason for not taking corrective measures.

9. The employer shall, without delay, submit a copy of the record referred to in paragraph 8 to the safety and health committee or the safety and health representative, if any.

## Minor injury reports

10. Every department shall keep a record of each minor injury of which the department is aware that affects employees in the course of employment. Such record shall contain

(1)  the date, time and location of the accident that resulted in the minor injury;
(2)  the name of the employee affected;
(3)  a brief description of the minor injury; and
(4)  the causes of the minor injury.

11. A record of all occupational injuries and illnesses that require first-aid treatment shall be maintained in accordance with chapter 2-5, First-aid directive.

## Written reports

12. Where an investigation referred to in paragraph 4 discloses that the accident resulted in

(1) a disabling injury to an employee;
(2) an electric shock, toxic atmosphere or oxygen deficient atmosphere that caused an employee to lose consciousness;
(3) the implementation of rescue, revival or other similar emergency procedures; or
(4) a fire or an explosion,

the results of the investigation shall, without delay, be reported in writing on the form and contain the information required by the form.

13. A copy of the report made in accordance with paragraph 12 shall be submitted by the department prescribed by appendix A:

(1) without delay, to the safety and health committee or the safety and health representative, if any; and
(2) within 14 days after the hazardous occurrence, to a safety officer at the regional office or district office.

14. Where an investigation referred to in paragraph 5 discloses that the accident resulted in a circumstance referred to in paragraph 12, the department shall, within 14 days after the receipt of the police report of the accident, submit a copy of the report to a safety officer at the regional office or district office.

## Annual report

15. Every department shall, not later than March 1 in each year, submit to the Minister of Human Resources Development, with a copy to the Deputy Secretary of the Human Resources Policy Branch of the Treasury Board, a written report setting out the number of accidents of which it is aware affecting any of its employees in the course of employment during the 12 month period ending December 31 in the preceding year.

16. The report referred to in paragraph 15 shall be in accordance with the relevant forms (see appendix A) and contain the information required by the forms.

17. In respect of the forms prescribed, departments may specify the use of different forms for individual departmental application if such forms provide, as a minimum, the same information.

## Reports and records

18. With regard to the content of those reports and records referred to in these procedures, where the word "employer" appears, it shall mean the Public Service department which is reporting the accident.

19. Subject to paragraph 20, every department shall keep a copy of each report and record referred to in these procedures for a period of two years after its submission to the regional safety officer, to the Minister of Human Resources Development or to the Treasury Board.

20. Every record referred to in paragraph 8 shall be kept by the department for a period of 10 years after the accident.

## Precedence

21. Where there may be a conflict between the meaning of, or the requirements of these procedures and those of the *Canada Occupational Safety and Health Regulations, Part XV, Hazardous Occurrence Investigation*, Recording and Reporting, the latter shall take precedence.

## Compensation

22. For purposes of Workers' Compensation, all work-related injuries and illnesses shall be reported in accordance with the requirements of the *Government Employees' Compensation Act*.

## Accident scene

23. Subject to paragraph 24, where an employee is killed or seriously injured in a workplace, no person shall, unless he or she is authorized to do so by a safety officer, remove or in any way interfere with or disturb any wreckage, article or thing related to the incident except to the extent necessary to

(1) save a life, prevent injury or relieve human suffering in the vicinity;
(2) maintain an essential public service; or
(3) prevent unnecessary damage to or loss of property.

24. No authorization referred to in paragraph 23 is required where an employee is killed or seriously injured by an accident or incident involving

(1) an aircraft where the accident or incident is being investigated under the *Aeronautics Act* or the *Canadian Aviation Safety Board Act*; or
(2) a motor vehicle on a public highway.

## Enquiries

Enquiries should be directed to the responsible officers in departmental headquarters who, in turn, may seek interpretation from the following:
Safety, Health and Employee Services Group
Staff Relations Division
Human Resources Policy Branch
Treasury Board Secretariat

*(section 15.8)*

| Title of Form | No. of Form | Initiated By | Available From |
|---|---|---|---|
| Employer's Annual Hazardous Occurrence Report | LAB/TRAV 393 (09/90) | HRDC | **Regional offices of HRDC** |

*(section 15.10)*

| Title of Form | No. of Form | Initiated By | Available From |
|---|---|---|---|
| Hazardous Occurrence Investigation Report | LAB/TRAV 369 (92/01) | HRDC | **Regional offices of HRDC** |

## CHAPTER 4-2 - PROCEDURES FOR OCCUPATIONAL HEALTH INVESTIGATIONS

### Introduction

1. Under the Public Service Occupational Safety and Health Policy, an important aspect of the Public Service health program concerns the provision of a program and procedures for the surveillance of work facilities, in order to identify and promptly correct hazards which may adversely affect the health of employees. Where a health hazard is suspected or deemed to exist, qualified personnel should be available to determine the type and extent of the hazard, if any, and, where necessary, to recommend measures that will eliminate or reduce the hazard.

2. Accordingly, the Treasury Board, pursuant to its authority under Section 7 of the *Financial Administration Act*, has authorized these procedures concerning the investigation and surveillance of occupational health hazards in the Public Service.

### Application

3. These procedures apply to all departments and other portions of the Public Service, as defined in Part I of Schedule I of the *Public Service Staff Relations Act*.

## Definitions

4. In these procedures:

*director* means the director, Occupational Medical Services, Medical Services Branch, Health Canada, Ottawa (directeur);

*environmental health officer* means an individual so designated by Health Canada, and whose responsibilities relative to these procedures are detailed in appendix A (agent d'hygiène du milieu);

*investigation* means the detailed examination of a work process, condition or facility, either in response to a specific request or incident, or as part of the overall program, in order to determine the existence, extent and type of a health hazard (enquête);

*manager* means the regional manager, Public Service Health, Medical Services Branch, Health Canada, or another official authorized to act on the Manager's behalf (gestionnaire);

*occupational health hazard* means an identifiable hazard to employee health which is capable of causing an occupational illness, and generally falls into one of the following categories:

(a) **chemical:** liquids, gases, dusts, fumes, mists and vapours;
(b) **physical:** ionizing and non-ionizing radiations, noise, vibration, sanitation, ventilation and extremes of temperature and pressure; and
(c) **biological:** insects, mites, moulds, yeasts, fungi, viruses and bacteria (risque pour la santé au travail);

*occupational illness* means any disease, abnormal health condition or disorder caused by exposure to environmental factors or substances associated with the work, and which includes acute and chronic illnesses or diseases which may result from inhalation, absorption, ingestion or direct contact with a substance (maladie professionnelle);

*occupational injury* means any bodily injury (such as a cut, fracture, sprain, amputation, etc.,) which results from a work accident or from an exposure involving a single incident in the work environment (blessure au travail);

*survey* means the pre-planned and systematic study or review of working conditions and facilities to identify and evaluate potential health hazards (étude).

## Responsibilities

5. Departments shall, in accordance with these procedures, be responsible for:

(1) initiating protective measures, as required, and maintaining ongoing conditions necessary to protect employees from exposure to health hazards at work;

425

(2)     formally requesting Health Canada to undertake occupational health investigations whenever health hazards are suspected;

(3)     affording environmental health officers access to work locations at all reasonable times; and

(4)     implementing recommendations made as a result of investigations and surveys.

6. Health Canada is responsible for:

(1)     ensuring the availability of qualified personnel to carry out the program;

(2)     undertaking, in liaison with the departments concerned, health investigations or surveys, and providing necessary direction to departments for the elimination or reduction of health hazards in the workplace; and

(3)     monitoring compliance with recommendations or directions issued.

**Criteria**

7. The criteria which govern general health conditions, types and levels of exposure, and the exposure thresholds pertaining to dangerous substances shall be as determined through reference to chapter 2-2, Dangerous substances directive and as contained in the *Canada Occupational Safety and Health Regulations* issued pursuant to the *Canada Labour Code*, Part II.

**Procedures for requesting an investigation**

8. The investigation and survey program is intended to augment departmental responsibilities and initiatives for the ongoing detection and elimination of hazardous or unhealthy conditions. Apart from individual responsibilities in this regard, such conditions may also be delineated through the activities of safety and health committees or safety and health representatives. Where a potential health hazard is suspected, departments should begin by consulting with local experts and specialists concerning the matter. Investigations should not be requested frivolously or carried out in respect of conditions which do not involve occupational health hazards as defined in these procedures.

9. In the event that an occupational health hazard is suspected, a written request, authorized and signed by a responsible departmental official, shall be forwarded to the applicable manager. Departmental requests shall include the following information:

(1)     the location and description of the facility to be investigated;

(2)     the general nature of the suspected health hazard; and

(3)     the name, address and telephone number of the departmental official to be contacted.

10. If an employee believes that he or she may be subject to hazardous working conditions that could cause an occupational illness, the employee may request that the person in charge arrange for an investigation of the hazard as soon as possible. Following confirmation that a formal investigation within the scope of these procedures is required, the person in charge shall make the necessary arrangements through the responsible departmental official.

11.  Upon receipt of a departmental request, the manager will, in liaison with the department concerned, schedule the investigation and advise the department accordingly.

## Distribution of reports

12.  Upon completion of an investigation or survey, a written report shall be prepared and forwarded under the authority of the manager to the departmental official concerned.  Directions concerning the implementation of recommendations should be specified, as required.

13.  One copy of each report shall be forwarded to the director.  The director will, where deemed necessary, forward copies of the report to the headquarters of the department concerned for attention and follow-up.

14.  The director will ensure that Human Resources Policy Branch of Treasury Board Secretariat is promptly notified of any serious case of non-compliance with a major recommendation or directive.

15.  It is the responsibility of each department to ensure appropriate internal distribution of each report, including the provision of a copy to departmental headquarters.  A copy of the report is to be made available also to the applicable safety and health committee or safety and health representative, if any, as soon as possible.  Pursuant to paragraph 10, an employee who requests an occupational health investigation shall be afforded the opportunity of reviewing the investigation report.

16.  If the report concerns the building systems, a copy shall be provided by the occupying department to the appropriate Public Works and Government Services Canada official or, where Public Works and Government Services Canada is not responsible, to the responsible building authority.  Where an investigation or survey has been requested by Public Works and Government Services Canada, each occupying department shall be notified of the results by Public Works and Government Services Canada.

## Implementation procedures

17.  Changes or measures recommended as a result of investigations or surveys shall be implemented by departments as soon as practicable.  The appropriate manager shall be advised when compliance has been effected, and also where an appreciable delay is foreseen in the implementation of recommendations.

## Procedures in the case of imminent and serious danger

18.  Where, in the opinion of an environmental health officer, a situation poses both an imminent and a serious danger to the health of employees, the department shall be directed to take the required immediate action to rectify or remove the imminent and serious danger.  If this danger cannot be immediately rectified or satisfactorily reduced, the environmental health officer shall contact the appropriate manager, who shall decide whether or not to order suspension

of the operations related to the imminent health danger until the condition has been rectified, or the degree of danger satisfactorily reduced.

19. Where a suspension order has been issued, it shall be the responsibility of Health Canada, Medical Services Branch, to immediately notify Human Resources Policy Branch of Treasury Board Secretariat and the Deputy head of the department concerned of the suspension order.

20. If a department considers that a suspension order is not warranted, it shall comply with the order, but may immediately request the manager to review the department's objections, and, if the matter is not mutually resolved, the department may appeal directly to the Safety, Health and Employee Services Group of Human Resources Policy Branch of Treasury Board Secretariat, which will rule on the matter.

## Ongoing surveillance program

21. In addition to investigations carried out in response to specific departmental requests, Health Canada will also undertake, under the authority of these procedures and the Public Service Occupational safety and health policy, an ongoing survey program of occupational health hazards at locations where such hazards are deemed to exist. Surveys shall be carried out in accordance with priorities and schedules established by the appropriate director.

22. The requirements of these procedures shall apply to this ongoing surveillance program (as outlined in paragraph 21) in all respects.

## Enquiries

Enquiries should be directed to the responsible officers in departmental headquarters who, in turn, may seek interpretation from the following:

> Safety, Health and Employee Services Group
> Staff Relations Division
> Human Resources Policy Branch
> Treasury Board Secretariat

## CHAPTER 4-2 - APPENDIX A - RESPONSIBILITIES OF THE ENVIRONMENTAL HEALTH OFFICER

1. The environmental health officer shall, with the assistance of other health specialists, as required:

(1) conduct, on behalf of the manager, occupational health investigations or surveys to determine:
  (a) the occupational health hazards which may prevail at work locations,
  (b) the scope and seriousness of these hazards, and their effect in impairing the health and well-being of employees, and
  (c) the measures required to eliminate or control these hazards, or to reduce their effects to an acceptable level;
(2) interpret and analyze the results of health investigations and surveys;

(3)    recommend control methods and procedures, where necessary, which will
       be suitable and effective for adequate employee protection;
(4)    prepare and forward, under the authority of the manager, a detailed report
       on each investigation or survey which includes:
    (a)    date or period conducted,
    (b)    location and description of facility,
    (c)    name of departmental official contacted,
    (d)    nature of problem,
    (e)    activities carried out,
    (f)    tests conducted,
    (g)    findings,
    (h)    formal recommendations,
    (i)    suggested improvements.

2.  Environmental health officers are not authorized to prepare or conduct
survey questionnaires without the prior approval of the director.

## CHAPTER 4-3 - PROCEDURES FOR OCCUPATIONAL EXPOSURE TO ASBESTOS

### Introduction

1.  Inhalation of asbestos fibres may result in adverse health effects to those
exposed.  These procedures, which are issued pursuant to the Dangerous
substances directive, chapter 2-2, are intended, therefore, to provide an outline
of basic measures which are required to be taken for the protection of
employees exposed to an asbestos process.

### Application

2.  These procedures apply to all Public Service departments and agencies, as
defined in Part I of Schedule I of the *Public Service Staff Relations Act.*

### Definitions

3.  In these procedures:

*asbestos* includes any of the minerals crocidolite, amosite, chrysolite,
anthophyllite, tremolite or actinolite (amiante);

*asbestos process* refers to any handling of materials which generates airborne
asbestos fibres, including (traitement de l'amiante)

(a)    the sawing, cutting, drilling or abrasion of asbestos materials;
(b)    the packing or unpacking of asbestos;
(c)    the installation or removal of asbestos insulation or coverings;
(d)    the mixing or application of asbestos cements, plasters, putties or similar
       compounds;
(e)    the cleaning of asbestos-contaminated clothing; or
(f)    the storage, conveyance or disposal of materials containing asbestos.

## General requirements

4. Departments and agencies shall:

(1) avoid the use or processing of asbestos if it is possible to substitute a less hazardous substance;
(2) ensure that every asbestos process under their jurisdiction is identified, and that each such process is carried out and controlled in compliance with the requirements of these procedures;
(3) notify the appropriate regional or zone director of the Medical Services Branch, Health Canada, concerning the details of every existing and new asbestos process under their jurisdiction;
(4) ensure that every person involved in an asbestos process is familiar with and complies with the requirements of these procedures.

## Control of airborne asbestos dust

5. No asbestos process shall proceed without appropriate control measures such as ventilation, separate enclosure or isolation of the process from workers, to minimize the hazardous dispersal of asbestos dust into the work environment. Advice concerning specific control measures may be obtained from the appropriate regional office of the Medical Services Branch, Health Canada.

6. Wet or damp processing of asbestos shall be instituted where practicable.

7. Ventilation equipment used for controlling and removing asbestos dust shall be maintained, operated and periodically tested by a competent person to ensure that it continues to operate at design performance in order that the asbestos content of the breathing zone shall not exceed the threshold limit value.

## Personal protective equipment

8. Where it is not practicable to control asbestos dust within the required threshold limit value as specified in paragraph 17, personal protective equipment and clothing shall be obtained and utilized. In such instances the department shall provide, and ensure that each exposed person uses, a respirator of a type recommended by Health Canada that is appropriate for the required degree of respiratory protection, and any special protective work clothing which may be required.

9. Every person who is required to wear personal protective equipment shall be fully instructed in the proper use, care and maintenance of that equipment.

10. Special protective work clothing shall be worn only in the workplaces or during operations for which such clothing is designated. A change room that is suitable for changing into and out of protective work clothing and for clean storage of street clothes shall be provided for the use of employees who work with asbestos. Protective clothing that has been exposed to asbestos shall not be taken away from the workplace by the employee.

11. Departments shall arrange for the laundering of such protective clothing, and this shall be done in a segregated manner. Clothing that is being laundered or sent for laundering shall be separated, identified and handled in a manner that does not expose any other persons to the asbestos hazard.

## Cleanliness of the workplace

12. Accumulations of asbestos waste or dust produced in any place of employment shall be removed at least once daily; heavy accumulations of such waste or dust shall be removed as frequently as is reasonably practicable during a work shift.

13. All cleaning to remove asbestos waste or dust shall be performed by vacuuming or wet cleaning methods to prevent the dispersal of asbestos dust into the environment. Vacuum equipment should be of a type acceptable to Health Canada for asbestos dust removal. Waste shall be collected in closed containers and disposed of in closed containers in compliance with applicable and local disposal regulations or requirements.

## Health surveillance

14. All employees, regardless of duties or mode of protection, who are regularly exposed to an asbestos process shall be medically examined through the facilities of Health Canada.

15. The examinations referred to in paragraph 14 shall be carried out in accordance with the Occupational health evaluation standard, chapter 2-13. Health Canada shall maintain detailed records of all employees whose health has been adversely affected through exposure to asbestos, and shall advise the employing department and the Treasury Board Secretariat of such cases.

16. Employees involved in an asbestos process shall be routinely informed by their department of all known asbestos hazards, and of the corresponding need to develop safe and healthful work and personal habits. Advice and information concerning such hazards may be obtained through Health Canada.

## Environmental surveillance

17. The following threshold limit values (TLVs) are the time-weighted average concentrations for a normal 8-hour work day or 40-hour work week, to which workers may be repeatedly exposed:

| Asbestos type | TLV |
| --- | --- |
| Crocidolite | 0.2 fibre longer than 5 micrometres per cubic centimetre of air. |
| Amosite | 0.5 fibre longer than 5 micrometres per cubic centimetre of air. |
| Other forms | 2 fibres longer than 5 micrometres per cubic centimetre of air. |

18. Air sampling shall be conducted in a manner to be specified by Health Canada. Samples shall be collected from within the breathing zone of the

employees, using membrane filters of 0.8 micrometer porosity, mounted in an open-face filter holder. Determinations shall be made at a magnification of 400 to 500 using phase contrast illumination.

19. All sampling and analyses shall be carried out according to a method approved by Health Canada.

## Enquiries

Enquiries should be directed to the responsible officers in departmental headquarters who, in turn, may seek interpretation from the following:

> Safety, Health and Employee Services Group
> Staff Relations Division
> Human Resources Policy Branch
> Treasury Board Secretariat

## CHAPTER 4-4 - PROCEDURES FOR TRACTOR SAFETY

### Application

1. These procedures are issued pursuant to the Materials handling directive, chapter 2-10, and are intended to provide a basic outline of the principal safety factors and requirements respecting tractors and their operation. They are to be applied by Public Service departments and agencies, as defined in Part I of Schedule I of the *Public Service Staff Relations Act*.

### Definition

2. In these procedures, a *tractor* (tracteur) means a vehicle designed for agricultural or industrial use which is used to pull, carry, propel or drive agricultural implements; and for landscaping, loading, digging, grounds keeping, highway maintenance and construction services. Tractors used in the servicing or towing of aircraft are not included.

### General responsibilities

3. Departments and agencies are, in accordance with the general principles set forth in these procedures, responsible for

(1) developing and enforcing detailed departmental rules and procedures for the safe operation of tractors, including instructions and procedures for the avoidance of roll-over;
(2) ensuring that every tractor is maintained in a safe operating condition;
(3) ensuring that every tractor operator is trained and qualified in all respects to operate the tractor to which the operator is assigned; and
(4) selecting tractors designed to perform safely under the most severe conditions of operation and use that may be encountered.

### Design and construction

4. Insofar as practicable, specifications for tractors should incorporate the following requirements:

(1)   Operator controls should conform to generally accepted design and safety directives and, when necessary, the function and direction of movement of controls shall be clearly marked.

(2)   Seats should incorporate a backrest, be adequately sprung or suspended, and be adjustable to the operator's height and weight.

(3)   Tractors not provided with a protective cab should be equipped with fenders over the rear wheels which extend beyond the full width of the wheels and are otherwise designed to protect the operator.

(4)   The exhaust from the engine should, where practicable, discharge vertically at least 40 inches (1 m) above the driver, or in such other manner or location whereby the driver is not exposed to the exhaust fumes. If the tractor is provided with a cab, the exhaust should discharge above the roof or in such other manner that exhaust gases are not drawn into the cab.

(5)   Batteries, fuel tanks, oil reservoirs and coolant systems shall, insofar as practicable, be constructed, located and maintained so that, in the event of upset, spillage will not come in contact with the operator.

(6)   Lighting shall be provided which is sufficient to ensure safe operation under all conditions.

(7)   Tractors operated on public roads and on thoroughfares used by other vehicles shall be equipped with lighting and other safety equipment required by the laws of the province or territory in which the tractor is operated.

**Guard devices**

5. All auxiliary equipment, implements and drive mechanisms including, where practicable, drive belts, shall be provided with suitable guarding devices.

6. Power take-off mechanisms shall be shrouded and, when not in use, enclosed by a guard of a strength sufficient to withstand potential load factors.

**Roll-over protection**

7. Where a tractor is likely to turn over under any circumstances of its operation, it shall be fitted with a roll-over protective structure (ROPS) that will prevent the operator from being trapped or crushed under the tractor. ROPS should be designed and installed so as to

(1)   tend to limit turn-over in any direction to 90°;

(2)   extend sufficiently to the rear to prevent the operator from coming in contact with attachments in case of a rear tip-over;

(3)   permit optimum visibility;

(4)   facilitate the operator's escape in the event of a roll-over;

(5)   permit convenient attachment and removal of the structure for maintenance repairs; and

(6)   meet, as a minimum, the test procedures and performance requirements set out in USA-OSHA Standards Part 1928, Sub-part C, 1928.51, 1928.52 and 1928.53, as amended from time to time.

8. Where a ROPS is fitted, a safety seat belt shall be provided and used when the tractor is being operated. Such seat belts shall be anchored in accordance

with USA-OSHA Standards, Part 1928, Sub-part C, 1928.51, and shall meet the requirements of SAE Standard J4C *Motor Vehicle Seat Belt Assemblies*.

9. Providing a tractor is not used in a manner and under such conditions that it could turn over, such tractor may be operated without a ROPS under the following circumstances

(1) inside a building or in other areas such as orchards, etc. where, normally, the vertical clearance is insufficient to allow a ROPS-equipped tractor to operate and where the use of the tractor is essential to the work being performed;
(2) with accessory equipment which is incompatible with a ROPS (e.g., pickers, harvesters, etc.);
(3) where, in the opinion of a Human Resources Development Canada safety officer, it is not technically feasible to install a ROPS.

## Protective cabs

10. Where a protective cab is installed on a tractor, it shall be constructed so as to provide

(1) the same degree of protection against roll-over as that prescribed for a ROPS;
(2) sufficient overhead protection from falling objects, where such hazard may be encountered;
(3) some means of emergency egress apart from the normal means of entry and exit; and
(4) noise attenuation to the extent practicable, in order to reduce dependence on personal hearing protection devices.

## Noise

11. Tractor operators subject to noise in excess of the levels and/or time-weighting outlined in the Noise control and hearing conservation directive, chapter 2-12, shall be provided with approved hearing protectors.

## Protective clothing

12. Personal protective equipment appropriate to the various hazards encountered shall be worn. This may include hard hats, eye protection, respirators, hearing protectors, reflector-type vests, etc. In this regard, reference should be made to the Personal protective equipment directive, chapter 2-14.

## Enquiries

Enquiries should be directed to the responsible officers in departmental headquarters who, in turn, may seek interpretation from the following:

Safety, Health and Employee Services Group
Staff Relations Division
Human Resources Policy Branch
Treasury Board Secretariat

## CHAPTER 4-5 - PROCEDURES FOR LIAISON WITH PRIVATE CONTRACTORS

### Introduction

1. The purpose of these procedures is to clarify the relationships which should exist between departments and private contractors, where the occupational health or safety of individuals may be adversely affected by activities of private contractors during construction, renovation, maintenance or any other operations, on or in federally owned or leased premises.

### Application

2. These procedures apply to all Public Service departments and agencies, as defined in Part I of Schedule I of the *Public Service Staff Relations Act*.

### Jurisdiction

3. As the work of private contractors and their employees is subject to the laws of the province or territory in which the work is being conducted, the appropriate provincial or territorial authorities have legal jurisdiction over health and safety conditions relative to such work.

### Departmental responsibility

4. If a departmental official becomes aware of a condition or situation arising out of the activity of a private contractor working on or in federally owned or leased premises, which could pose a hazard to the health or safety of Public Service employees or the general public, that official shall ensure that the appropriate details concerning the hazard are relayed immediately to the manager responsible for the letting and/or control of the contract.

5. The manager responsible for the letting and/or control of the contract, after receiving the details of the dangerous condition, shall:

(1) make direct arrangements with the contractor to effect the necessary changes to assure the health and safety of those exposed; or
(2) where resolution of the situation is not achieved to the satisfaction of the manager or responsible departmental official, advise the appropriate provincial or territorial authority of the matter; or
(3) in extreme cases, request the assistance of federal safety officers by contacting the nearest district or regional office of Human Resources Development Canada.

### Enquiries

Enquiries should be directed to the responsible officers in departmental headquarters who, in turn, may seek interpretation from the following:

Safety, Health and Employee Services Group
Staff Relations Division
Human Resources Policy Branch
Treasury Board Secretariat

## CHAPTER 4-6 - PROCEDURES FOR CORRECTION OF PHYSICAL SAFETY AND HEALTH HAZARDS

### Application

1. These procedures apply to all Public Service departments and agencies, as defined in Part I of Schedule I of the *Public Service Staff Relations Act*.

### General

2. The prompt correction of physical hazards which could affect the occupational safety or health of Public Service employees is of vital importance because delays might result in occurrences which lead to injury or illness. Whenever hazardous situations are identified through internally-organized departmental programs, or through the services of outside inspection authorities, it is essential that such hazards be rectified with the least possible delay.

### Hazard correction

3. Departments that are responsible for the maintenance of their own property should establish internal procedures to ensure that identified safety or health hazards are given priority attention. If, however, a department must call upon the services of Public Works and Government Services Canada to effect hazard correction, the following special procedure is to be used:

(1) The request shall be made using form DPW 337, Tenant Service Request for Estimate or Work (CGS Cat. No. 7540-21-879-7571).

(2) In Section 1 of the form, under "Work (Description)", preface the detail with the term SAFETY/HEALTH HAZARD, and indicate the priority of the request by completing the "Date Work Required" space. If the request is of an urgent nature, the form should be tagged accordingly.

(3) In an emergency situation, the work may be requested by telephone, with form DPW 337 being completed later.

(4) If the request is the result of a direction issued by a safety officer under the *Canada Labour Code*, Part II, or by an environmental health officer of Health Canada, attach a copy of such direction.

4. Requisitions concerning health or safety hazards shall be actioned by Public Works and Government Services Canada on a priority basis, and the necessary work will be carried out as soon as practicable. If Public Works and Government Services Canada is unable to meet the required date, the requesting department will be provided with an estimated completion date.

5. Departments and agencies are responsible for ensuring that this special procedure is used only in respect of matters pertaining to employee health and safety, and that all follow-up actions are carried out, as required.

Enquiries should be directed to the responsible officers in departmental headquarters who, in turn, may seek interpretation from the following:

Safety, Health and Employee Services Group
Staff Relations Division
Human Resources Policy Branch
Treasury Board Secretariat

## CHAPTER 4-7 - PROCEDURES FOR SAFETY OFFICERS AND SPECIAL SAFETY MEASURES

### Application

1. These procedures apply to all departments and other portions of the Public Service, as defined in Part I of Schedule I of the *Public Service Staff Relations Act*.

### Definitions

2. In these procedures:

*board* means the Public Service Staff Relations Board (commission);

*danger* means any hazard or condition that could reasonably be expected to cause injury or illness to a person exposed thereto before the hazard or condition can be corrected (danger);

*person in charge* means a qualified person appointed by management to ensure the safe and proper conduct of an operation or the work of employees (personne responsable);

*regional safety officer* means a person designated as a regional safety officer pursuant to the *Canada Labour Code*, Part II (agent régional de sécurité);

*safety and health committee* means a committee established pursuant to Part II of the *Canada Labour Code* and chapter 2-20, Committees and representatives directive (comité de sécurité et de santé);

*safety and health representative* means a person appointed as a representative pursuant to Part II of the *Canada Labour Code* and chapter 2-20, Committees and representatives directive (représentant en sécurité et en santé);

*safety officer* means a person designated as a safety officer pursuant to the *Canada Labour Code*, Part II, and includes a regional safety officer (agent de sécurité);

*workplace* means any place where an employee is engaged in work for the employee's department (lieu de travail).

## Introduction

3. Human Resources Development Canada safety officers are, subject to government security requirements, authorized to enter Public Service establishments and carry out a range of duties intended to enforce the requirements of the *Canada Labour Code*, Part II, and regulations issued thereunder. The following procedures, based upon the *Canada Labour Code*, shall be applicable to such activities.

## Powers of safety officers

4. A safety officer may, in the performance of his or her duties and at any reasonable time, enter any workplace controlled by a department and in respect of any workplace, may:

(1)  conduct examinations, tests, inquiries and inspections or direct the department to conduct such examinations, tests, inquiries and inspections;
(2)  take or remove for analysis, samples of any material or substance or any biological, chemical or physical agent;
(3)  be accompanied and assisted by such persons and bring such equipment as deemed necessary to carry out his or her duties;
(4)  take photographs and make sketches;
(5)  direct the department to ensure that any place or thing specified by the safety officer not be disturbed for a reasonable period of time pending an examination, test, inquiry or inspection in relation thereto;
(6)  direct the department to produce documents and information relating to the safety and health of departmental employees or the safety of the workplace, and permit the examination and making of copies of extracts of such documents and information by the safety officer; and
(7)  direct the department to make or provide statements, in such form and manner as the safety officer may specify, respecting working conditions and material and equipment that affect the safety or health of employees.

## Certificate of authority

5. A safety officer shall be furnished by the Minister of Human Resources Development with a certificate of his or her authority and on entering any workplace shall, if so required, produce the certificate to the person in charge thereof.

## Duty to assist

6. The person in charge of any workplace and every person employed thereat or in connection therewith shall give a safety officer all reasonable assistance in his or her power to enable the safety officer to carry out the duties.

7. No person shall obstruct or hinder a safety officer engaged in carrying out his or her duties.

8. No person shall make a false or misleading statement either orally or in writing to a safety officer engaged in carrying out his or her duties.

## Special safety measures

9. Where a safety officer is of the opinion that any provisions under the *Canada Labour Code*, Part II, are being contravened, he or she may direct the department or employee concerned to terminate the contravention within such time as the safety officer may specify and the safety officer shall, if requested by the department or employee concerned, confirm the direction in writing if the direction was given orally.

10. Where a safety officer considers that the use or operation of a machine or thing or a condition in any place constitutes a danger to an employee while at work:

(1)  the safety officer shall notify the department of the danger and issue directions in writing to the department directing the department immediately or within such period of time as the safety officer specifies:
 (a) to take measures for guarding the source of danger, or
 (b) to protect any person from the danger; and
(2)  the safety officer may, if he or she considers that the danger cannot otherwise be guarded or protected against immediately, issue a direction in writing to the department directing that the place, machine or thing in question shall not be used or operated until the directions are complied with.  However, nothing in this paragraph prevents the doing of anything necessary for the proper compliance with the direction.

11. Where a safety officer issues a direction under paragraph 10(2), he or she shall affix to or near the place, machine or thing in respect of which the direction is made, a notice in such form and containing such information as the Minister of Human Resources Development may specify, and no person shall remove the notice unless authorized by a safety officer.

12. Where a safety officer issues a direction under paragraph 10(2) in respect of any place, machine or thing, the department shall discontinue the use or operation of the place, machine or thing and no person shall use or operate it until the measures directed by the safety officer have been taken.

13. Where a safety officer issues a direction in writing under paragraph 10 or makes a report in writing to a department on any matter under the *Canada Labour Code*, Part II, the department shall forthwith:

(1)  cause a copy or copies of the direction or report to be posted in such manner as the safety officer may specify; and
(2)  give a copy of the direction or report to the safety and health committee, if any, for the workplace affected or the safety and health representative, if any, for that workplace.

14. Where a safety officer issues a direction in writing under paragraphs 9 or 10 or makes a report referred to in paragraph 13 in respect of an investigation made by him or her pursuant to a complaint, the safety officer shall forthwith give a copy of the direction or report to each person, if any, whose complaint led to the investigation.

**Review of a safety officer's direction**

15.  Any department, employee or union that is aggrieved by any direction issued by a safety officer under the *Canada Labour Code*, Part II, may, within fourteen days of the date of the direction, request that a regional safety officer for the region in which the place, machine or thing in respect of which the direction was issued is situated, review the direction.

16.  The regional safety officer may require that an oral request for a review under paragraph 15 be made, as well, in writing.

17.  The regional safety officer shall in a summary way inquire into the circumstances of the direction to be reviewed and the need therefor, and may vary, rescind or confirm the direction and thereupon shall in writing notify the employee, department or union concerned of the decision.

18.  A request for a review of a direction shall not operate as a stay of direction.

19.  Paragraph 15 does not apply in respect of a direction of a safety officer that is based on a decision of the safety officer that has been referred to the Board pursuant to Part II of the *Canada Labour Code* or to Refusal to work directive, chapter 2-19.

**Enquiries**

Enquiries should be directed to the responsible officers in departmental headquarters who, in turn, may seek interpretation from the following:

> Safety, Health and Employee Services Group
> Staff Relations Division
> Human Resources Policy Branch
> Treasury Board Secretariat

**CHAPTER 4-8 - PROCEDURES FOR SAFETY RESPONSIBILITIES – MANAGERS AND SUPERVISORS**

**Introduction**

1.  The Public Service Occupational safety policy outlines a requirement for departments and agencies to develop and publish statements of internal safety policy which include specific and general safety responsibilities.  While it is expected that this requirement will result in the broad identification of program responsibilities, it will also be necessary to assign a degree of direct safety responsibility to individual managers and supervisors.  As the effectiveness of safety programs is directly related to the leadership and participation of management and supervisory personnel, it is important that there be an awareness of individual responsibilities and accountability in this regard.

**Development**

2.  The most appropriate time to formally consider the identification of occupational safety and health responsibilities of managers and supervisors is

during the initial assignment of duties and responsibilities in a position, or at the time of periodic review of job content. As the position description is the basic document to describe the work performed, it must accurately reflect all related functions. Accordingly, the extent of safety responsibility can be determined by a careful analysis of the work in relation to the inherent duties and responsibilities of a position.

## Implementation

3. Departments and agencies should therefore, identify positions which require specific delineation of ongoing safety and health program responsibilities and, notwithstanding any lack of reference in classification directives, include appropriate statements of such responsibility in the applicable position descriptions.

## Enquiries

Enquiries should be directed to the responsible officers in departmental headquarters who, in turn, may seek interpretation from the following:

> Safety, Health and Employee Services Group
> Staff Relations Division
> Human Resources Policy Branch
> Treasury Board Secretariat

## CHAPTER 4-9 - PROCEDURES FOR NATIONAL AND REGIONAL SAFETY AND HEALTH COMMITTEES

### Application

1. These procedures apply to all departments and other portions of the Public Service, as defined in Part I of Schedule I of the *Public Service Staff Relations Act*.

### Requirement

2. To assist in the overall administration of the safety and health program, each department shall develop, in consultation with employee representatives, internal policy directives and instructions governing the establishment and operation of a national safety and health committee and, where applicable, regional safety and health committees.

### Operating principles

3. Departmental program directives and instructions shall be developed to incorporate the following, as a minimum:

(1)  the detailed operating rules, terms of reference, functions and composition of such committees, including the provision for representation from management and employees;

(2)  the promotion of all aspects of the Safety and Health Program, including related research, as required;

(3)     the convening of committee meetings on a regular basis or as determined by the parties concerned;

(4)     the availability of all records, reports and documents required for use by the committees, and the distribution of notices, minutes and records of meetings to the committee members; and

(5)     the provision for review, at a higher level, of any matter which cannot be resolved by a regional committee.

4. Where the formation or use of a national safety and health committee or a regional safety and health committee is deemed by both the department and the employee representatives as impracticable or unnecessary, other existing operational joint consultation arrangements may be used to fulfil the intent of this directive.

## Workplace committees

5. The establishment of safety and health committees for departmental workplaces, as required in Part II of the *Canada Labour Code*, is governed by the Committees and representatives directive (chapter 2-20), and departments shall ensure that the requirements of the directive are applied as necessary.

## Enquiries

Enquiries should be directed to the responsible officers in departmental headquarters who, in turn, may seek interpretation from the following:

    Safety, Health and Employee Services Group
    Staff Relations Division
    Human Resources Policy Branch
    Treasury Board Secretariat

## CHAPTER 4-10 - PROCEDURES FOR TRAINING IN OCCUPATIONAL SAFETY AND HEALTH

### Introduction

Both the *Canada Labour Code*, Part II and the Treasury Board Occupational safety and health (OSH) policy require that employers (departments) provide the information, instruction and training necessary to ensure safety and health at work. The advent of the Workplace Hazardous Materials Information System (WHMIS) on October 31, 1988 has re-emphasized this need.

A few departments are already doing a creditable job of OSH training. Most of the others, however, have not yet instituted comprehensive and consistent training programs in this area. Treasury Board Secretariat (TBS), therefore, commissioned a training needs study of the issue through the OSH Interdepartmental Committee (OSH/IDC). The study was carried out by an OSH/IDC sub-committee with representation from several departments. The results were reviewed by both the Professional and Technical Training Board and the Management Training Board. They were then approved by the Staff Training Council.

**Application**

This procedure applies to departments and other Public Service organizations (herein referred to as "departments") as listed in Part I of Schedule I of the *Public Service Staff Relations Act*.

**General requirements**

The study identified four programs to meet the minimum training and information requirements in the OSH area, as follows:

- an OSH management overview;
- an OSH management and supervisory training course;
- a training course for safety and health representatives and committee members;
- an information program for all employees.

Additional details on these programs are given in appendix A.

The Public Service Commission (PSC) will provide the first three programs interdepartmentally at various locations throughout Canada, charging the normal course fees. Departments have the option of sending representatives to these sessions or of purchasing the PSC material when it is available and conducting their own courses. The PSC is also prepared to "tailor-make" courses for individual departments.

Departments are responsible for providing the fourth program, namely the information program for all employees. There are two elements here: meeting the needs of present employees and meeting those of new employees. See appendix A for details.

Departments may continue to use existing OSH training and information programs as long as they meet, as a minimum, the content of the programs described in this procedure. Departments may also adjust their OSH training and information initiatives to reflect individual and organizational needs; but the intent of the OSH legislation, of the Treasury Board policy and of this procedure must be fulfilled.

With regard to WHMIS, note that each of the four programs covers the basic elements of this initiative but that may not be enough to meet the need in all situations. Departments must meet all of the education requirements mandated in the WHMIS legislation. These requirements are specified in the *Canada Occupational Safety and Health Regulations*, Part X, Hazardous Substances. In general, they include the development and implementation of an employee education program on hazard prevention and control for the workplace in consultation with the safety and health committee or representative. Through these programs, employees must be made aware of the product identifiers of hazardous substances used in their workplaces; of the information disclosed on labels and material safety data sheets (and how to read and interpret it); of procedures for the safe storage, use, handling and disposal of hazardous substances; of pipes, valves, controls and safety devices; and of procedures for emergencies.

The WHMIS education program must be reviewed in consultation with the safety and health committee or representative at least once each year or more often if conditions change or if new products are introduced into the workplace.

For details on the WHMIS education requirements see Sections 10.17 to 10.19 in the regulations referred to above.

## Reporting

Departments are required to report annually on OSH training in the Multi-Year Human Resource Plan. For those departments that have completed Memoranda of Understanding with Treasury Board under Increased Managerial Authority and Accountability, information on OSH training is to be included in the *Annual Management Report*.

## Monitoring

As noted at the beginning of this chapter, TBS monitors the effectiveness and application of the various procedures. This will be done in this case by:

- reviewing the annual reports from departments, i.e. MYHRP or AMR;
- reviewing audit and evaluation reports, both internal and external.

## References

*Canada Labour Code*, Part II

*Treasury Board Manual*, "Occupational Safety and Health" volume, OSH policy.

WHMIS legislation

## Enquiries

Enquiries about this procedure should be directed to responsible officers in departmental headquarters.

For interpretations of specific provisions, direct questions to the following, whichever is appropriate:

## General questions on training or information in the OSH area

Safety, Health and Employee Services Group
Human Resources Policy Branch
Treasury Board Secretariat

## Information on PSC courses, course design and materials of OSH training programs

Training and Development Canada
Public Service Commission

## CHAPTER 4-10 - APPENDIX A - SYNOPSIS OF THE OSH TRAINING AND INFORMATION PROGRAMS

Following is a synopsis of the OSH training and information programs.

### 1. OSH management overview

The objective of this one-day overview is to provide senior managers and others with the basic knowledge needed to manage the OSH programs in their organizations.

The seminar modules are:

(a)  Introduction to management and supervisory roles and responsibilities in OSH (including WHMIS)
(b)  Job safety analysis
(c)  Workplace inspections
(d)  Accident/hazardous occurrence investigation
(e)  Refusal to work
(f)  Roles and responsibilities of safety and health committees and representatives.

### 2. OSH management and supervisory training course

The objective of this three-day course is to provide managers and supervisors with the knowledge and skills to effectively organize and implement an OSH program at their work sites.

The seminar modules are:

(a)  Management and supervisory roles and responsibilities in OSH
(b)  Job safety analysis
(c)  Workplace inspections
(d)  Accident/hazardous occurrence investigation
(e)  Refusal to work
(f)  Roles and responsibilities of safety and health committees and representatives
(g)  WHMIS.

### 3. Training course for safety and health committee members

The objective of this two-day program is to provide selected employees with the knowledge and skills to function effectively as members of safety and health committees or as safety and health representatives.

The target population is all groups and levels of employees who are or are going to be members of safety and health committees or safety and health representatives (including both management and employee members of safety and health committees).

The course modules are:

(a) The role of a safety and health committee
(b) Effective committee membership
(c) Workplace hazards
(d) Accident investigation
(e) Improving safety and health performance
(f) WHMIS.

## 4. Information program for all employees

Departments are responsible for preparing the detailed information internally and for deciding how it will be disseminated.

Following are the subjects that need to be covered:

(a) *Canada Labour Code*, Part II
  – employee rights including right to know, right to participate and right to refuse to work;
  – role, authority and responsibility of Human Resources Development Canada;
  – role of Health Canada;
  – role of the Treasury Board Secretariat.
(b) Departmental roles and responsibilities
  – Safety and health policy;
  – safety and health programs;
  – role of the safety and health officer.
(c) Safety and health committees and representatives
  – mandate — roles and responsibilities
  – make-up of committees
  – employees' right to participate.
(d) Employees' responsibilities

To contribute to a safe work environment by:
  – learning and applying safe work practices
  – using safety equipment (e.g. saw guard)
  – reporting faulty conditions and unsafe acts
  – reporting accidents and incidents requiring first aid
  – cooperating with safety and health committees
  – complying with directions given by the designated safety and health officer
  – complying with employers' instructions.
(e) WHMIS
  – what the system is
  – how it works
  – general information on material safety data sheets and labels.

This information can be imparted to employees in a variety of ways, for example:

(a)  to existing employees through brochures, leaflets, posters, articles in departmental newsletters or newspapers, or in meetings led by supervisors, etc.;

(b)  for new employees, each one should be given the information as part of the normal orientation process.

## CHAPTER 5 - GUIDES

### Introduction

Occupational safety and health (OSH) in the Public Service is governed by policies, directives and standards, and procedures which are mandatory for departments to follow. Certain activities, however, require additional guidance to assist departments in meeting their responsibilities. OSH guides provide more detailed direction in the form of generally accepted, good OSH practices.

## CHAPTER 5-1 - SAFETY GUIDE FOR LABORATORY OPERATIONS

### 1. Introduction

1.1  The operation of laboratories exposes employees to a wide range of occupational risks and hazards. Due to the complex and highly variable nature of these operations, these guidelines provide only a basic outline of the principal safe practices and procedures applicable thereto. All applicable and authorized directives or special instructions should be available for reference by laboratory staff.

1.2  Departments and agencies which carry out laboratory work should develop and issue their own detailed safety directives, based on the general requirements of this guide and good industrial safety practices. The directives should indicate, where necessary, when compliance is mandatory for safe operations. Such directives should also include specific references to the use of personal protective equipment in relation to the hazard or exposure encountered.

1.3  While all those employed in laboratories have a responsibility to follow safe practices and support the safety program, the principal responsibility and authority for ensuring the safety of laboratory operations rests with the person in charge.

1.4  Where chemicals are being used, all materials and procedures should be reviewed regularly for hazards by a professional chemist, biochemist or chemical engineer; changes in materials and procedures should be made to reduce any hazards observed.

1.5  Effective communications between all levels of management and between managers and employees is essential for the promotion of safe and healthful working conditions. It is recommended, therefore, that wherever appropriate, safety committees be organized to meet as required to discuss, plan and review

447

the safety of laboratory operations, and where necessary make recommendations to management relative to occupational safety and health.

1.6  A program of preventive maintenance of facilities and equipment is also viewed as an important aspect of laboratory safety. In this regard, laboratory personnel should be encouraged to report faulty equipment or procedures and to draw attention to preventive maintenance requirements where applicable.

## 2.  General laboratory design criteria

### 2.1  General

This section sets out certain general information and suggested criteria for the design of safe laboratory facilities, intended to be of assistance to those responsible for laboratory design. The guidelines outlined in this chapter are not intended to cover all facets of laboratory design.

2.1.1  Application — It is recommended that these criteria be considered for application or incorporation, where reasonably practicable, into the design of any new or renovated laboratory facilities.

2.1.2  Fire protection — Information on fire protection directives may be obtained from the Office of the Fire Commissioner of Canada in Human Resources Development Canada.* Design aspects related to fire protection or fire safety are excluded from this section except where reference to fire protection directives is required.

2.1.3  Interpretation — Persons requiring assistance or advice concerning the interpretation or application of these criteria on laboratory design, should contact the Buildings Directorate, Architectural and Engineering Services, Public Works and Government Services Canada, Ottawa.

### 2.2  General considerations for design

2.2.1  The design of laboratories, and their furnishings, should aid the implementation of good safety practice yet remain economical. The key to a successful design is a thorough understanding by all design disciplines of the nature and type of work to be conducted in the proposed laboratory.

2.2.2  Benching — Continuous tops are needed in wet laboratory areas to prevent the entry of contaminants and liquids between counter units. In dry laboratory areas, joints are acceptable between benching unit tops but these should be secured and tight. Generally, uninterrupted benching should be provided with a cleansweep for housekeeping and without hazard to glassware caused by service outlets.

2.2.3  Service outlets — Deck-mounted services with outlets installed behind the plane of a raised front are recommended. When services themselves could be hazardous due to heat or pressure (steam, compressed gases), the outlet tips should, if possible, be angled down toward the bench top, and should be equipped with reliable and safe shut-off and operating devices. Controls for services should be easily reached without the need for the operator to lean forward over benching or to pass the hand among glassware, equipment, etc. to reach the control.

2.2.4  Storage — Corrosive chemicals require to be stored on low shelves or cupboards for safety; chemicals of hazardous combination require separate storage.  Extremely strong oxidizers (e.g. perchlorates) or potentially unstable compounds require individual and separate storage.

2.2.5  Floor strength — Since occupancy and use can change quite frequently, introducing appreciable change in floor loading requirements, care should be taken to ensure that the building design is adequate with respect to loading, and that the building structure remains safe after other, changed, operations begin.  Final drawings should show permissible floor loads, which should be carefully observed.

2.2.6  Pressure vessels — Designs and installations should be in accordance with related directives.  All fittings should meet requirements of the inspecting authority.

2.2.7  Environmental directives — Spaces for employee occupancy in laboratories should be maintained at the established comfort level.  Spaces requiring special functional or operational conditions — temperature, humidity, air purity or other — should be served by separate environmental engineering systems.

2.2.8  Fume hoods — Contamination from laboratory work and chemical testing procedures should not be permitted to reach the breathing zones of laboratory personnel.  This can be achieved by restricting the contaminant-producing procedures to an enclosure or hood which is exhausted to the outside.  A laboratory fume hood is a ventilated enclosed work space consisting of side, back and top enclosure panels, a work surface or deck, a work opening called the face, and an exhaust plenum equipped with horizontal adjustable slots for the regulation of air flow distribution.  The work opening may be restricted or may be equipped with operable glass doors for observation and shielding purposes.  In the design of a fume hood and its exhaust system the following factors should be considered:

(a)    effective capture velocities to remove contaminants;
(b)    a balanced air supply;
(c)    even air distribution;
(d)    safe construction materials for hood, ducts and fans;
(e)    hood located away from corridors or doors;
(f)    exhaust air control — air foils, adjustable air slots, baffles and bypass systems;
(g)    exhaust dispersal to atmosphere and/or exhaust air treatment;
(h)    width of hood — as small as possible for maximum safety;
(i)    hazards of specific laboratory operations.

2.2.8.1  The protection that the fume hood gives the operator is influenced by factors such as:

449

(a)    replacement air provisions;
(b)    the proximity of breathing zones to the contaminant emission point;
(c)    the frequency, duration and intensity of contaminant emission;
(d)    room air current disturbances;
(e)    the direct handling and process manipulation required;
(f)    the operator attendance time.

Advice on the degree of hazard associated with specific contaminants may be obtained from regional offices of the Medical Services Branch, Health Canada.

2.2.8.2  Since it is impossible to provide one type of hood for all uses, it is imperative that laboratory managers and laboratory supervisors ensure that hoods are restricted to the uses for which they have been designed. A general purpose fume hood for use in chemical and associated laboratories should be capable of being used for a broad spectrum of material. During its expected working life both the materials used in it and the staff associated with it are likely to change. It is therefore important that any limits imposed on the hood use be clearly posted and maintained.

2.2.8.3  Thermal convection currents within the hood, mechanical agitation and aspirating action by cross currents of air outside the hood can adversely affect safe operation and allow contaminants to enter the breathing zone. Hoods should therefore be kept free of unnecessary apparatus and be operated with the sash at the smallest practical opening. When the hood is not in use, the sash should be closed for maximum safety.

2.2.8.4  All fume hoods, including perchloric acid fume hoods, should be operated at an average face velocity in the range of 0.40 — 0.50 m/s (80-100 fpm) with a typical operating sash opening that is normally, but not invariably, 30 cm (12"). For perchloric acid hoods, the method and materials of construction are of paramount importance.

2.2.8.5  Special-use fume hoods, for example those that contain centrifuges, high-heat loads, ultrasonic equipment, etc., should be evaluated on an individual basis.

2.2.8.6  Each hood should be checked for adequate face velocity. Tests should be performed upon installation, and annually or more frequently as required. Appendix A provides these assessment procedures.

**Note**: While determining an adequate face velocity is necessary to ensure the proper performance of a fume hood, a test result in the recommended range does not, by itself, guarantee containment and elimination of contaminants.

2.2.8.7  Correct sash opening and face velocity should be prominently and permanently displayed on the fume hood.

2.2.8.8  Fume hoods should be constructed of suitable fire and corrosion-resistant materials and with glazing components appropriate to the hazard involved.

450

2.2.8.9 Piped services with outlets inside a fume hood should be remotely controlled from accessible locations outside the hood enclosure. Fittings and installations should meet all applicable codes.

2.2.8.10 The hood exhaust fan motor and all wiring and equipment should bear the approval label of, and be installed in accordance with, the *Canadian Electrical Code* CSA 22.1. Electrical equipment should be of a type suitable to the hazards of the location as defined in the *Canadian Electrical Code*, Section 18.

2.2.9 Biological safety cabinets — The selection, use and maintenance of biological safety cabinets should be in accordance with Health Canada and Medical Research Council of Canada Laboratory Biosafety Guidelines.

2.2.9.1 Biological safety cabinets should be tested when newly installed or moved, and annually thereafter, in accordance with CSA Standard Z316.3-M87 *Biological Containment Cabinets*: Installation and Field Testing.

### 2.3 Laboratory design codes, standards and references

2.3.1 General requirements — The following should be followed in the design and construction of laboratories:

– *National Building Code of Canada;*
– fire protection engineering directives — and other fire prevention or protection directives issued by the Fire Commissioner of Canada;
– all applicable regulations and regulatory guides issued by the Atomic Energy Control Board under the authority of the Atomic Energy Control Act;

   **Note**: The purchase, possession and use of radioactive materials may require a licence issued by the Atomic Energy Control Board (AECB), design approval for a laboratory or device using radioactive material, specific training and a dosimetry program. Advice should be obtained from the AECB.

– all applicable codes issued by the Radiation Protection Bureau of Health Canada;
– Laboratory Biosafety Guidelines issued by Health Canada and the Medical Research Council of Canada;
– Guide to the care and use of experimental animals (Volume 1) issued by the Canadian Council on Animal Care;
– *Hazardous Substances Code* — issued by the National Fire Protection Association, and the Chlorine Institute Inc.;
– all health and safety directives approved by the Treasury Board for application in the Public Service of Canada.

2.3.2 Boilers and pressure vessels — Boilers and pressure vessels used in connection with building systems should comply in all respects with the Treasury Board directive and procedures respecting the safe operation of boiler and pressure vessels.

2.3.3 Illumination — Lighting systems should comply with the *Canada Occupational Safety and Health Regulations*, Part II, Levels of Lighting, made under Part II of the *Canada Labour Code*. Minimum recommended levels of illumination on the task should be as shown in the following chart:

### Table I

### Minimum recommended levels of illumination

| Area/Operation | Footcandles | (lux) |
|---|---|---|
| Reading instruments, gauges, etc., where errors could be the cause of a hazardous condition | 75 | (750) |
| Working with hazardous substances of severe or moderate hazard | 70 | (750) |
| General laboratory work of low hazard: | | |
|     Medium or fine work | 50 | (500) |
|     Rough work | 30 | (300) |
|     Emergency shower locations | 5 | (50) |
|     Emergency lighting | 1 | (10) |

2.3.4 Floors — Floor coverings or finishes should resist liquid penetration, be easy to clean and resistant to slipping.

2.3.5 Doors — Doors should be sufficient in both number and size and so located as to allow quick emergency evacuation. Conspicuous marking on glass doors or panels, and vision panels on free-swinging solid doors, may eliminate certain traffic hazards.

### 2.3.6 Ladders and floor and wall openings

2.3.6.1 Ladder installations should comply with the *Canada Occupational Safety and Health Regulations*, Part II, Building Safety.

2.3.6.2 Floor and wall openings and holes should be guarded as required by the *Canada Occupational Safety and Health Regulations*, Part II, Building Safety.

### 2.3.7 Utilities

2.3.7.1 All piping systems should be clearly marked and available to inspect, with identification in accordance with CSA Standard B53, *Code for Identification of Piping Systems*.

2.3.7.2 Potable water systems should not be endangered by laboratory work. Great care is necessary to ensure that flexible or other temporary connections to potable systems do not prevent the proper function of protection devices such as vacuum breakers and backflow preventers in the permanent system.

2.3.7.3  Central vacuum systems should not be connected to zones which are potential sources of contamination and from which the system could spread the hazard.  Radio-isotope and microbiology areas are such zones; vacuum requirements in these areas should be created by a separate vacuum pump in each room requiring the service.

2.3.7.4  All electrical installations and facilities in a laboratory or workplace should comply with the recommendations of the Canadian Standards Association Standard C22.1 and amendments thereto, and be approved for use in accordance with that directive, for the classification of the hazard in the workplace.

2.3.7.5  Standby electrical power should be available for use in the event of a commercial power failure and should have capacity to supply at least emergency lighting plus any hazardous facilities such as ventilated animal cages, some incinerators and refrigerators, etc.

2.3.8  Emergency equipment — Equipment as indicated hereunder should be provided to deal with emergency situations involving hazardous materials:

(a)  Emergency shower and/or eye wash equipment should be provided wherever there is a significant exposure to hazardous materials and a risk of skin or eye injury due to accidental splashes of such materials.  The temperature of any connected water supply should not exceed 100°F (38°C).
(b)  Emergency power facilities should be provided wherever, in hazardous areas, a failure of power supply would cause dangerous conditions.

2.3.9  Radiation emitting equipment — The design, construction, functioning, installation, maintenance, operation and use of x-ray, laser, microwave, ultrasound and ultraviolet radiation emitting equipment must comply with recommendations, safety codes and regulations issued by the Radiation Protection Bureau of Health Canada (see Chapter 4).

2.3.10  **Laboratory furniture**

2.3.10.1  All laboratory benches, tables, and cupboards should be secure against upset.  Stools, if not fixed, should be solid and stable and adequate knee-space should be provided.

2.3.10.2  Bench widths should be such that utility controls located at the back can be reached safely.  Gangways between benches should be wide enough to permit safe movement in normal working conditions and quick escape in an emergency.

2.3.10.3  Laboratory furnishings should be constructed of materials to suit the functional needs of the laboratory.

2.3.11  Equipment safeguards — To the extent that it is reasonably practicable, all machines purchased for laboratory use should be so designed and constructed as to be safe. If an apparatus is a significant work hazard it should be fitted with a guard that will effectively prevent injury to any person.

2.3.12 Protective shields — Suitable shields and barricades should be provided to protect laboratory personnel from the hazards of explosion, rupture of apparatus and systems from over-pressure, implosion due to vacuum, sprays or emission of toxic or corrosive materials, or flash ignition of escaping vapours.

### 2.3.13 Safety containers

2.3.13.1 Designs of chemical laboratories should incorporate provision of functional space and facilities for temporary storage and use of appropriate safety containers to facilitate safe disposal of waste liquids or materials for which discharge to a building drain is not acceptable (e.g. solvents, radioactive materials).

2.3.14 Static grounding — Static bonding and grounding should be provided wherever flammable liquids are dispensed from metal containers of over 5-gallon (22.5 liter) capacity.

### 2.3.15 Warning signs

2.3.15.1 Particular systems and facilities designed into a laboratory may require warning signs, or colour symbols, to represent hazard (e.g. chemical stores, radioactive and microbial filtration, perchloric exhaust components in penthouses and on a roof).

2.3.15.2 The design of any warning sign should stipulate the location and angle of disposition of the sign for clearest display, and call for such signs to be painted according to the established colour coding.

### 2.3.16 Biological considerations for design

2.3.16.1 Laboratories handling pathogenic micro-organisms should be designed in accordance with Health Canada and Medical Research Council of Canada Laboratory Biosafety Guidelines.

### 2.3.17 Design of animal facilities

2.3.17.1 The laboratory animal area typically includes rooms for inoculation and autopsy, as well as rooms for holding infected animals. In all cases, safe working conditions for personnel should be provided to prevent any potential hazard possible from undesired animal cross-infection. Design of facilities should be in accordance with guidelines issued by the Canadian Council on Animal Care.

### 2.4 References

*Handbook of Laboratory Safety*,
The Chemical Rubber Company,
18901 Cranwood Parkway,
Cleveland, Ohio 44128

*Laboratory Biosafety Guidelines*,
Office of Biosafety,
Laboratory Centre for Disease Control,
Health Canada,
Ottawa, Ontario
Canada
K1A 0L2
(ISBN 0-662-17695-2)

*Guide to the care and use of experimental animals (Volume 1)*,
Canadian Council on Animal Care,
1000-151 Slater Street,
Ottawa, Ontario
K1P 5H3

*Microbial Contamination Control Facilities*,
Van Nostrand Reinhold Publishing Company,
1410 Birchmount Road,
Scarborough, Ontario

*Accident Prevention Manual for Industrial Operations*,
National Safety Council,
425 North Michigan Avenue,
Chicago, Illinois 60611

*Industrial Ventilation Manual*,
American Conference of Governmental and Industrial Hygienists,
1014 East Broadway,
Cincinnati, Ohio 45202

*Threshold Limit Values*, and Amendments,
American Conference of Governmental and Industrial Hygienists.

*Industrial Hygiene Practices Guide for Laboratory Hood Ventilation*,
American Conference of Governmental and Industrial Hygienists.

*The Tuberculosis Diagnostic Laboratory*,
Canadian Journal of Public Health,
Canadian Public Health Association,
125 Yonge Street,
Toronto, Ontario
M5C 1W5

*Chemical Safety Data Sheets*,
Manufacturing Chemists Association,
1825 Connecticut Avenue, N.W.,
Washington, D.C. 20009

Office of the Fire Commissioner of Canada,
Technical Information Bulletins, Human Resources Development Canada.

*Canadian Electrical Code*, Part I, Safety Standard For Electrical Installations
(CSA C22-1-1986),
Canadian Standards Association,
178 Rexdale Blvd.,
Rexdale, Ontario
M9W 1R3

*ASHRAE Handbooks*,
American Society of Heating, Refrigerating and Air Conditioning Engineers, Inc.,
1791 Tullie Circle, N.E.,
Atlanta, Georgia 30329

## 3. General safety practices

This section sets out general safety practices for several areas of principal concern. It does not attempt to list all of the precautions and safety measures which may be required. More detailed information concerning these and other specific procedures and guidelines may be obtained from the appropriate references listed in this chapter.

### 3.1 Cryogenics

#### 3.1.1 General

3.1.1.1 A cryogen is any refrigerant used to obtain a temperature lower than -58°F (-50°C). Potential hazards occur because cryogenic fluids are extremely cold, and in some processes very small amounts of liquids are converted into large volumes of gas. Liquids which boil at very low temperatures will condense oxygen from the air, which in turn creates an explosion or combustion hazard.

3.1.1.2 The user of cryogenic fluids should have a thorough knowledge of the characteristics of the gas at temperatures and pressures being used, and the appropriate safety precautions for the handling of the individual liquid. Particularly, users should know how to recognize and eliminate leaks, and what to do in the event of an explosion or implosion.

3.1.1.3 Only authorized and qualified personnel should have access to the storage area for cryogenic fluids. Gaseous or liquid oxygen should be kept in its own particular area, without any other gases being allowed except gaseous nitrogen and gaseous carbon dioxide. Liquid nitrogen should not be stored with helium, hydrogen or oxygen. Walls and floors of such storage areas should be concrete or reinforced concrete.

3.1.2 Hazards — Danger of fire and explosion exists with escaping cryogens such as oxygen and hydrogen. The danger is such that even materials normally non-combustible will ignite if allowed to become coated with an oxygen-rich condensate. Thermal shock to containers, or a gas pressure greater than the containers are designed to hold, may cause explosions. Implosions may result from pressures produced by cryo-pumping which are nearly equal to the existing atmospheric pressure on the equipment being used, unless this equipment is designed to withstand such pressure changes.

Structural or other material coming into contact with cryogenic fluids may become combustible, explosive, or subject to failure from strain or impact due to altered physical characteristics. Direct or indirect uninsulated contact with cryogenic fluids causes cold burns (frostbite); delicate surfaces, such as eyes, can be damaged by a brief exposure. With the exception of oxygen, rapid expansion of cryogenic fluids results in an oxygen deficient atmosphere if the immediate environment is inadequately ventilated; this can lead to asphyxiation.

### 3.1.3 **Special safety precautions**

3.1.3.1 Warning signs should be posted where cryogens are stored or being used, and the name of the cryogens should be shown.

3.1.3.2 Proper ventilation where cryogens are stored or being used is required, to reduce the danger of explosion, fire or asphyxiation.

3.1.3.3 Vessels containing cryogens should be thermally isolated from sources of heat. When not in use, containers used for the transport of cryogenic fluids should be secured with chains or straps to a substantial support, such as a wall or a fixed bench, to protect against upsets.

3.1.3.4 All vessels containing cryogens should be provided with a vent or other approved safety device which permits the escape of excess pressure and vapors.

3.1.3.5 Containers should be filled only with the liquids that they were designed to hold. Each container should be labelled as to which cryogenic fluid it contains.

3.1.3.6 When pouring a liquid cryogen into a Dewar flask or other container, a metal funnel should be used, and a face shield and insulated gloves should be worn.

3.1.3.7 Personnel should always stand clear of boiling or splashing cryogen, and perform operations slowly to minimize any boiling or splashing.

3.1.3.8 Vessels designed and constructed as containers for cryogens should not be welded or heated while the vessel contains a cryogen.

3.1.3.9 Liquid nitrogen heavily contaminated with oxygen should be handled with precautions applicable to liquid oxygen. The appearance of a blue tint in liquid nitrogen is a direct indication of oxygen contamination.

### 3.1.4 **Personal protection**

3.1.4.1 Eyes and face should be protected with a face shield whenever there is a danger of injury from physical or chemical agents.

3.1.4.2 Gloves should be worn when handling anything that is or may have been in contact with a cryogenic liquid. Asbestos gloves are preferable, but leather gloves may be used. The gloves should fit loosely, so that they can be removed quickly (thrown off) if liquid spills or splashes into them

457

3.1.4.3 A knee length laboratory coat with cuffless long sleeves, or a full length apron of non-porous material which fastens at the back, should be worn. Coats and aprons should have neither pockets nor cuffs.

3.1.4.4 Boots with tops sufficiently high to be covered by the trouser leg (which should be cuffless) should be worn.

3.1.4.5 An appropriate eye wash fountain or eye wash bottle and safety shower should be available.

3.1.4.6 Watches, rings, bracelets or other jewellery should not be worn by personnel handling cryogens.

### 3.1.5 Cold traps

3.1.5.1 Cold traps improperly employed can impair accuracy, destroy instrumentation and systems, and be a physical hazard. In addition, the slush mixtures frequently used in cold traps should be handled with care, because many are toxic and present explosive hazards that are not necessarily referred to in the literature.

3.1.5.2 Operations should always be performed slowly, to minimize boiling and splashing, when charging a warm condenser or inserting objects into a cryogenic liquid.

3.1.5.3 If liquid nitrogen is the coolant, the trap should be charged only after the system is pumped down. This is because liquid air containing oxygen can condense in the trap, and a considerable amount of the liquid oxygen in it·may also condense, creating a major hazard.

3.1.5.4 The trap should be suitably vented or exhausted, if the cooling bath is removed, so that any condensate which will then evaporate from the trap will not pressurize the system.

### 3.1.6 References

*Handbook of Laboratory Safety*,
Chemical Rubber Company,
18901 Cranwood Parkway,
Cleveland, Ohio 44128

*Liquefied Atmospheric Gases*,
(Precautions and Safe Practices for Handling),
February 1971, Form 9888-L,
Union Carbide Corporation,
123 Eglington Street East,
Toronto, Ontario

*Handling Cryogenic Fluids*,
Neary, R.M. (1960),
Union Carbide Corporation

*Liquefied Helium*,
(Precautions and Safe Practices for Handling, 1962),
Union Carbide Corporation

*Liquefied Nitrogen*,
(Technical Data, 1960),
Union Carbide Corporation

*Gaseous Oxygen*,
National Safety News,
Data Sheet D-472, December 1958,
National Safety Council,
425 North Michigan Avenue,
Chicago, Illinois 60611

*Handbook of Compressed Gases*,
Compressed Gas Association,
Van Nostrand Publishing Company,
1410 Birchmount Road,
Scarborough, Ontario

*Treasury Board Manual*, "Occupational safety and health" volume, Dangerous substances directive (Chapter 2-2), and Personal protective equipment directive, (Chapter 2-14),
Treasury Board of Canada

*Canada Occupational Safety and Health Regulations*, Part X and Part XII, Human Resources Development Canada.

## 3.2 Compressed gas

3.2.1 General — A gas having a pressure in the container in excess of 40 psia (280 kPa) at room temperature, and a flammable liquid having a Reid vapor pressure exceeding 40 psia (280 kPa) at 100°F (38°C), are classified as compressed gases. They must be handled with care in all phases: storage, transportation, connection, use and disposal. All personnel who handle, manipulate or work in the presence of compressed gases should be made aware of the dangers of exploding cylinders, and the projectile-like behaviour of a cylinder under circumstances of a sudden pressure release. All personnel should be made aware of the physical, chemical, and toxic properties of any gas used in the laboratory operation. Contracts and purchase orders should show the required CGA connection number for the cylinder to be supplied.

3.2.2 Cylinder contents — Each cylinder should be logged in and out of the laboratory operation, recording the serial number stamped on the cylinder and the contents as received. Under no circumstances should attempts be made to alter the chemical composition of the cylinder contents by introduction of impurities. If the contents of a cylinder are not definitely known, the cylinder should be clearly labelled "contents unknown" and the supplier notified for disposal. Coding of cylinders varies among suppliers. Depending on the grade of the product contents, some suppliers choose to code by purity rather than chemical content. Colour codes may not be reliable indicators of cylinder contents.

3.2.3 Properties of contents — Each person should be aware of the chemical, physical and physiological properties of gases used. Charts, listing characteristics of each gas and toxic thresholds, should be posted in a prominent location for reference and emergency situations. All toxic gases should be either used in, or vented into, a fume hood. All fume hoods so used should be marked accordingly, specifying the hazard. Cylinders should not be exposed to temperatures in excess of 120°F (50°C). Cylinders containing flammable or oxidizing gases should be used in locations having good ventilation.

3.2.4 Cylinder handling — Misuse, abuse or mishandling of compressed gas cylinders may result in serious accidents. Observance of the following will help reduce the hazards:

3.2.4.1 Cylinders should be protected from cuts or abrasions and not allowed to drop or strike each other violently.

3.2.4.2 Cylinders weighing in excess of 40 lbs (18 kg) (total) should be transported by cart, properly retained in a vertical position.

3.2.4.3 Cylinders may be rolled on the bottom edge for short distances but should not be dragged.

3.2.4.4 Cylinders should not be transported without the valve protection cap securely in place.

3.2.4.5 Always assume cylinders to be full and handle them accordingly.

3.2.4.6 Oil and grease should never be permitted to come in contact with oxygen cylinder valves, regulators, hoses or associated equipment, nor should combustible substances be used as lubricants. The operator should ensure that there is no oil or grease on the hands, gloves or clothing.

### 3.2.5 Cylinder storage

3.2.5.1 Cylinders should be stored in a secure, dry, well-ventilated area, clear of exit routes and fire exits, heat or ignition sources, and with valve protection caps securely in place.

3.2.5.2 Segregated areas should be defined according to cylinder content: flammable, oxidizing and inert.

3.2.5.3 Indoor storage areas for oxidizing gases should be separated from flammable gases and highly combustible materials by at least 20 feet (6.0 m), and by an approved fire-resistant partition.

3.2.5.4 Indoor ventilation should be provided at both floor and ceiling levels and conform to fire regulations.

3.2.5.5 Cylinders containing gases such as acetylene, liquified propane and liquified carbon dioxide should be stored upright.

3.2.5.6 Storage areas should be fitted with cylinder racks securely anchored to the wall at a height appropriate for the cylinder to be stored.

3.2.5.7 Cylinders should be individually secured to the storage rack, not more than two rows deep, using chains, straps or bars.

3.2.5.8 Full and empty cylinders should be stored separately, with the latter clearly identified as such.

3.2.6 Leaks — Leaking cylinders are hazardous and wasteful.

3.2.6.1 A contingency plan for leaking cylinders should be defined and known to all laboratory personnel in areas where compressed gases are used.

3.2.6.2 Where available, the faulty cylinder should be placed in a walk-in fume hood, the fume hood identified, and the supplier notified. Alternatively, the cylinder should be moved out of doors to a secure area to await disposal by supplier.

3.2.6.3 Some stem valves used on cylinders for low molecular weight gases such as hydrogen will leak when fully opened. Under no circumstances should any adjustment to the stem packing nut or pressure relief safety nut be even considered. Such actions are extremely hazardous, and are the responsibility of the supplier alone.

3.2.6.4 Leaks resulting from improper plumbing or worn fittings should be identified, using approved liquids or detection instrumentation. If wear is the reason for the problem, the components should be replaced.

3.2.6.5 Fittings should not be tightened beyond the manufacturer's specifications.

3.2.7 Regulators — Each person expected to use compressed gases should be instructed on the proper installation and use of regulators.

3.2.7.1 Cylinders of compressed gas should only be connected to regulators specified for use with the contents of the cylinder. C.G.A. regulations should be adhered to at all times.

3.2.7.2 The seat of the cylinder stem valve should be cleaned before coupling with the regulator. The stem valve should not be used to blow out the regulator fitting seat.

3.2.7.3 The regulator should be closed before coupling. Do not over-tighten the coupling nut from the regulator stem.

3.2.7.4 Once the regulator is installed, and before use or further connection to apparatus, the regulator-to-cylinder connections should be checked for leaks.

3.2.7.5 In addition to the use of liquids and detection instruments, the regulator can be used to detect leaks. Open the cylinder stem valve and note the pressure. Close the stem valve and wait 15 minutes. There will have been no pressure drop if the regulator/cylinder connection is leak-free.

3.2.7.6 A similar procedure can be applied to other portions of the gas plumbing to verify the integrity of the system. Should a leak be detected, close the cylinder stem valve.

Breathing apparatus — Laboratories handling toxic gases must have appropriate personal protective equipment on hand in case of leaks or accidents. Self-contained breathing apparatus is preferred, but respirators with suitable fresh canisters may be acceptable, bearing in mind the nature of the gas and the possible concentrations.

### 3.2.8 References

*Handbook of Laboratory Safety*,
The Chemical Rubber Company,
18901 Cranwood Parkway,
Cleveland, Ohio 44128

*Handbook of Compressed Gases*,
Compressed Gas Association,
Van Nostrand Reinhold Publishing Company,
1410 Birchmount Road,
Scarborough, Ontario

*The Use and Handling of Compressed Gases*,
Bulletin 259, revised 1969,
U.S. Department of Labour,
U.S. Government Printing Office,
Washington, D.C. 20402

*Production, Storage and Handling of Liquid Natural Gas*, (Z-276),
Canadian Standards Association,
178 Rexdale Blvd.,
Rexdale, Ontario

*Matheson Gas Data Book*,
Matheson of Canada Limited,
Whitby, Ontario

### 3.3 Glass

3.3.1 General — Directive safe practices, including the wearing of eye and face protection, are necessary in the handling and use of glass to prevent injuries or illnesses from explosions, ruptures from overpressure, implosion due to vacuum, spills, sprays or emission of toxic or corrosive materials, or flash ignition of escaping vapours.

3.3.2 Disposal — Broken or cracked glass should not be placed in waste bins designed to receive paper and other laboratory waste. A separate metal container, appropriately labelled, should be available for such use in each laboratory.

3.3.3 Storage — Glass and glassware should be stored on shelves no higher than a person of average height can reach from floor level. Delicate pieces should be protected by storing in cartons clearly marked for easy identification. No item of glassware should protrude over the edge of shelving.

3.3.4 Selection of glassware — When selecting a piece of glassware for use, care should be taken to ensure that it is designed for the type of work planned. For pressures even slightly above normal, pressure bottles should be specifically chosen, and vacuum flasks should be used for filtration with the aid of suction. Types of glass rods and tubing can be identified by refraction and comparison with approved directives. Where caustics are used, glass-to-teflon connections and stoppers (or suitable alternatives) may reduce hazards, especially in the reduction of seized joinings.

3.3.5 Setting up apparatus — Apparatus (a combination of two or more units) should be set up with units adequately supported by clamps on stands. Laminated safety-glass protective shields should be placed around the apparatus to protect workers on both sides of the bench, if necessary.

3.3.6 Cutting tubing and rods — The ends of any glass piece cut in the laboratory should be squared and fire-polished prior to its employment. Protective hand covering should be used when working with glass rods and tubing.

3.3.7 Glass and rubber or cork connections — The correct bore should be selected, so that the insertion can be made without undue force. The glass and stopper should be wet (water or glycerine). Appropriate hand and eye protection should be used. Extreme care should be exercised when removing a glass rod or tube that is stuck to a rubber stopper.

3.3.8 Heating of glassware — Care should be taken to ensure that the type of glass to be used will withstand the heat to be applied. The heat source, and the method of heating to be used, should be selected carefully in relation to the liquid or material to be heated.

3.3.9 Glassware under pressure or vacuum — Heated pressure vessels should be shielded in case an accident occurs. Pressure should not be applied internally to a liquid in glassware to expel the contents. Personnel should be protected against implosion of evacuated glassware, using guards of wire screen or perforated metal.

3.3.10 Seized glass-to-glass surfaces — During attempts to separate, extreme care and patience should be exercised and hands must be protected. Glass-to-teflon or other suitable alternatives may reduce hazards when caustics are used, as noted in 3.3.4.

3.3.11  Cleaning glassware — Before cleaning glassware the user should ensure that each piece is free of any material that might present a hazard.  The use of mild cleaners is preferred to strong acids or caustics.  Should the latter be used, the glassware should be well rinsed and dried afterwards.  Hand and eye protection should be stressed, and procedures devised which will reduce the hazards from possible breakage.

3.3.12  Transporting glassware — To reduce the hazards from breakage during transport, special chemical-resistant metal or plastic containers of adequate size should be used to transport all bottles containing acids, alkalines, or other corrosive or flammable liquids.  Desiccants under vacuum may be transported in a wooden box or metal shield; in such instances appropriate carrying tongs should be used when handling breakers and other glassware.

3.3.13  Labelling bottles — All reagent bottles and other containers of laboratory chemicals should be clearly labelled and dated.  A coat of clear lacquer applied to the label will protect it.

3.3.14  Ullage in bottles — Bottles should be filled to not more than three-fourths of their capacity at room temperature.

3.3.15  Special hazards — The handling of hazardous products in glass containers should be controlled by local laboratory directives.

### 3.3.16  References

*Guide for Safety in the Chemical Laboratory*,
Manufacturing Chemists Association,
1825 Connecticut Avenue, N.W.,
Washington, D.C.  20009

*Laboratory Glassware*,
Safety Education Data Sheet No. 23,
National Safety Council,
425 North Michigan Avenue,
Chicago, Illinois  60611

*Bottles and Broken Glass*,
Safety Education Data Sheet No. 355,
National Safety Council

*Safety Measures in Chemical Laboratories*,
Third Edition — 1964,
National Chemical Laboratory,
Teddington, Middlesex, England

*Handbook of Laboratory Safety*,
The Chemical Rubber Company,
18901 Cranwood Parkway,
Cleveland, Ohio  44128

*Treasury Board Manual,* "Occupational Safety and health" volume, Dangerous substances directive, (Chapter 2-2), and Personal protective equipment directive, (Chapter 2-14), Treasury Board of Canada.

*Canada Occupational Safety and Health Regulations,* Part X and Part XII, Human Resources Development Canada.

### 3.4 Instruments and other equipment

3.4.1 General — All instruments, and associated electrical equipment, should be inspected periodically for defects and replaced as necessary. The following review of the more common equipment will aid in determining which safety items to check on various types of equipment. Where good industrial laboratory practice prescribes it, personal protective equipment including hand, eye and face protection should be worn, as is appropriate to the hazard associated with the use of the various types of equipment.

3.4.2 Autoclaves — All autoclaves should be provided with interlocks to prevent the opening of the charging door until all pressure has been relieved. High pressure types should have integral explosion protection and control for safe operation. Autoclaves should be tested after installation, and on a regular basis, by the use of biological indicators.

3.4.3 Calorimeter bombs — Adequate shielding as a protection against explosion should be used.

3.4.4 Centrifuges — Centrifuges should be of double-walled construction to prevent fly-aways, and equipped with a disconnect switch on the lid. The centrifuge should be located where vibration will not cause items to fall off nearby shelves. Exhaust ventilation should be provided, especially if flammables are to be centrifuged.

3.4.5 Chromatography equipment — Insulation for radiation, and for ventilation to contain and remove hazardous vapor, should be provided.

3.4.6 Distillation apparatus — Fail-safe devices should be used to guard against possible fluctuations or failure in water pressure and electrical power.

3.4.7 Fraction collectors — Fraction collectors should be isolated from sources of ignition. Adequate ventilation is necessary and construction should be explosion proof.

3.4.8 Microtomes — A lock to prevent inadvertent operation, and a guard to protect the operator against the cutters, should be provided.

3.4.9 Paraffin dispensers and vacuum infiltrators — An automatic over-temperature shut-off should be in series with the thermostatic control.

3.4.10 Ovens — Ovens used in service with explosive materials should be equipped with blow-out panels or magnetic latches which open at pressures slightly above one atmosphere. Forced draft ventilation, inert gas purging, exhaust ventilation or other appropriate means should be used, to prevent a build-up of explosive concentrations of vapour in ovens. Reliable, well-maintained and accurately calibrated thermostatic controls, with units clearly marked, should be used to prevent excessive heating. All controls should be designed to be fail safe. Where an oven is supported by a table or counter, the counter or table top should be constructed of non-combustible material, or adequate ventilation should be provided between the supporting surface and the bottom of the oven.

3.4.11 Electrical and electronic instruments and equipment — These should be inspected periodically for hazardous leakage currents, and repaired or replaced as necessary. The use of ground-fault circuit interruptors should be considered where electrical/electronic equipment is used in locations which increase the possibility of shock hazard.

3.4.12 Mercury vapour — Mercury diffusion pumps, and any other equipment that produces mercury vapour, should be provided with exhaust ventilation.

## 3.5 Storage of chemicals

3.5.1 General — Due to the wide range of chemicals and materials used in laboratories, good storage practice and reliable current inventory control is important. Chemicals should not be stored in alphabetical order because of the danger of incompatibility. Neglect of the physical and chemical properties of stored materials may result in fires, explosions, emission of toxic gases, vapours, dusts or radiation, and various combinations of these effects. Care should be taken to provide separate storage or other special conditions, where required, for certain materials and chemicals including pesticides and herbicides. Since refrigerators are often used for the storage of highly volatile or reactive materials, it is essential that all controls, switches, etc. be explosion proof. A current inventory list of all chemicals in stock should be maintained, and each item should be identified as to its hazard in accordance with NFPA booklet, (National Fire Protection Association), 704.

3.5.2 Flammable materials — Such materials should be stored in places that are cool and adequately ventilated. Continued liaison should be maintained with the local fire prevention authorities regarding the type and disposition of such materials.

3.5.3 Oxidizing materials — These are not usually combustible, but will produce oxygen for accelerated burning of combustible material, and should be stored separately. Examples of classes of such compounds are organic and inorganic peroxides, oxides, permanganates, perchlorates, and chlorates.

3.5.4 Water sensitive materials — These are materials which react with water, steam or water solutions; examples are lithium, sodium, potassium, acid anhydrides, and concentrated acids or alkalies. Because many of these materials are flammable, it is essential that the advice of the Office of the Fire Commissioner of Canada be obtained regarding the installation of automatic sprinkler systems in the storage area which houses them.

3.5.5 Acid sensitive materials — Such materials react with acid and acid fumes; examples are lithium, sodium, arsenic, selenium and cyanides. Acids should not be stored close to these materials.

3.5.6 Toxic hazards — These are materials which under either normal or disaster conditions, or both, can be dangerous, e.g. carbon tetrachloride and materials which are toxic because of their radioactivity. In general, materials which are toxic as stored, or which can decompose into toxic components due to contact with heat, moisture, acids or acid fumes, should be stored in a cool, well-ventilated place, out of direct sunlight, away from areas of high fire hazard. Examples of toxic materials are mercury, benzene, carbon tetrachloride and other hydro-carbons, organic nitro compounds, and organic phosphate compounds.

**Caution**: Appropriate respiratory protection and training should be provided where dangerous levels of noxious gases or vapours may be given off or created by stored chemicals.

### 3.5.7 **References**

*Handbook of Compressed Gases*,
Compressed Gas Association,
Van Nostrand Reinhold Publishing Company,
1410 Birchmount Road,
Scarborough, Ontario

*Construction Safety*,
Construction Safety Association of Ontario,
74 Victoria Street,
Toronto, Ontario

*Accident Prevention Manual for Industrial Operations*,
National Safety Council,
Chicago, Illinois 60611

*Dangerous Properties of Industrial Materials*,
Van Nostrand Reinhold Publishing Company,
1410 Birchmount Road,
Scarborough, Ontario

*Handbook of Laboratory Safety*,
The Chemical Rubber Company,
18901 Connecticut Avenue, N.W.,
Cleveland, Ohio 44128

*Chemical Safety Data Sheets*,
Manufacturing Chemists Association,
1825 Connecticut Avenue, N.W.,
Washington, D.C. 20009

*Treasury Board Manual*, "Occupational safety and health" volume,
Dangerous substances directive (Chapter 2-2),
Treasury Board of Canada.

*Canada Occupational Safety and Health Regulations,*
Part X,
Human Resources Development Canada.

*Fire Protection Engineering Standards*,
Office of the Fire Commissioner of Canada,
Human Resources Development Canada.

## 3.6 Environmental chambers

3.6.1 The hazards of environmental chambers are related to exposure to heat and cold (heat exhaustion, heat stroke, frostbite and skin burns, eye damage, respiratory tract damage), and exposure to toxic gases and fumes arising from test equipment within the chamber or from escaping refrigeration gases.

3.6.2 Safety precautions — The following safety precautions should be followed with respect to environmental chambers.

3.6.2.1 Maintain outside surveillance of personnel working in environmental chambers, particularly those working alone. Continuous two-way communication should be provided where practicable.

3.6.2.2 Personnel should be advised of temperature ranges before entering, and provided with appropriate personal protective clothing and equipment.

3.6.2.3 An outside warning light should indicate when someone is in the chamber.

3.6.2.4 An emergency alarm system, audio and visual, that can be triggered off either inside or outside the chamber, should be available and tested periodically.

3.6.2.5 Gasket heaters should be installed where required, and used to prevent doors from freezing shut in low temperature rooms.

3.6.2.6 Safety devices such as exits and alarms, break-out tools, fire emergency equipment, emergency resuscitation, and first-aid equipment should be provided, and personnel should be trained in their use.

3.6.2.7 Adequate ventilation should be provided.

### 3.6.3 References

*Cold Room Testing of Gasoline and Diesel Engines*,
Data Sheet No. 465 of 1958,
National Safety Council,
425 North Michigan Avenue,
Chicago, Illinois 60611

*Engineering Environmental Simulation Facilities*, by T.R. Ringer,
National Research Council of Canada,
Ottawa, Ontario
K1A 0R6

*The New Canadian Laboratory for Arctic Testing*, by J.L. Orr and
D.G. Henshaw,
National Research Council of Canada

## 4. Radiation

**Note**: The purchase, possesion, use, transportation and disposal of radioactive
materials and radioisotopes is subject to the provisions of the Atomic Energy
Control Act and Regulations. Information and advice should be obtained from
the Atomic Energy Control Board.

4.1 This section sets out certain requirements concerning the design, operation
and maintenance of laboratory and other facilities involved with the use of
radioactive materials, X-ray emitting equipment, microwave, and ultrasonic and
laser radiating devices. (Refer also to Chapter 2.)

4.2 All new facilities, and all modifications and additions to existing facilities,
should meet the specifications described in the following documents and their
subsequent amendments, which have been produced by the Radiation
Protection Bureau, Health Canada:

*Safety code 6*: Recommended safety procedures for the installation and use
of radiofrequency and microwave devices in the frequency range 10MHz —
300GHz.
— 79 — EHD — 30

*Safety code 14*: Policy respecting the use of radionuclides in humans.
— 79 — EHD — 44

*Safety code 20A*: X-ray equipment in medical diagnosis Part A:
Recommended safety procedures for installation and use.
— 80 — EHD — 65

*Safety code 21*: Recommended safety procedures for the selection,
installation and use of baggage inspection X-ray equipment.
— 78 — EHD — 20

*Safety code 22*: Radiation protection in dental practice recommended safety
procedures for installation and use of dental X-ray equipment.
— 80 — EHD — 66

*Safety code 23*: Guidelines for the safe use of ultrasound Part I: Medical and
paramedical applications.
— 80 — EHD — 59

**Safety code 24**: Guidelines for the safe use of ultrasound Part II: Industrial and commercial applications.

— 80 — EHD — 60

**Safety code 25**: Shortwave diathermy Guidelines for limited radiofrequency exposure.

— 80 — EHD — 98

**Note**: The foregoing documents do not cover all conceivable eventualities. Where assistance is required in interpretation or measurement, design or working techniques, departments should consult the Radiation Protection Bureau, Health Canada in Ottawa.

### 4.3 X-ray equipment and facilities

Recommendations concerning the installation facilities, shielding and mode of use of X-ray equipment, are detailed in publications 78-EHD-30, 80-EHD-66 and 80-EHD-65 with data included to aid the design of facilities of an acceptable directive. The X-ray equipment itself should conform to at least the minimum directives of design, construction and functioning detailed in regulations proclaimed under the *Radiation Emitting Devices Act*.

### 4.4 Microwave, radiofrequency, ultrasound, laser and electro-optic equipment and facilities

Publication 79-EHD-30 provides requirements for microwave and radiofrequency equipment radiating >10MHz, facilities and mode of use, similar in scope to those of item 4.3. Recommendations regarding the safe use of ultrasound, shortwave diathermy and demonstration laser devices are contained in publications 80-EHD-59, 80-EHD-60, 80-EHD-90 and 77-EHD-5. In addition, the following devices should conform with regulations proclaimed under the *Radiation Emitting Devices Act*, specifying minimum directives of design, construction and functioning: — microwave ovens, laser scanners, ultraviolet sunlamps, demonstration laser devices, high intensity mercury vapour discharge lamps, ultrasound therapy devices.

### 4.5 Safety codes

The safety codes are constantly being updated and current information on their availability can be requested from the Radiation Protection Bureau, Health Canada.

## 5. Laboratory work in microbiology

Basic safety practices, as well as safety practices related to specific risk groups and containment levels, should conform with Health Canada and Medical Research Council of Canada Laboratory Biosafety Guidelines.

## 6. Control, handling, and disposal of laboratory wastes

### 6.1 Procedures and regulations

6.1.1  All waste should be controlled, handled and disposed of in a manner which will not cause injury to employees or the public, or damage to property, and in compliance with applicable municipal, provincial and federal regulations or requirements.

6.1.2  Any procedures, directives, regulations or directives issued by a department or other authority having specific jurisdiction in respect to the control, handling or disposal of laboratory wastes (Environment Canada, Atomic Energy Control Board, etc.) shall take precedence over these guidelines.

6.1.3  Practices for disposing of infectious or contaminated biomedical waste should conform with Health Canada and Medical Research Council of Canada Laboratory Biosafety Guidelines.

### 6.2  Identification of disposals

6.2.1  The level of hazard of disposable materials should be identified on a label in four categories of health, fire, reactivity and environment, by reference to Environment Canada *Code of Good Practice for Management of Hazardous and Toxic Waste at Federal Establishments*. Abbreviated definitions of the degree of hazard in each category follow:

#### Health

4.  Short exposure may cause death or major injury.
3.  Prolonged or repeated exposure may cause serious injury.
2.  Concentrations may be toxic.
1.  No known health hazard.

#### Fire

4.  Very flammable gases or volatile liquids.
3.  Liquids and solids which will burn readily under normal conditions.
2.  Substances which must be heated before ignition can occur.
1.  Substances which will not readily burn.

#### Reactivity (stability)

4.  Readily detonates or explodes.
3.  Can detonate or explode but requires strong initiating force or heating under confinement.
2.  Mild reaction, unlikely to be hazardous.
1.  Normally stable.

**Environment**

4. Substances which cause major residual damage.
3. Substances which could cause serious damage.
2. Intense or continuous application could cause residual damage.
1. No environmental hazard.

6.2.2 When labelling a material with respect to hazard, consideration should be given to hazards arising from contact with other substances during disposal, and hazards produced by the disposal procedure. The material should be labelled according to the highest hazard which might be encountered from it during disposal.

## 6.3 Storage and disposal

6.3.1 The collection and segregation for disposal of all waste originating within a department's laboratories, offices, and workshops, and the disposal of all unidentifiable waste, remains the responsibility of that department, through the person in charge.

6.3.2 The individual scientist or laboratory technician is responsible for rendering safe what are considered hazardous materials before placing them in the collection area for pickup or disposal.

6.3.3 When explosive or poisonous materials are synthesized in a laboratory, the product should be identified prior to disposal.

6.3.4 Where the disposal of wastes presents special problems (e.g. emission of poisonous gases when being burned), detailed procedures and instructions on their disposal should be specified for the person responsible for the actual disposal.

6.3.5 Waste materials should not be accumulated in laboratories or storage areas. Particular attention should be given to those materials that tend to develop explosive properties over a period of time, and to those materials bearing a date beyond which the material should not be retained.

6.3.6 Combustible waste — This waste should be kept in a storage locker which is not adjacent to buildings. The storage locker should be of fire resistant material, well ventilated, and marked as follows "Flammable Material — Danger — Keep Away", and/or in accordance with applicable directives provided by the Office of the Fire Commissioner of Canada. The storage lockers should be equipped with suitable locks.

6.3.7 Solid waste — This is best divided into three classes:

(a) Glass
(b) Combustibles
(c) Non-combustibles (excluding glass)

Suitably lettered and colour-coded containers are a convenient method of providing receptacles for solid waste. However, care should be taken not to introduce heated or unstable materials without removing the hazard, i.e. they should be cooled or decomposed.

6.3.8 Venting of gaseous waste — The emptying of gas cylinders can be very hazardous, and in most cases the advice of a scientist, a laboratory technician or a safety officer should be sought beforehand. In venting gaseous waste products from a reaction, arrangements should be made for the gas either to enter the fume extraction system or to be led directly outside the building. Care should be taken to ensure that the system can deal with the waste products, and produce an effluent that is toxicologically acceptable.

6.3.9 Liquid wastes — Materials immiscible with water, flammable liquids or solutions containing cyanides and chromates, should never be discarded into drains or ditches. Incompatible materials should be kept separately and disposed of separately. Random bulking of waste liquids can be very dangerous.

Some wastes may be disposed of by diluting with sufficient water and flushing into the sewage system. Materials that can be disposed of in this way should be so designated by the person in charge. Used oils and hydrocarbons may have commercial value and, if so, should be stored in suitable receptacles for ultimate disposal. Oil that is highly contaminated (i.e. more than thirty per cent by volume) with solvents or other chemicals, or with extremely hazardous materials at any concentration, should be classified as chemical waste and handled accordingly.

6.3.10 Burning — The burning of material for disposal should be carried out in an approved incinerator. The Environmental Protection Service (EPS) will provide assistance to individual facilities in developing handling procedures for, and disposal of, hazardous and toxic waste by open pit burning.

6.4 **Additional information**

For further information on the hazards of a specific chemical substance and recommendation for its disposal as a laboratory waste, reference should be made to the Environmental Protection Service of Environment Canada, and to the *Laboratory Waste Disposal Manual*, Second Edition (1969) published by the Manufacturing Chemists Association, Washington, D.C.

6.5 **References**

Laboratory Biosafety Guidelines,
Office of Biosafety,
Laboratory Centre for Disease Control,
Health Canada,
Ottawa, Ontario
Canada
K1A 0L2
(ISBN 0-662-17695-2)

*Laboratory Waste Disposal Manual,*
Manufacturing Chemists Association,
1825 Connecticut Ave. N.W.,
Washington, D.C. 20009

*Precautionary Labeling of Hazardous Chemicals,*
Chemical Manufacturers Association

*Chemical Safety Sheets,*
Manufacturing Chemists Association

*Poisons: Properties, Chemical Identification,*
*Symptoms and Emergency Treatment,*
Vincent J. Brookes and Morris B. Jacobs,
2nd Edition, 1958,
Van Nostrand Reinhold Publishing Company,
1410 Birchmount Road,
Scarborough, Ontario

*Dangerous Properties of Industrial Materials,*
N. Irving Sax, 3rd Edition,
Van Nostrand Reinhold Publishing Company

*National Fire Codes* (Vol. 15-1975),
National Fire Prevention Association,
470 Atlantic Avenue,
Boston, Mass. 02210

*Code of Good Practice for Management of Hazardous and Toxic Waste at*
*Federal Establishments,*
Environmental Protection Service,
Environment Canada

*Code of Good Practice for Handling Solid Waste at Federal Establishments,*
Environmental Protection Service,
Environment Canada

## 7. **Medical surveillance**

7.1  Laboratory employees are subject to periodic health evaluations where required, in accordance with the Occupational health evaluation standard (chapter 2-13).

## 8. **First-aid — Health and medical services**

8.1  The provisions of first-aid facilities, and the training of employees in first-aid, should follow the requirements outlined in the First-aid directive (chapter 2-5).  All other applicable and authorized standards, or special instructions concerning first-aid equipment and procedures particular to each laboratory operation, should be available for reference by laboratory staff.

8.2 Where it has been determined that special first-aid facilities, supplies and training are required, advice and arrangements for such supplies or training should be obtained through the nearest Regional Medical Services office of Health Canada.

8.3 Suspected or potential health or environmental hazards in laboratories should be investigated in accordance with the procedures outlined in chapter 4-2, Occupational health investigations and surveys.

## 9. Fire prevention and emergency procedures

9.1 Fire prevention and protection in the Public Service is under the jurisdiction of the Office of the Fire Commissioner of Canada.

9.2 Measures concerning fire prevention and protection should be in compliance with the *National Fire Codes* of the National Fire Prevention Association, Fire Protection Technical Information Bulletins and the Fire Protection Engineering Standards issued by the Office of the Fire Commissioner of Canada. It is the responsibility of the person in charge to ensure that all possible precautions are taken to prevent fires and explosions.

9.3 Where specific hazards may require emergency evacuation of staff, or other special safety measures, a disaster plan should be developed and approved by or in cooperation with the person in charge. This plan should be updated as necessary, and evacuation and/or disaster procedures should be rehearsed on a regular basis.

### Enquiries

Enquiries should be directed to the responsible officers in departmental headquarters who, in turn, may seek interpretation from the following:

> Safety, Health and Employee Services Group
> Staff Relations Division
> Human Resources Policy Branch
> Treasury Board Secretariat

## CHAPTER 5-1 - APPENDIX A - ASSESSMENT PROCEDURES

### 1. Apparatus

#### 1.1 To measure air velocity at sash opening of fume hood

Anemometer with specifications as follows:
Range: 0 — 1.5 m/s (0 — 300 fpm). Maximum.
Accuracy: plus or minus 0.03 m/s (6 fpm).
Calibration: regularly, as recommended by manufacturer. Normally this will be every six (6) months.

**Note**: Vane anemometers are not acceptable.

### 1.2 **To measure air volume in exhaust duct**

Pitot tube and manometer with specifications as follows:
Range: to suit requirements.
Accuracy: plus or minus 1% of full range span.
Calibration: as recommended by manufacturer.

### 1.3 **To measure speed of exhaust fan impeller and motor**

Non-contact type tachometer with specifications as follows:
Range: to suit requirements.
Accuracy: plus or minus 2% of full scale.
Calibration: as recommended by manufacturer.

### 1.4 **To measure voltage and current to exhaust fan motor**

Volt-ammeter with specifications as follows:
Type: clamp-on.
Range: to suit requirements.
Accuracy: plus or minus 3% of full scale.
Calibration: as recommended by manufacturer.

**Note**: Calibration shall be performed by an approved testing laboratory. The National Research Council will calibrate the tachometer and the volt-ammeter. The manufacturer of the anemometer should have facilities for instrument re-calibration. For a partial listing of approved testing laboratories in the private sector, consult the Occupational Health Unit, Medical Services Branch, Tunney's Pasture, Ottawa.

## 2. **Procedures**

### 2.1 **Face velocity**

2.1.1  Fume hoods shall normally be tested empty. However, in exceptional circumstances, it may be essential for the safety of the operator to test a hood with fixed equipment in place.

2.1.2  With the fume hood and the normal air supply operating, raise the sash to its typical operating level which is normally, but not invariably, 30 cm (12").

2.1.3  If the hood is an auxiliary air supplied hood, it shall be tested in both modes — with auxiliary air "ON", and with auxiliary air "OFF".

2.1.4  Divide the sash opening into equal areas calculated as follows (see also Appendix B):

width = not greater than 30 cm (12").

height = not greater than 15 cm (6"). There must be at least three (3) rows.

2.1.5 Measure the air velocity in the centre of each area described in 2.1.4 in the plane of the sash. Avoid false reading caused by vibration and movement of the anemometer sensor.

To avoid disturbing air flow patterns, the operator shall stand at least 50 cm (20") back from, and to one side of, the face of the hood.

**Note**: Readings that fluctuate indicate an unacceptable degree of turbulence which must be investigated.

2.1.6 Record these readings on an appropriate report form.

**Note**: Where face velocity has changed significantly since the previous tests, ventilation system performance must be reviewed and recorded.

## 2.2 Exhaust air volume

2.2.1 With the fume hood and the normal air supply operating, raise the sash to its typical operating level which is normally, but not invariably, 30 cm (12").

2.2.2 Determine exhaust air volume using duct traverse procedures as described in ASHRAE Handbooks (current editions).

2.2.3 Record these readings on an appropriate report form.

## 2.3 Fan performance

2.3.1 With the fume hood and the normal air supply operating, raise the sash to its typical operating level which is normally, but not invariably, 30 cm (12").

2.3.2 Determine the rotational speed of the fan impeller in accordance with procedures described in ASHRAE Handbooks (current editions).

2.3.3 Measure fan motor current and voltage.

2.3.4 Record these readings on an appropriate report form.

## 2.4 Smoke tests

2.4.1 With the fume hood and the normal air supply operating, raise the sash to its typical operating level which is normally, but not invariably, 30 cm (12").

2.4.2 Using a smoke source such as Draeger CH216 or MSA5645, observe the behaviour of smoke released just outside the centre of the sash opening. The smoke should flow smoothly into and through the hood without looping, eddying or mixing, which are signs of turbulence and indications of possible problems.

**Note**: The operator must not allow his hand or arm to interfere with air flow patterns.

The smoke may be corrosive and should be handled with care.

2.4.3  Repeat step 2.4.2 in the corners of the fume hood opening.  Watch for signs of reverse flow from the hood.

2.4.4  Draw the smoke source from the fume hood and note the distance at which the smoke is no longer reliably captured.  Watch for signs of turbulence in front of the hood.

2.4.5  Release smoke along the walls and floor of the fume hood.  Watch for eddies flowing forward especially along the sides of the base, carrying smoke to the front lip of the hood.

2.4.6  Release smoke around any equipment, apparatus, tanks or sinks in the fume hood.  Watch for eddies bringing smoke to the face of the fume hood.

2.4.7  Release smoke behind and close to the sash at several points across the width of the opening.  Watch for eddies developing under and sometimes in front of the sash bar, fed from inside the fume hood.  This can sometimes be a serious problem with bulky sash bars, and requires modifying the hood to correct it.

2.4.8  If the air flow pattern in the room is causing a problem, it may be necessary to release smoke at various points in the open room both before and after any remedial action is taken.

2.4.9  Record the results on an appropriate report form.

3. **Report**

3.1  Complete all information required by the report forms.

3.2  The report shall be accompanied by a sketch of the room showing the location of the hood, windows and doors, all major furniture, air supply, return and/or exhaust outlets and a schematic of each exhaust system showing all components.  Any sources of significant cross-drafts should be noted.

4. **Assessment**

4.1  As stated in 2.2.8.4, the average face velocity shall be between 0.40 and 0.50 m/s (80 and 100 fpm).  The optimum recommended face velocity is 0.50 m/s (100 fpm).  If any individual reading falls below 0.30 m/s (60 fpm) or rises above 0.60 m/s (120 fpm), contaminant containment could be jeopardized, especially when the additional turbulence created by the operator moving in, or passing through, the working zone is taken into consideration.  In such cases, the fume hood shall be taken out of service, the problem investigated, all faults corrected and the hood re-tested before being returned to use.

4.2  If comparisons with previous similar test records indicate a declining average face velocity (and thus declining exhaust volume), this may indicate problems such as corrosion of the exhaust fan, loose belts, duct failure, etc.  In such cases, the fume hood may have to be taken out of service, the problem investigated, all faults corrected and the hood re-tested before being returned to use.

4.3 All turbulence shall be investigated and remedial action taken before the hood is returned to use.

4.4 Cross-drafts should be reduced to less than 0.10 m/s (20 fpm).

## 5. Documentation

5.1 A copy of all test records for each fume hood should be provided to laboratory supervisory personnel and retained on site.

5.2 A label should be applied to the hood indicating the date of most recent testing. This label should also indicate the usual working level of the hood sash opening.

## CHAPTER 5-1 - APPENDIX B - SAFETY GUIDE FOR LABORATORY OPERATIONS

## LABORATORY FUME HOOD TEST RECORD FUME HOOD CHARACTERISTICS AND FACE VELOCITY

Laboratory organization: . . . . . . . . . .     Responsible officer: . . . . . . . . . .
Building: . . . . . . . . . . . . . . . . . . . . . . .     Room designation: . . . . . . . . . . .

Fume hood type: . . . . . . . . . . . . . . . . .     Manufacturer: . . . . . . . . . . . . . .
Fume hood identifier: . . . . . . . . . . . . . .     Size of fume hood: . . . . . . . . .
Slot/baffle positions: . . . . . . . . . . . . . .     Sash height: . . . . . . . . . . . . . .

### FUME HOOD FACE VELOCITY

W

H

Average face velocity: . . . . . . . . . . . . . .
Previous average face velocity: . . . . . . .     Date of previous test: . . . . . . . . .
Instrument: . . . . . . . . . . . . . . . . . . . . .     Serial no: . . . . . . . . . . . . . . . .
Tests performed by: . . . . . . . . . . . . . . .     Date: . . . . . . . . . . . . . . . . . . . . .

## SMOKE TEST

| Clause | Test | Results | | Comments |
|--------|------|---------|---|----------|
| | | Acceptable | Not Acceptable | |
| 2.4.2 | Turbulence in front of hood at centre | | | |
| 2.4.3 | Turbulence in front of hood at corners | | | |
| 2.4.4 | Capture distance from face of hood | | | |
| 2.4.5 | Reverse flow inside hood along walls | | | |
| 2.4.5 | Reverse flow inside hood along floor | | | |
| 2.4.6 | Eddies inside fume hood | | | |
| 2.4.7 | Eddies under sash bar | | | |

Smoke source: . . . . . . . . . . . . . . . . . . . . . . . . . . . . . . . . . . . . . . . . . . . . .

Tests performed by: . . . . . . . .          Date: . . . . . . . . . . . . . . . . . . . .

## LABORATORY FUME HOOD TEST RECORD EXHAUST FAN PERFORMANCE AND EXHAUST DUCT TRAVERSE

| | |
|---|---|
| Laboratory organization: . . . . . . . . . . . | Responsible officer: . . . . . . . . . |
| Building: . . . . . . . . . . . . . . . . . . . . . . | Room designation: . . . . . . . . . . . |
| Fume hood type: . . . . . . . . . . . . . . . . | Manufacturer: . . . . . . . . . . . . . . |
| Fume hood identifier: . . . . . . . . . . . . . | Size of fume hood: . . . . . . . . . |
| Slot/baffle positions: . . . . . . . . . . . . . . | Sash height: . . . . . . . . . . . . . . . |

### EXHAUST FAN PERFORMANCE

| Motor: | Rated | Tested | Previous test |
|---|---|---|---|
| Voltage: | .... Volts | .... Volts | .... Volts |
| Current: | .... Amps | .... Amps | .... Amps |
| Speed: | .... RPM | .... RPM | .... RPM |
| **Exhaust fan:** | **Rated** | **Tested** | **Previous test** |
| Static pressure at suction: | .... Pa/ln. WG | .... PA/ln. WG | .... Pa/ln. WG |
| Static pressure at discharge | .... Pa/ln. WG | .... Pa/ln. WG | .... Pa/ln. WG |
| Impeller speed: | .... RPM | .... RPM | .... RPM |

| Instrument: | Make & Model number: | Serial No: |
|---|---|---|
| Volt-ammeter: | . . . . . . . . . . . . . . . . . . . . . . | . . . . . . . . . . . |
| Tachometer: | . . . . . . . . . . . . . . . . . . . . . . | . . . . . . . . . . . |
| Tests performed by: | . . . . . . . . . . . . . . . . . . . . . . | Date: . . . . . . . |

## DUCT TRAVERSE

Duct size: . . . . . . . . . . . . . . . . .        Test location: . . . . . . . . . . . . . .

| Vertical position | | Horizontal position | | | | |
| No. Distance | 1 V.P. Vel | 2 V.P. Vel | Distance 3 V.P. Vel | 4 V.P. Vel | 5 V.P. Vel |
| --- | --- | --- | --- | --- | --- |
| 1 | | | | | |
| 2 | | | | | |
| 3 | | | | | |
| 4 | | | | | |
| 5 | | | | | |

Calculated air volume: ... (L/s)(CFM)   Air volume (previous test): ... (L/s)(CFM)

| Instrument: | Make and Model No.: | Serial No: |
| --- | --- | --- |
| Manometer: | . . . . . . . . . . . . . . . . . . . . . . | . . . . . . . . . . |
| Tests performed by: | . . . . . . . . . . . . . . . . . . . . . . | Date: . . . . . . . |

## LABORATORY HOOD TEST RECORD LABORATORY LAYOUT

Laboratory organization: . . . . . . . . . . . .     Responsible officer: . . . . . . . . . .
Building: . . . . . . . . . . . . . . . . . . . . . . .     Room designation: . . . . . . . . . . .

Fume hood type: . . . . . . . . . . . . . . . . .     Manufacturer: . . . . . . . . . . . . .
Fume hood identifier: . . . . . . . . . . . . . .     Size of fume hood: . . . . . . . . . .
Slot/baffle positions: . . . . . . . . . . . . . .     Sash height: . . . . . . . . . . . . . .

Include a sketch of the laboratory showing all doors, windows, furniture, air
supply outlets, return and/or exhaust inlets, canopy hoods, exhaust terminals,
fume hoods, biological safety cabinets, window air conditioners, etc. See
APPENDIX A, 3. REPORT.

Normal position of laboratory door(s)
during normal laboratory operations:                          Closed/Open

Normal position of windows during
normal laboratory operations:                                 Closed/Open

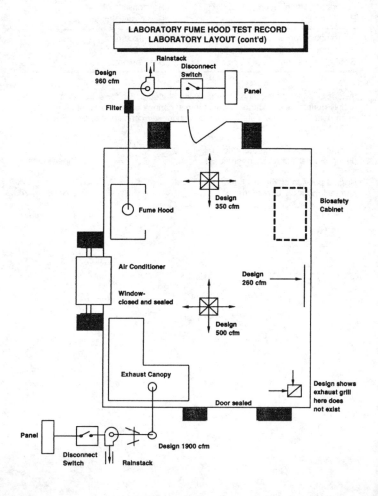

LABORATORY FUME HOOD TEST RECORD
LABORATORY LAYOUT (cont'd)

# CHAPTER 5-2 - A GUIDE TO ACCIDENT INVESTIGATION

## Introduction

An effective accident-investigation, record-keeping and reporting system is the heart of an occupational safety and health program. It facilitates the identification of health and safety hazards, enables the design and provision of preventive measures, and helps determine overall safety and health program priorities. Most importantly, an accident investigation that is properly and efficiently carried out and is followed by prompt remedial action is one of the most effective methods of reducing accidents.

An accident investigation has two purposes: to determine why the accident happened by identifying all work-related facts associated with it; and, to subsequently modify work conditions and procedures to prevent a similar occurrence. The supervisor responsible for the work must be totally committed to the accident investigation process. These guidelines are based on that premise.

## Definition

An **accident investigation** is the determination of the facts of an accident by inquiry, observation, and examination and an analysis of these facts to establish the causes of the accident and the measures that must be adopted to prevent its recurrence (enquête sur un accident).

## Departmental responsibilities

Each department is responsible for ensuring that work accidents occurring within its jurisdiction are investigated, recorded and reported, the causes determined and appropriate measures taken to prevent similar occurrences.

The procedures contained in chapter 4-1, Accident investigation and reporting, are to be followed whenever an accident occurs, and this guide should be used in carrying out the accident investigation itself.

## The Investigator

## The Supervisor

The supervisor, by the nature of his or her position and its inherent responsibilities, possesses a detailed knowledge of the work and the conditions under which it is done and is the appropriate person to undertake the accident investigation. Primary responsibility for investigation should, therefore, rest with the supervisor. Acceptance of this responsibility by each supervisor, combined with a personal commitment to the time and effort involved in such investigations, is required if the program is to operate effectively. The supervisor's responsibility extends beyond determining the cause of the accident and includes exercising supervisory responsibilities to ensure that proper remedial action is promptly taken.

## Technical advisers and specialists

When investigating accidents of a serious nature, or those which involve highly technical processes, it may be appropriate to engage the services of a technical adviser or other specialist with specialized knowledge of an operation or equipment to assist in the investigation. An "outside" specialist may also provide an additional degree of objectivity to the investigation. When such a person is authorized to conduct or assist in an investigation, it should be carried out in collaboration with the responsible supervisor.

## Safety and health officer

Where appropriate, the departmental safety and health officer can offer guidance in coordinating an accident investigation. The officer cannot be expected to provide technical advice on every operation or to take the place of the responsible supervisor concerning the detailed work operation or procedures. The officer can, however, often help to determine the cause of an accident, as a result of general knowledge and experience with similar accidents and their causes.

## Safety and health committee or representative

Where an accident is reported at a workplace that is represented by a safety and health representative, the committee or representative must take part in the accident investigation. A committee's involvement in the investigation is to include the participation of committee members as appointed by the committee chairpersons.

## Team approach

The investigation should be carried out by the supervisor directly responsible for the work at the time of the accident. Such investigation may be all that is required to establish the causes and to prescribe corrective action. However, in the event of a serious accident, and particularly when the causes are not readily determinable, a team approach may be advisable. The team would include the supervisor in every case, the safety and health committee or representative, if any, and other qualified personnel such as the safety and health officer, technical specialists, and advisers. It is the supervisor's responsibility to assess his or her own ability to investigate an accident and decide on the extent of backup required.

## The accident investigation

## Purpose of investigation

The primary purpose of the accident investigation is to establish the causes as quickly as possible through the identification and examination of all information associated with the accident. The ultimate purpose is to make the required changes in the work conditions and procedures that will eliminate or reduce the risk of a similar occurrence.

## Definition of accident

For purposes of this guide, an *accident* is defined as an event that results in a fatality, work injury, property damage or material loss arising out of, linked with or occurring in the course of employment.

## Fact-finding only

The investigation should be undertaken in a constructive spirit. It is not a fault-finding exercise, and irrespective of the causes determined or the involvement of various personnel, the occasion should not be used for apportioning blame. In establishing the existence of human error, such actions should be dealt with objectively.

## Consideration of information

As soon as possible after the accident, all information relating to the accident should be obtained and a conclusion reached concerning the causes. The investigator should, however, be cautious about accepting incomplete conclusions as the basis for establishing the cause. A normal tendency to reinforce preconceived thoughts as to the probable cause should also be resisted. When original factual evidence is not available, conclusions should be based only on substantiated facts or on the best possible logical assumptions. Circumstantial evidence should only be considered, with extreme caution, when no other evidence exists.

## Planned approach to investigations

Each department should institute a planned approach to accident investigations supported by general internal procedures, as necessary, to guide supervisors. Fundamentally, such a planned approach should incorporate the following sequence of actions:

- prevent the removal of evidence or the change of conditions at the work scene;
- determine the specified work procedure;
- verify evidence of the witnesses and, if possible, the injured employee;
- record results of special tests or re-enactments;
- review and select findings and establish causes;
- recommend appropriate changes based on the conclusions; and
- carry out the changes recommended to prevent a recurrence of the accident.

A planned system of approach incorporating these actions will help to insure that each accident investigation is carried out in an orderly and effective manner. More detailed guidelines respecting each phase of the investigation follow.

## Action following accident

Depending on the seriousness of the accident, the scene of the accident should be isolated as much as possible. If someone is injured, immediate medical attention should be given. Ensure that the scene of the accident remains undisturbed and that the requirements of Section 84 of the *Canada Labour Code*, Part II, are complied with.

When the exact physical situation cannot be maintained for the investigation, a sketch should be prepared and, if possible, photographs taken as soon as possible.

The names and addresses of all witnesses should be obtained immediately.

If there is a possibility that any circumstances surrounding the accident may constitute danger to anyone, action should be taken (without awaiting the final outcome of the investigation) to remove the apparent hazard or temporarily discontinue the work under investigation.

The investigation should proceed as soon as possible after the accident.

### Investigation interviews

Personnel involved in the accident, including witnesses and, if possible, the injured employee, should be interviewed as soon as possible following the accident, while the events are still clear in their minds, and the resulting information and statements should be recorded in writing. In many cases the injured employee will be the principal source of information.

Before interviews take place, however, the investigator should ensure that he or she is thoroughly familiar with the approved procedures governing the work being performed at the time of the accident.

Any person directly involved should be interviewed first, to determine what was being done before the accident; this should be followed by interviews with co-workers involved or associated in the operation, or in the immediate vicinity. In most circumstances, it is advantageous to interview each person separately.

Before interviewing the injured employee, the investigator should obtain assurance that the interview is medically permissible. Basic information relative to the employee, his or her injury, or other circumstances, should be obtained before the interview. (If the employee's answers vary considerably from known information, the person's physical condition may be such that questioning is inappropriate at that time.)

Proper interview techniques are important. Normally, for example, it is well to remind the person being interviewed of the constructive purpose of the investigation. The investigator should do everything reasonable to put the person at ease and should never appear hostile. Ask the person what happened and try not to interrupt. When more information is needed, it is usually better not to pose direct questions but to ask for clarification of key points. Also ask any pertinent questions required to complete the accident investigation report.

## Cause categories

The two principal cause categories consist of personal actions by the injured employee or by someone else, and environmental conditions surrounding the work.

In many accidents a combination of the effects of both cause categories may be evident, and a careful appraisal of the sources of such factors should disclose all of the underlying contributory causes.

A considerable number of potential causes of accidents, arising both from unsafe acts and unsafe conditions, have been identified and catalogued. A list of these is provided in Appendix A of this guide, as an aid to investigators. (Refer also to Section 5 of this guide "Contributory causes and factors", for additional information.)

## Cause determination

The specific work procedure (whether right or wrong) that was being performed by the employee at the time of the accident should be determined. The employee's actions immediately before the accident should be compared with the approved procedure for that job. One should attempt to establish and record (in written form) whether the employee lacked skill, knowledge, training, or awareness. Did the employee take a shortcut in an attempt to avoid inconvenience or discomfort? Was the job being performed covered by a safety rule or directive and, if so, was there a deviation from the directive?

It should be determined whether the employee's work environment contributed to the accident. Were there any defects or deviation from approved conditions in respect to tools, equipment, vehicles, or the surrounding work area? If a faulty condition is detected, it should be determined whether the condition was caused by normal deterioration, excessive use, abuse, or faulty design.

## No substantiated causes

Occasionally, investigators may be unable to decide on the cause. For example, key facts or supporting details may be absent in relation to the type of accident and the injury, or to the employee's version of these. In some instances it may be found also that the type, location, or severity of injury cannot be related to the circumstances of the work or to the accident. In such situations the supervisor should extend the investigation to verify that the accident and the injury occurred as described. If such verification cannot be made, the investigation report should indicate this. (Supervisors should also bring such cases to the attention of the person or persons responsible for reporting the injury for purposes of workers' compensation (*Government Employees' Compensation Act*), and provide the essential information concerning lack of verification for inclusion in the applicable Workers' Compensation Accident or Injury Report or other form specified for this purpose.)

## Caution

Occasionally it may be useful to re-enact certain elements of the operating procedure, or test equipment under similar conditions of use. In such an event, proceedings should be carried out with extreme caution, briefing the participants fully on the relative hazards and on the safe procedures to be followed. It should be verified that in the event of an accidental deviation from the work procedure or the failure of some equipment, there would be no possibility of damage or injury. If there is any doubt about the safety of a re-enactment it should not be carried out.

## Final recommendations

Once the error that may have contributed to the cause of the accident has been identified, it is the responsibility of the supervisor, supported where necessary by the investigation team, to prescribe the action that must be taken, based on the findings of the investigation, to rectify the hazard or reduce the risk of a similar accident (see "Corrective action", below). Once action is recommended, it is the responsibility of local management to review and to change the work procedures or equipment appropriately, and to provide for ongoing monitoring and inspection systems to maintain the work in a safe and healthful manner.

## Corrective action

### Physical conditions

If the investigation revealed that the cause of the accident was related to physical or environmental conditions, action should be taken, as appropriate, to:

* modify or change facilities or personal equipment or other physical elements at the work location to eliminate or minimize the hazards concerned;
* undertake special technical studies, tests, or analyses, or initiate manufacturing or design inquiries.

### Personal action or inaction

If the accident investigation revealed that someone's action or inaction contributed to the accident, the following steps may be appropriate:

* revise the job procedure;
* undertake the safety training or instruction of all employees involved in similar operations;
* undertake campaigns to ensure employees' compliance with safety procedures or directives;
* seek professional medical advice when an underlying mental or physical problem of the employee is suspected;
* publicize facts and causes of the accident among other supervisors and employees who may be subject to similar hazards.

## Supervisor's responsibilities

In addition to the primary responsibility for ensuring that the accident investigation is carried out, the supervisor is responsible for taking whatever immediate corrective actions are required within the scope of his or her own authority.

Should any changes in procedures or conditions be required that are beyond the supervisor's assigned authority to approve, recommendations concerning them should be made to higher management in accordance with local departmental procedures. The supervisor should clearly state what is being recommended and give reasons in support of the recommendations to prevent recurrence of the accident.

## Unit or division head's responsibilities

The term unit head or division head signifies the level of management to which the first-line supervisor generally reports. The unit head should review the corrective action recommended by the supervisor, as recorded on the accident investigation report form and other supporting documents.

It is usually the unit or division head's responsibility to decide what action is to be taken with respect to any changes recommended by the investigating supervisor. If further study is indicated regarding a proposed change, personal responsibility should be assigned to someone for conducting the study, and a completion date should be established.

The unit or division head should indicate approval or concurrence in writing, with the recommendations made and action to be taken. This practice ensures that this level of management is completely familiar with the accident and its details. It also enables the unit or division head to assess the supervisor's acceptance of responsibility and commitment, to concur with the supervisor's findings, and to add any comments.

## The accident investigation report

### General

The accident investigation report form is the basic vehicle for providing and summarizing all the facts relating to the accident. Its systematic use is essential in giving line supervisors and managers the opportunity to propose corrective action and to formally indicate their commitment to follow up the corrective measures.

The supervisor in charge of the work is responsible for the completion of the accident investigation report, and the data ultimately recorded on the report form are the end result of the completed investigation process. The report should not be regarded as final until the investigation has been completed, the results (including the corrective action to be taken) recorded, and the report signed by the supervisor. (The accident investigation report should not be confused with the accident report form used for reporting an injury for purposes of Workers' Compensation pursuant to the *Government Employees' Compensation Act*.)

## Public Service mandatory reporting requirements

In accordance with chapter 4-1, Accident investigation and reporting, an investigation form shall be completed for every accident that results in:

- a disabling injury to an employee;
- an electric shock, toxic atmosphere or oxygen deficient atmosphere that caused an employee to lose consciousness;
- the implementation of rescue, revival or other similar emergency procedures; or
- a fire or an explosion.

A copy of the accident investigation report shall, within 14 days after the accident, be submitted to the regional safety officer at Human Resources Development Canada's regional office.

## Departmental option

In the case of other types of work accidents and injuries, detailed procedures concerning the supervisor's investigation shall be determined by each department according to its own operating requirements.

## Distribution of reports

The appropriate distribution of the completed accident report form is the principal method of publicizing and disseminating the accident investigation information to those involved in the operation, coordination, maintenance, or monitoring of the safety and health program. Distribution should take place as follows:

- copies of the accident investigation report should be forwarded to appropriate levels of management, safety and health committees or representatives, safety and health officers, and elsewhere, according to the safety and health program requirements of the department.

## The Public Service report form

The standard form recommended for the Public Service is the Hazardous Occurrence Investigation Report, (see appendix B), which has been designed and specified for use in support of the accident prevention program and is available through Human Resources Development Canada's regional offices.

## Completion of report form

It is essential that each section of the report form be fully completed and that details be recorded accurately. If this is not done, the appropriate use of the report may be prejudiced, and the value of the entire investigation negated. The following general guidelines for completing the report form may be helpful:

### Section A – Identification data

For use by Human Resources Development Canada only.

### Section B – Related information

Basic information concerning location of the accident and personnel involved. Weather conditions should be specified where applicable.

### Section C – Accident description

A clear and concise statement that describes the sequence of events that led to the accident. If extensive additional notes and comments have been made during the investigation, they should be condensed and may be attached to the report.

In most situations, it is helpful to provide a sketch, diagram or photograph.

### Section D – Property damage

The type and extent of damage to property, material or facilities. The actual cost or an estimate of the cost to repair damaged property and restore it to its original condition should be indicated.

If there is no property damage, this should be stated.

### Section E – Injured employee

Information respecting the employee injured as a result of the accident. The injury or illness sustained should be clearly described. The direct cause of the injury (e.g., struck by, tripped over, etc.) should be determined and specified.

### Section F – Accident cause

The unsafe conditions or acts which caused the accident (refer to appendix A for guidance). Subsequent corrective measures should be based on these facts.

### Section G – Corrective action

The remedial action taken to correct the unsafe condition or act. State completion date where required. Indicate further preventive measures considered necessary.

### Section H – Investigator

Details concerning the person who has been designated to investigate the accident.

### Section I – Committee/representative

Information regarding involvement of the safety and health committee or representative in the accident investigation.

## Management comment

The responsible senior manager at each work location should review the form, provide comments, and sign it. These comments should state or confirm the action to be undertaken to prevent a similar accident. Where necessary, the manager should assign specific responsibility to the appropriate person to ensure that the required changes will be made, and should follow up such action personally.

## Use of report by safety personnel

Departmental safety and health officers should receive or have access to copies of all accident investigation reports. Such reports provide the officer with a good deal of valuable information concerning the general status of the safety and health program and, more specifically, are useful in the following applications:

- to provide information concerning the efficacy of the accident investigation program, thereby evaluating safety performance and progress at various locations;
- to evaluate the types of accidents and their severity, thus providing information to the safety office for use in the monitoring and advisory role;
- to provide a basis for the completion of statistics, identifying accident trends, and denoting possible deficiencies in the safe supervision of the work;
- to indicate when special safety and health surveys and inspections may be required, or to identify the requirements for a review of job procedures, or to initiate special studies.

## Contributory causes and factors

As previously referred to in this guide, it is often determined that accidents are caused by more than one unsafe act or condition, and in combination with a number of underlying, less evident contributory factors. These contributory factors or causes may arise from particular defects in the organization or from personnel actions or both. The following accident description may help to demonstrate this:

"The operator of a circular saw reached over the running saw to pick up a piece of scrap which was lodged near the saw blade. His hand touched the blade, which was not guarded, and his thumb was amputated."

The unsafe condition — an unguarded saw.

The unsafe act — cleaning or clearing a moving machine.

Possible contributory factors:

- Existence and tolerance of an unsafe condition (unguarded saw blade).
- Failure to establish and enforce safe operating rules (rules should prohibit cleaning a running or moving machine).

- Disregard of job safety instructions (if an instruction existed stating that the machine must be stopped before cleaning).
- Lack of knowledge or training (worker unaware of safe practice, i.e. the need to stop machine before cleaning).
- Lack of routine safety inspection (which would have identified the unsafe condition and action).

It is apparent from this example that the elementary identification of the unsafe condition or act will not suffice for corrective action. The underlying contributory factors or reasons for the unsafe condition and the unsafe act should be established in order to determine and provide the appropriate corrective measures.

In the example outlined the recommendations might include the following:

- that guards be designed and maintained in place on this and other similar equipment;
- that job safety training be undertaken to stress that moving machinery must be stopped and safely secured before cleaning or other maintenance;
- that supervisors enforce observance of the above requirement;
- that appropriate signs or directives be prepared and posted where the employees may be reminded of the safe procedures to be followed; and
- that routine inspections be arranged to ensure maintenance of required safety conditions and procedures.

An investigation is neither successful nor complete until all the possible causes and contributory factors are considered, and the actual causes identified and acted upon.

### Enquiries

Enquiries should be directed to the responsible officers in departmental headquarters who, in turn, may seek interpretation from the following:

Safety, Health and Employee Services Group
Staff Relations Division
Human Resources Policy Branch
Treasury Board Secretariat

### CHAPTER 5-2 - APPENDIX A - LIST OF FACTORS CONTRIBUTING TO ACCIDENTS

#### Unsafe conditions

#### Inadequate guarding

– Guard weak, defective, poorly designed
– Inadequately guarded
– Improper shoring in mining, construction, excavating.

### Defective

- Tough
- Slippery
- Sharp-edged
- Poorly designed
- Low material strength
- Poorly constructed
- Inferior composition
- Decayed, aged, worn, frayed, cracked

### Hazardous arrangement or procedure

- Unsafely stored or piled tools or material
- Congestion of working space
- Inadequate aisle space or exits
- Unsafe planning or layout of traffic or process operations
- Unsafe processes
- Overloading
- Misaligning
- Inadequate drainage

### Improper illumination

- Insufficient light
- Glare
- Unsuitable location or arrangement (producing shadow or contrast)
- No light

### Improper ventilation

- Insufficient air changes
- Unsuitable capacity, location, or arrangement of system
- Impure air source
- Abnormal temperature and humidity (confined area)

### Unsafe dress or apparel

- No goggles or face shields
- Goggles or face shields defective, unsafe, or unsuited for work
- No gloves or mitts
- Gloves or mitts defective, unsafe, or unsuited for work
- No apron
- Apron defective, unsafe, or unsuited for work
- No shoes
- Shoes defective, unsafe, or unsuited for work
- No respirator
- Respirator defective, unsafe, or unsuited for work
- High heels
- Loose hair
- Loose clothing
- Inadequately clothed
- No leggings

- Leggings defective, unsafe, or unsuited for work
- Lack of protective headgear, or hard hat unsafe or unsuited for work
- No welder's helmet or hand shields
- Welder's helmet or hand shields defective, unsafe, or unsuited for work
- No welder's protective clothing (spats, capes, sleeves, jackets, and other) or protective clothing defective, unsafe, or unsuited for work
- No babbitting mask
- Babbitting mask defective, unsafe, or unsuited for work
- No safety belts
- Safety belts defective, unsafe, or unsuited for work

## Unguarded

- Lack of guard, screen, enclosure, barricade, fence, insulation, railing, rope (as opposed to inadequate guarding)

## Unsafe design or construction

- Hazard built into new equipment or structures
- Faulty architecture, design, or engineering
- Faulty assembly, manufacture, or construction (as opposed to defective through wear and tear or abuse)

## Unsafe acts

### Operation without authority, failure to secure or warn

- Starting, stopping, using, operating, firing, moving, without authority or without giving proper signal
- Failing to lock block, or secure vehicles, switches, valves, press rams, other tools, materials, and equipment against unexpected motion, flow of electric current, steam
- Failing to shut off equipment not in use
- Releasing or moving leads without giving warning
- Failing to place warning signs, signals, tags
- Failure of crane signaller to give proper signal

### Operating or working at unsafe speed

- Running
- Feeding or supplying too rapidly
- Driving too quickly
- Driving too slowly
- Throwing material instead of carrying or passing it
- Jumping from vehicles or platforms
- Walking backwards
- Working too fast or too slowly (endangering self and others)

### Making safety devices inoperative

- Removing safety devices
- Blocking, plugging, tying of safety devices

- Replacing safety devices with those of improper capacity (higher amperage electric fuses, low-capacity safety valves)
- Misadjusting safety devices
- Disconnecting safety devices
- Failing to secure safety devices

## Using unsafe equipment, using hands instead of equipment, or using equipment unsafely

- Using defective equipment (mushroom head chisels)
- Unsafe use of equipment (e.g. using iron bars for tamping explosives, operating pressure valves at unsafe pressures or volume)
- Gripping objects insecurely or improperly

## Unsafe loading, placing, mixing, combining

- Overloading
- Crowding or unsafe piling
- Lifting or carrying too heavy loads
- Arranging or placing objects or material unsafely (parking, placing, stopping, or leaving vehicles, elevators, and conveying apparatus in unsafe position for loading and unloading)
- Injecting, mixing, or combining one substance with another so that explosion, fire, or other hazard is created (injecting cold water into hot boiler, pouring water into acid)
- Introducing objects or materials unsafely
- Portable electric lights inside boilers or in spaces containing inflammables or explosives
- Moving equipment in congested workplaces
- Smoking where explosives or inflammables are kept
- Placing or leaving on working surfaces (tools, materials, debris, rope, chain, hose, electrical leads)
- Oil, water, grease, paint on working surfaces

## Taking unsafe position or posture

- Exposure under suspended loads (fixed or moving)
- Putting body or parts of body into shaftways or openings
- Standing too close to openings
- Walking on girders, beams, or edges of working surfaces when not necessary
- Not using proper methods of ascending and descending
- Entering vessel or enclosure when unsafe because of temperature, gases, electric, or other exposures
- Working on high-tension conductors from above instead of below
- Lifting with bent back or while in awkward position
- Riding in unsafe position on platforms, tailboards, and running boards of vehicles (tailing on or stealing rides, riding on apparatus designed only for materials)
- Exposure on vehicular right of way
- Passing on grades and curves, cutting in and out, road hogging
- Exposure to falling or sliding objects

## Working on moving or dangerous equipment

- Getting on and off equipment such as vehicles, conveyors, elevators
- Cleaning, oiling, adjusting of moving equipment
- Caulking or packing of equipment under pressure (pressure vessels, valves, joints, pipes, fittings)
- Working on electrically charged equipment (motors, generators, lines, or other electrical equipment)
- Welding or repairing of equipment containing dangerous chemical substances

## Distracting, teasing, abusing, startling (horseplay)

- Calling, talking, or making unnecessary noise
- Throwing material
- Teasing, abusing, startling, horseplay
- Practical joking
- Quarrelling or fighting

## Failure to use safe attire or personal protective devices

- Failing to wear goggles, gloves, masks, aprons, shoes, leggings, protective hats
- Wearing high heels, loose hair, long sleeves, loose clothing
- Failure to report defective safety apparel

## Reasons for some unsafe acts and conditions

## Possible personal defects

- Improper attitude
- Conflicting motivations
- Violent temper
- Absentmindedness
- Wilful intent to injure
- Nervous, excitable
- Failure to understand instructions, regulations, and rules
- Wilful disregard of instructions, regulations, and rules.

## Lack of knowledge

- Unaware of safe practice
- Unpractised or unskilled

## Bodily defects

- Defective eyesight
- Defective hearing
- Muscular weakness
- Fatigue
- Existing hernia
- Crippled
- Existing heart disease or other organic weakness

- Intoxicated
- General physical condition not adapted to job
- Existing injury (cut, laceration, bruise)

**Possible organizational defects**

- Lack of safe job procedures
- Inadequate training
- Failure to establish and enforce safety rules
- Inadequate supervisory training
- Tolerance of unsafe conditions
- Inadequate design or layout (engineering)
- Inadequate inspection program
- Inadequate preventive maintenance program
- Inadequate safety standards for purchasing

## APPENDIX B - RELEVANT FORM

| Title of Form | No. of Form | Initiated By | Available From |
|---|---|---|---|
| Hazardous Occurrence Investigation Report | LAB/TRAV 369 (92/01) | HRDC | **Regional offices of HRDC** |

## CHAPTER 5-3 - SAFETY GUIDE FOR OPERATIONS OVER ICE

### 1. Introduction

#### 1.1 General

1.1.1 Ice covers are used for transportation routes, as a surface on which structures can be erected, and for the temporary storage of materials.

1.1.2 This guide is concerned primarily with fresh water ice bridges, which are intended to support a gross vehicle weight of no more than 25 tons (22.5 tonnes). An ice bridge can be a natural untouched ice cover, a built-up, or a combined reinforced and built-up crossing route.

1.1.3 When loads are expected to exceed 25 tons (22.5 tonnes) or when operations will be conducted over salt water ice covers, advice should be sought from the Geotechnical Section, Division of Building Research, National Research Council of Canada, Ottawa, Ontario, K1A 0R6.

1.1.4 Information on the safe use of ice covers for aircraft operations is available from Transport Canada.

## 1.2 Purpose

1.2.1 The purpose of this safety guide is to:

(a) specify rules of good safety practice for all Public Service employees engaged in operations on ice covers;
(b) provide information on the thickness of ice required to support moving and stationary loads;
(c) specify methods for determining ice thickness and quality; and
(d) outline approved methods for the preparation and maintenance of ice bridges.

## 2. Properties of ice covers

### 2.1 Ice formation

2.1.1 Ice forms on fresh water when the surface temperature falls to zero degrees Celsius, or at lower temperatures if dissolved impurities are present. While the underside of the ice cover in contact with the water will remain close to the melting temperature, the upper surface will be nearer to the surrounding air temperature.

2.1.2 The date of annual freeze-up, the rate of ice growth, and the quality of the ice cover depend on various factors such as air temperature, solar radiation, wind speed, snow cover, wave action, currents, and the size and depth of the water body. Generally, small lakes and slow-moving streams freeze over earlier than larger lakes or fast moving streams.

2.1.3 While there are many different types of ice, the two types of major concern are:

(a) clear ice — formed by the freezing of water;
(b) snow ice — formed when water-saturated snow freezes on top of ice, making an opaque white ice which is not as strong as clear ice.

### 2.2 Ice colour

2.2.1 The colour of ice, which may range from blue to white to grey, provides an indication of its quality and strength:

(a) clear blue ice is generally the strongest;
(b) white opaque ice (snow ice) has a relatively high air content, and its strength depends on the density: the lower the density the weaker the ice; but high density white ice has a strength approaching that of blue ice;
(c) grey ice generally indicates the presence of water as a result of thawing, and must be considered highly suspect as a load-bearing surface.

### 2.3 Ice thickness

2.3.1 The other major factor determining the bearing capability of ice is its thickness. Care must be taken when determining the thickness of ice covers to ensure that the readings are properly taken and are an accurate representation of the area under consideration.

2.3.2 Currents have a distinct bearing on the temperature required to form ice. Rivers and channels with strong currents may remain open all winter despite low air temperatures. Springs can cause currents, and also be the source of warmer water; currents can also cause variations in ice thickness without changing the uniform surface characteristics.

2.3.3 When selecting the site of an ice bridge, currents and springs should be located and avoided. Frequent checks of the ice thickness should be made in areas suspected of being affected by currents.

2.3.4 Ice under an insulating snow blanket thickens very slowly even in low temperatures. A heavy snow cover, before significant ice growth, may cause the ice to remain unsafe throughout the winter.

## 3. Bearing capability of ice

### 3.1 General

3.1.1 The load bearing capacity of ice covers depends on the quality of ice, its thickness, ice and air temperatures, temperature changes and solar radiation.

3.1.2 Clear blue ice is the standard of quality against which other types of ice are compared. White opaque ice, or snow ice, is normally considered to be only half as strong.

3.1.3 Ice covers may consist of alternate layers of clear ice and snow ice, and each layer should be measured so that the effective thickness may be calculated. For example, an ice cover with a total thickness of 8 inches (20 cm) consisting of a 4 inch (10 cm) layer of clear ice and a 4 inch (10 cm) layer of snow ice would have an effective thickness of 6 inches (15 cm).

3.1.4 The strength of ice is generally increased by low temperatures. The increase is progressive from zero to minus eighteen degrees Celsius and remains fairly constant below this point. However, a marked drop in temperature can temporarily cause internal stress in an ice cover and reduce its bearing capacity. This can often occur during overnight periods when the temperature is much lower than the preceding average for the day.

3.1.5 The removal of snow from an ice cover during periods of low temperature has an effect similar to a marked temperature drop. The bearing capacity of ice should be considered to be reduced by 50 per cent for 24 hours after these conditions.

### 3.2 Determining ice thickness

3.2.1 Prior to use, the ice should be measured to determine whether its effective thickness is adequate to support the expected load. The graph presented in Appendix A should be used as a guide to the required thickness for the loads involved.

3.2.2  To initially determine effective ice thickness, the rule of thumb "one inch (2.5 cm) of clear blue ice for every thousand pounds (450 kg)" may be used.

**Caution**

Ice that is less than six inches (15 cm) thick should not be used for any crossing.  Because of natural variations, thickness may be less than 2 inches (5 cm) in some areas.

3.2.3  The effective thickness can vary considerably in an ice cover.  In particular, dangerously thin areas can occur due to currents in the covers of rivers and estuaries, and on lakes near the inlet or outlet of rivers and streams. Careful attention should be given to reduced ice thickness close to shorelines and around ridges and leads.

3.2.4  The thickness can be determined by drilling test holes spaced at a maximum of 50 feet (15 m) apart in rivers, and 100 feet (30 m) apart on a lake.

3.2.5  Crossings should be checked for ice thickness once a week when average air temperatures vary between -15 and -5 degrees Celsius; and daily when the temperature is above -5 degrees Celsius.  Checks can be less frequent when ice thickness substantially exceeds requirements.  A new hole should be drilled for each ice measurement.

3.2.6  Ice that is no longer supported by water, due to lowering water levels, may be too weak to support the loads to be applied; conversely, a rising water level can result in the formation of two ice layers with an intervening water layer.  Ice thickness tests will reveal these conditions.

### 3.3  **Parked and stationary loads**

3.3.1  Ice behaves elastically under moving loads; that is, the ice is depressed while loaded but recovers its original position after the load has passed.

3.3.2  With a stationary load the ice surface will sag continuously and may fail, depending on the strength of the ice cover.  The safe bearing capability for stationary loads should be considered to be 50 per cent less than that for moving loads.

3.3.3  The sequence of failure for stationary loads is as follows:

(a)  radiating cracks form at the bottom of the cover immediately beneath the load (and ultimately propagate through the cover);
(b)  circular cracks form at the upper surface of the cover at some distance from the load (noticeable sagging of the ice may occur);
(c)  the ice shears in a circle immediately adjacent to the loaded surface (failure may be imminent).

3.3.4 The initial radial cracks may not be of immediate concern if the load bearing capacity of the ice is substantially higher than the load. However, prolonged application of the load should cause concern about possible ice failure.

3.3.5 Stationary loads should be moved under any of the following conditions:

(a) when radial cracks develop;
(b) if noticeable sagging is observed;
(c) if the rate of sagging increases;
(d) if continuous cracking is heard or observed;
(e) if water appears on the surface of the cover.

3.3.6 The accumulation of drifted snow, often caused by stationary loads, may mask the indicators listed in paragraph 3.3.5 as well as increase the static load on the ice. Vehicles should be parked at least 5 lengths apart and in such a way that snow drifts do not interfere with other vehicles.

## 3.4 **Effects of speed**

3.4.1 When a vehicle travels over an ice cover, a hydrodynamic or resonance wave is set up in the underlying water. This wave travels at a speed that depends upon the depth of the water, the thickness of the cover and the degree of elasticity of the ice. If the speed of the vehicle coincides with that of the hydrodynamic wave, the stress on the cover due to the wave reinforces that due to the vehicle, and can increase the maximum stress in the ice to the point of failure. The wave action tends to crack the ice in a checkerboard pattern.

3.4.2 Particular care should be exercised when approaching or travelling close to shore, or over shallow water, because of more severe stressing of the cover due to reflection of the hydrodynamic wave. Roads and vehicle approaches should meet the shoreline at an angle of not less than 45 degrees.

3.4.3 If the weight of a loaded vehicle is one-half or less than that determined from Figure 1 as safe for the thickness of the ice being used, speed is not critical. When the weight is greater, and for ice thickness less than 30 inches (75 cm), speed should be carefully controlled and in general be kept below 10 m/h (15 km/h).

## 3.5 **Cracks**

3.5.1 The ice usually has many cracks made by thermal contraction or movements of the ice cover. Except at the thaw period cracks do not necessarily indicate a reduction in the load-bearing capability of the cover.

3.5.2 A dry crack with an opening of less than 1/8 inch (0.32 cm), which does not penetrate very deeply into the ice cover, will not cause serious weakening. Where a single dry crack in excess of one inch (2.5 cm) is noted, loads should be reduced by one third; for intersecting cracks of this size the loads should be reduced by two thirds. Dry cracks should be repaired by filling with water or slush.

3.5.3 A wet crack indicates that the crack penetrates completely through the ice cover and therefore affects the load-bearing capacity, which should be reduced by one-half in the case of a single wet crack. If two wet cracks meet at right angles the reduction is to one-quarter of that for a good cover. Most wet cracks refreeze as strong as the original ice cover; however a core sample should be taken to ascertain the depth of healing.

3.5.4 Due to normal thermal contraction, cracks sometimes form at the middle of a road in the direction of travel; but these do not seriously reduce the bearing capability if they remain dry. If cracks form parallel to the road, at the sides, they do indicate over-stressing (perhaps by snow deposits from clearing operations) and possible fatigue due to excessive traffic. If such cracks develop, particularly if they are wet, road use should cease at once, and not be recommenced until the cracks are healed.

3.5.5 Fluctuating water levels may produce cracks near and generally parallel to the shoreline. These cracks are often accompanied by a difference in the levels of the floating and the grounded ice. If these cracks are wet, loads should be reduced accordingly. With extreme level differences, appropriate bridging repair (flooding, reinforcing) may be necessary.

## 3.6 Spring thaw

3.6.1 Ice covers will begin to decay in the spring as the ice warms and begins to melt. The ice will thaw in the sunlight, but in the early spring may refreeze at night. Intensive thawing begins only in atmospheric temperatures above freezing.

3.6.2 Snow is a poorer thermal conductor than ice. A covering of 3 to 4 inches (7.5 to 10 cm) of clean snow on an ice bridge will reduce significantly the solar radiation penetrating the cover, thus prolonging the period of use.

3.6.3 Travel over an ice bridge displaying water on the surface should be executed with great caution and only if absolutely necessary. If mild weather continues and the water disappears, it may indicate that the ice is honey-combed, in which case the use of the area as an ice bridge should be discontinued immediately.

3.6.4 If the average air temperature has been above zero degrees Celsius for three days or more, then use of an ice-bridge should cease.

## 4. Preparation of ice bridges

### 4.1 Building techniques

4.1.1 A marked route over a natural ice cover can be utilized as an ice bridge, but since this may not provide sufficient strength for repetitive use, various techniques may be used to increase the safe load-bearing capability.

4.1.2  When temperatures are low and early winter use is not required, ice thickness can be increased by keeping the intended crossing snow-free, or by compacting the snow so that its normal insulating qualities are diminished.  The natural rate of ice growth will thus be accelerated and the required thickness will eventually be reached.

4.1.3  If there is a need for a bridge when temperatures are not low enough to obtain the necessary natural thickness by the time of required use, the ice thickness can be increased by flooding:  adding water on top of the existing ice cover.

## 4.2  Flooding

4.2.1  The flooding operation is normally carried out with small lightweight pumps, rather than larger pumps which are less portable.

4.2.2  Flooding may be started as soon as the natural ice is about 3 inches (7.5 cm) thick and strong enough to bear the weight of persons and pumps. The initial flooding should be limited to a depth of about one inch (2.5 cm).

4.2.3  Subsequent floodings for "lifts" should be limited to that depth of water that will freeze within 12 hours.  As a rule of thumb, an average air temperature of -18 degrees Celsius will freeze 2 inches (5 cm) of water overnight.  With average temperatures of -31 degrees Celsius or lower, lifts may be increased to 3 1/2 inches (9 cm) the freezing rate respectively.  Wind or snow on the surface will increase or decrease the freezing rate respectively.

4.2.4  Thicker lifts can lead to a layer of water between the old ice surface and the new layer of ice.  When covered by succeeding lifts of warm water, this layer may not freeze until well after the bridge has been completed.  Such lifts may also overload and crack the existing ice cover.

4.2.5  To achieve maximum strength in the bridge, any snow cover should, if possible, be removed before each flooding operation.  However, dragging or packing the snow to an even thickness and then flooding — "slushing" — provides a thicker sheet in less time but the resulting ice is not as strong.

4.2.6  If banks of snow are constructed on each side of the bridge to contain the flooding, they should be at least 150 feet (45 metres) apart; however, a 200 foot (60 metre) wide bridge is preferable.

4.2.7  Snow banks may leak after freezing has begun so that a crust of ice is formed with an air-filled void between it and the initial ice cover.

4.2.8  Flooding should take place from the bridge centre line, letting the water feather out to seek its own level.  This method also provides a wider bridge surface.

4.2.9  Ice formed by the flooding process will be stress-free if each lift is allowed to become completely frozen before the next flooding.

### 4.3 Reinforcement

4.3.1 An ice bridge built in more temperate climates or intended for repeated use may be reinforced with grasses, brush or logs. Such a bridge can then take a greater load for the same thickness, being held together by the reinforcing inclusions. It can heal itself more easily after cracking and is less likely to fail catastrophically.

4.3.2 One disadvantage to reinforcement is the added time and effort required for construction. Another is the effect of local radiational heating of the reinforcing inclusions, particularly during the spring thaw, which will increase the rate of decay of the bridge.

4.3.3 It is preferable to locate the reinforcing items in the bottom portion of the final ice bridge; they should be placed and frozen in as early as possible.

4.3.4 Reinforcing logs, properly placed in an ice bridge, will make possible a reduction of ice thickness of up to 25 per cent.

### 4.4 Maintenance

4.4.1 On completion, the following rules should be observed in order to increase the safety and life of the ice bridge:

(a) The bridge must be kept clear of excessive snow, and the snow banks kept well back, with slopes of no more than a ratio of 1 to 5. The weight of snow banks can weaken the ice underneath and form relatively deep ditches by slow sagging, and therefore should be levelled out if higher than 3 feet (1 metre) or two thirds of the ice thickness, whichever is the larger.

(b) A covering of 3 to 4 inches (7.5 to 10 cm) of compacted snow will give good traction and will also provide a cushion. Glare or snow-free ice breaks up rapidly under traffic in extreme cold.

(c) The surface should be kept clear of dirt or other dark material, such as oil spots, which will absorb solar radiation and melt into the ice. Puddles of water also absorb heat from the sun and should be "repaired" by filling with snow.

(d) The ice bridge should be checked for cracks daily and on foot, and its thickness measured as outlined in article 3.2. A longitudinal crack more or less down the centre line may occur, particularly if the ice thickness has been increased by flooding. If dry, this crack is not serious. Wet cracks should be repaired immediately and loads reduced until the refreezing process is completed (see article 3.5).

### 4.5 Operating precautions

4.5.1 Following are a number of general precautions which should be taken when testing for ice thickness or crossing ice covers:

(a) All persons involved in operations over ice covers should be familiar with the hazards involved, the precautions to be taken and the basic rescue techniques required in case of a breakthrough.

(b) Single persons or single vehicles should not venture onto an ice cover when there is no help at hand.

(c) When testing, persons on foot should carry long poles, to be used as an aid to rescue in case of a breakthrough, or alternatively be securely roped together, with minimum spacing of 50 feet (15 m).

(d) Light vehicles used during test periods and initial build-up should be equipped with an extended frame of logs to provide support if the vehicles break through the ice cover.

(e) A rope at least 50 feet (15 m) long, or equivalent to water depth, with a float, may be attached to test vehicles as an aid to marking and recovery.

(f) Vehicle doors and cab hatches should be removed or lashed open; seat belts must NOT be worn.

(g) Adequate spacing must be maintained between vehicles; it is recommended that an interval of at least 100 feet (30 m) be observed.

(h) Vehicle speed should not normally exceed 10 m/h (15 km/h) in order to avoid the effects of the hydrodynamic wave, nor should speed be less than 1 m/h (1.5 km/h) in order to avoid the effects of stationary load.

(i) Where practicable, precautionary and speed limit signs should be erected at each end of the ice bridge, and the route across the ice cover clearly marked.

(j) Traffic lanes should alternate across the width of the ice bridge, working gradually from one side to the other before starting over again. This reduces the danger of deterioration of the ice and makes possible a choice of routes if dangerously cracked areas develop or breakthrough occurs.

(k) Equipment required for rescue operations, such as "mats" (chained or wire-linked small logs or heavy planks as a platform for rescue vehicles) jacks, hoists, etc., should be available near by.

(l) Frequently it is the second vehicle in a convoy which encounters ice failure problems. Before a second heavily loaded vehicle proceeds along the ice bridge, it is advisable to have it preceded by a more lightly loaded vehicle to check the route.

(m) For a period of 24 hours after a marked drop in temperature, or following the removal of snow from the ice cover during periods of low temperature, loads should be reduced by 50 per cent and night-time travel should be discouraged.

## 5. The use of snowmobiles on ice covers

### 5.1 General

5.1.1 Drownings resulting from snowmobiles going through ice are the greatest single cause of fatalities arising out of the use of these machines. However, snowmobile operations over ice covers can be conducted safely by using common sense and observing the basic precautions.

5.1.2 As the total load — machine, operator and ancillary gear — may weigh approximately 500 pounds (225 kg) or more, a substantial thickness of ice is required for support.

5.1.3 Difficulties in control, steering and stopping are increased on snow-free ice, particularly at higher speeds.

### 5.2 Operation precautions

5.2.1 The following is an outline of some of the basic precautions:

(a) Where there is an alternative, single machines should not be operated unaccompanied over ice covers.

(b) Should single machine operation be unavoidable, the shore base should be notified of the route to be taken, the destination and probable time of return.

(c) Operations should not be conducted over ice covers less than 6 inches (15 cm) thick.

(d) Operators should know of and avoid locations where currents or springs may cause dangerous thinning of the ice cover.

(e) Fog may indicate the proximity of open water; speed should be reduced and great care taken.

(f) When unexpectedly encountering open water normal action is to slow down, brake gently and turn away; otherwise, turn as sharply as possible. If a turn cannot be made in time or a skid results, the operator should roll off the machine.

(g) Glare from the sun and ice may obscure obstacles or dangerous areas; anti-glare sun glasses should be worn under these conditions.

(h) Operations at night or at high speeds should be restricted to well-marked and known safe trails or crossings.

(i) Unless essential, snowmobiles should not be operated on ice bridges or roads with other types of traffic.

(j) Avoid operating over slush or water-covered ice; but if unavoidable, ensure that the tracks are cleared of ice and slush.

## References

Additional technical information concerning ice formation and its use is available in the following publications:

Publication CL1-7-71
*Freeze-up and Break-up Dates of Water Bodies in Canada*
Information Section
Central Service Directorate
Atmospheric Environment Services
Environment Canada

Technical Memorandum No. 56
*The Bearing Strength of Ice*
National Research Council

Research Paper No. 469, NRCC 11806
*Use of Ice Covers for Transportation*
National Research Council

Information and advice may be obtained also from the National Research Council of Canada, Division of Building Research, Geotechnical Section, Ottawa, Ontario K1A 0R6.

Enquiries should be directed to the responsible officers in departmental headquarters who, in turn, may seek interpretation from the following:

Safety, Health and Employee Services Group
Staff Relations Division
Human Resources Policy Branch
Treasury Board Secretariat

THICKNESS OF GOOD QUALITY FRESH WATER ICE, cm

ÉPAISSEUR DE LA GLACE D'EAU DOUCE DE BONNE QUALITÉ, en cm

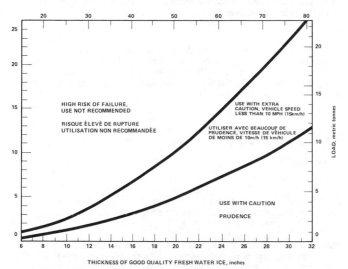

HIGH RISK OF FAILURE, USE NOT RECOMMENDED

RISQUE ÉLEVÉ DE RUPTURE UTILISATION NON RECOMMANDÉE

USE WITH EXTRA CAUTION, VEHICLE SPEED LESS THAN 10 MPH (15km/h)

UTILISER AVEC BEAUCOUP DE PRUDENCE, VITESSE DE VÉHICULE DE MOINS DE 10m/h (15 km/h)

USE WITH CAUTION

PRUDENCE

LOAD, metric tonnes

THICKNESS OF GOOD QUALITY FRESH WATER ICE, inches

The allowable load should be reduced by one half for operations on white opaque ice.

RECOMMENDED BEARING CAPACITY BASED ON EXPERIENCE

ÉPAISSEUR DE LA GLACE D'EAU DOUCE DE BONNE QUALITÉ, en pouces

La charge permise doit être réduite de moitié dans le cas des travaux effectués sur glace blanche opaque.

PORTANCE RECOMMANDÉE D'APRÈS L'EXPÉRIENCE

## CHAPTER 5-4 - SAFETY GUIDE FOR FIELD OPERATIONS

### Foreword

A wide range of activities, environmental conditions and hazards are common to field operations, requiring special care and attention on the part of those involved directly, and on the part of supervisory staffs responsible for the control of such operations.

This guide has been prepared accordingly, to focus attention on some major areas of concern, and to serve as a basis for the development of more detailed departmental safety rules and procedures. Particular note should be made of action to be taken before personnel are dispatched on any field operation: thorough planning and careful preparation are essential to the continued safety and good health of those involved.

### Purpose

Field operations, particularly those undertaken in isolated areas, expose personnel to a wide range of unique occupational risks and hazards. The purpose of this guide, therefore, is to provide an outline of basic occupational health and safety practices and procedures which may be applied and further developed as required by departments.

### Application

This guide applies to all Public Service departments and agencies, as defined in Part I of Schedule I of the *Public Service Staff Relations Act*.

### Definitions

In this guide:

*field officer* means a party chief or other officer to whom, during field operations, is delegated the responsibility to manage a project, or a part thereof (agent des opérations sur le terrain);

*field operations* means those operations and activities conducted by individuals or groups of persons away from the department's premises, such as surveys of an engineering and research nature, forest fire fighting, rescue operations or work parties (opérations sur le terrain).

### General

Departments undertaking field operations should, based on the general requirements of this guide, issue detailed safety directives governing the conduct of field operations, appropriate to the particular risks and hazards which may be foreseen. In this regard, the following general practices are recommended:

(1) one member of each field party (normally the field officer in charge) should be appointed as a safety officer;

(2) where a safety officer is not the field officer in charge, a clear definition of the safety officer's authority and responsibility should be provided;

(3) an appropriate official or authority in the area should be advised of the geographical location of an isolated field operation, its estimated duration, the normal and emergency methods of communication, and the names or the number of personnel in the party. Provincial forest services should be advised as a courtesy;

(4) all relevant safety and health directives and standards applicable to the Public Service should be reviewed prior to departure (see the section entitled "Other references" for list);

(5) inexperienced personnel who will be working in isolated areas should be provided, where required, with basic survival training or information.

Members of the field party should be briefed by the designated safety officer on the basic safety and health rules to be observed according to the type of field operation and the expected environment, including the following:

(1) the location of the nearest available emergency medical facility, police station, military or forestry establishment;

(2) the procedures to be followed in the event personnel become lost, or in the evacuation of seriously ill or injured personnel;

(3) the location and method of operating of any emergency equipment provided or available in the area;

(4) the procedure to be followed for carrying out regular field operating reviews for the purpose of identifying and eliminating unsafe and unhealthy conditions and practices; and

(5) procedures to be followed in the event of encounters with wildlife, particularly bears.

## Equipment

Departments should ensure that all field and safety equipment is checked for suitability and serviceability prior to issue, and re-checked by field and/or safety officers on receipt. Personnel should be instructed on the proper use, care and maintenance of field and safety equipment.

## Boat and water safety

Comprehensive safety guidelines, embodied in the Transport Canada publication *Boating Safety Guide*, concerning the use of boats, both powered and unpowered, should be followed. The wearing of approved type life jackets should be enforced in accordance with the provisions of the Personal protective equipment directive, chapter 2-14.

## Snowmobiles

Safety rules concerning the operation of snowmobiles, such as those outlined in the booklet *Play Safe with Snowmobiles* (available from the Canada Safety Council), should be followed. In addition, snowmobiles should be operated in compliance with local regulations governing their use. Additional safe operating procedures are contained in the Safety guide for operations over ice, chapter 5-3.

## Motor vehicles and trailers

Departmental safety rules and procedures concerning the operation and use of government-owned or leased motor vehicles and trailers, encompassing the applicable requirements of the Motor vehicle operations directive, chapter 2-11, should be developed and enforced.

## Diving operations

Personnel performing diving duties should be physically and mentally fit to perform each task, and be in possession of a valid certificate of qualification satisfactory to the department.

## Sanitation and hygiene

The field officer in charge should ensure, as far as is practicable, that personnel involved in the preparation and serving of food are free from any communicable disease, and that sanitation and shelter facilities are maintained in a manner that does not constitute a health or ecological hazard.

## Vehicle traffic hazards

All field operation crews exposed to hazards from vehicular traffic should wear a high visibility vest or other similar clothing, and use appropriate warning signs or be protected by a high visibility barricade in accordance with the Personal protective equipment directive, chapter 2-14.

## Tent heaters, gasoline stoves and lanterns

All heaters, gasoline stoves and lanterns should be carefully fuelled and lighted, and care should be taken to keep all open-flame models away from combustible materials. During the use of such equipment in tents, shelters, or any confined enclosure, adequate ventilation should be assured to eliminate the possibility of carbon monoxide poisoning or oxygen deficiency.

## Firearms

No person should be allowed to handle or use a firearm unless the department is satisfied as to that person's sense of responsibility, competence and demonstrated knowledge of accepted safety practices in the use of firearms.

## Use of explosives

Good industrial safety practices and departmental regulations, where applicable, should be followed with respect to the handling, preparation or firing of explosive charges. Information in this regard is available from Human Resources Development Canada in Technical Data Sheet "The Safe Use of Explosives in Federal Enterprises". The *Explosives Act of Canada* should also be observed.

## Fire prevention and fire fighting

Basic fire fighting rules and procedures should be developed and enforced for specific field operations consistent with the general requirements and standards prescribed by the Fire Commissioner of Canada. Provincial forestry requirements should be observed where applicable. Fire fighting should be controlled by experienced crew leaders, and appropriate safety clothing and footwear worn.

## Air transport operations

The safety of personnel travelling by air is a prime consideration when making arrangements for transport by charter aircraft. Departments should conclude contracts only with those air carriers who have demonstrated compliance with Transport Canada directives relating to air carriers. These directives are found in the Air Navigation Orders (ANOs) available from Public Works and Government Services Canada, which contain safety-related regulations; designated safety officers should carefully review the following ANOs, as appropriate, before the operation begins:

(1) ANO VII, 3 — Standards and Procedures for Air Carriers Using Small Airplanes in Air Transport Operations.
(2) ANO VII, 2 — Standards and Procedures for Air Carriers Using Large Airplanes in Air Transport Operations.
(3) ANO VII, 6 — Standards and Procedures for Air Carriers Using Rotorcraft in Air Transport Operations.

Non-compliance with these safety procedures, or any other unsafe practices or conditions, should be reported to Transport Canada, ASP, Transport Canada Building, Ottawa, K1A 0N8.

## Emergency equipment and exits

A seatbelt is required for every passenger. The seatbelt should be secured during takeoff and landing and whenever considered necessary by the flight crew. Each passenger seat shall be provided with printed information listing the emergency equipment carried, and the location and operation of emergency exits. The emergency equipment and rations detailed in ANO V, 12, shall be carried on all flights conducted within the "sparsely settled area" which is defined in the order. There should be a readily-accessible life jacket or flotation device for each person on board a floatplane, and the location of these made known to the passengers by the pilot or aircrew.

## Personal clothing

The probable temperature in the area of the flight should be known, and appropriate footwear (not street oxfords) and clothing worn or carried, including clothing to protect against insects in summer months.

## Emergency locator transmitter (ELT)

All aircraft are required to carry a compact radio, which transmits a distinctive signal on the emergency frequency of 121.5 mHz for the detection and location of downed aircraft. It will normally be triggered "on" automatically during a forced landing. If the ELT is not automatically triggered, it can be turned on manually. The battery life of an ELT is at least 100 hours, and signals can be heard up to 100 miles (160 km) away by high-flying aircraft. The ELT provides a homing signal to pinpoint location and greatly reduces time-to-rescue.

Information on the location and operation of the ELT is placarded in the cabin, and its location is marked externally on the aircraft. It is usually mounted behind the cabin or to the rear of the aircraft. Before boarding the aircraft the crew should describe the location and operation of the ELT.

More information on the ELT and the search and rescue system is given in a colour slide/sound presentation available on loan from any Regional Office of Transport Canada.

### Flightplans

The purpose of the flightplan is to ensure that a record is available if an emergency develops. For every flight, a flightplan form should be filed by the pilot through the company (or an agent) for transmission to air traffic control. If this is not possible, the pilot is required to notify a responsible person of his proposed flight by means of a flight notification or flight itinerary. This should specify the estimated duration of the flight or series of flights, the estimated time of return, the route or the area boundaries of the flight operation, and the location of any overnight stops.

The pilot is responsible for the safe conduct of the flight, and should not be unnecessarily distracted from the task of flying. Passengers should not request changes from the flightplan for personal reasons such as sightseeing, photography, low flying, etc.

### Weather

It is the pilot's responsibility to determine if weather conditions are suitable for a safe flight, and passengers should not attempt to influence the pilot's decision in this regard. Normally, safe visual flight, i.e. flight with visual reference to the ground, requires a 1000-ft (300 m) cloud ceiling and 3 miles (4.8 km) forward visibility. However, minimum requirements permit flight with a 700-ft (210 m) cloud ceiling and 1 mile (1.6 km) forward visibility. If weather falls below these limits, visual flight must be discontinued. Visual flight above a cloud layer is not permitted.

### Cargo

Personal baggage and equipment should be properly secured. When cargo is carried in the cabin with passengers, it should be secured by nets, strapping or other tied down to prevent shifting in flight. Cargo should not be placed so as to restrict the use of emergency or regular exits. It is the pilot's responsibility not to exceed the aircraft's total "maximum gross weight", and to ensure that the load is distributed so that the aircraft is within its centre of gravity limits. Passengers carrying their own baggage and equipment should ask the pilot where it is to be placed; the pilot should not be pressed to put on extra items that might overload the aircraft.

**Propellers and rotors**

Every year rotating propellers and rotors cause fatal and serious injuries because they are difficult to see when in motion. Passengers should not board, leave or work around aircraft when propellers or rotors are in motion. The helicopter is often an exception to this rule, when rotors must be kept in motion at remote landing sites. Passengers should receive a thorough briefing from the pilot or crew member before boarding or leaving a running helicopter. Most accidents occur when persons walk into the tail rotor. The safe procedure is to crouch low and approach or depart the helicopter from the side or the front but never near the tail rotor area. Never walk downslope toward the helicopter and never walk upslope away from the helicopter.

A passenger should not normally perform any crew function unless safety is otherwise in jeopardy. In any case, a passenger must receive a thorough briefing, including safety procedures, from the pilot or flight crew member. An inexperienced floatplane passenger attempting to assist the pilot to dock is exposed to extreme danger from the rotating propeller. Passengers have been struck by propellers while walking forward to the front end of the float to tie the aircraft to a dock or mooring point. After docking, twin-engine floatplanes create a hazard to persons when rotating propellers overhang the dock. Passengers should not disembark from these aircraft until the propellers are stopped.

Additional safety information is given in a colour slide/sound presentation entitled "Safety Around Small Aircraft", available from any Regional Office of Human Resources Development Canada.

**Other references**

In addition to other applicable safety standards, procedures and guides approved by the Treasury Board for the Public Service, the following specific documents should be reviewed in conjunction with this guide, and applied as appropriate:

(1)  First-aid and health directive, chapter 2-5.
(2)  Tools and machinery directive, chapter 2-9.
(3)  Personal protective equipment directive, chapter 2-14.
(4)  Motor vehicle operations directive, chapter 2-11.
(5)  Occupational health evaluation standard, chapter 2-13.
(6)  Safety guide for operations over ice, chapter 5-3.

**Enquiries**

Enquiries should be directed to the responsible officers in departmental headquarters who, in turn, may seek interpretation from the following:

> Safety, Health and Employee Services Group
> Staff Relations Division
> Human Resources Policy Branch
> Treasury Board Secretariat

## CHAPTER 5-5 - A GUIDE ON VIDEO DISPLAY TERMINALS (VDTS)

### Introduction

New office technology, in particular the video display terminal (VDT), has given rise to several health concerns for employees and their managers.

### Purpose

This guide is intended to provide general information about these concerns and the generally accepted ways of dealing with them. It also briefly addresses the desirable features of office workplaces and work organization. Consult the references at the end of this guide for more information.

### Health concerns

### Radiation

The concerns over radiation hazards of VDTs have attracted a great deal of media attention in recent years. Nevertheless, the evidence of reputable studies all over the world indicate that VDTs do not emit x-radiation (ionizing radiation) above background levels.

These findings are the same as those published by the Bureau of Radiation and Medical Devices of Health Canada, the agency responsible for enforcing the *Radiation Emitting Devices Act..*

The Department reports that the Bureau tested VDT models beyond standard radiation testing methods with the help of a specialized counting facility for low-level emissions. The result was that no x-ray emission was detected regardless of whether the sets were on or off.

With respect to non-ionizing radiation, (radiofrequency, microwave, infrared, visible light and ultraviolet), the Department and the National Institute for Occupational Safety and Health (NIOSH) of the United States of America have found the levels of emissions to be either non-detectable or well within the national standards set to protect people.

Extremely low frequency (ELF) emissions were found to be of very low intensity that were comparable to those from common electrical and electronic devices in homes and offices.

Consequently, the Department of Health reports that VDTs do not emit radiation or emit so little that no standard in the world classifies them as hazardous.

### Radiation and pregnancy

Some epidemiological studies investigating the risk of spontaneous abortions (miscarriages), stillbirths, prematurity and congenital malformations for women exposed to VDTs during pregnancy have found some spontaneous abortion risk related to extended work with VDTs. However, the nature of these retrospective studies and their methodologies raised questions about their scientific validity.

According to experts in the field, the scientific investigations of the VDTs themselves rule out allegations that radiation emissions from VDTs may have adverse effects, including spontaneous abortions, on pregnant employees.

Additionally, the prospective epidemiological study of pregnant VDT operators carried out by NIOSH in 1991 confirmed that the use of VDTs and exposure to the accompanying electromagnetic fields were not associated with an increased risk of spontaneous abortion.

Since VDTs do not emit x-radiation, leaded aprons for pregnant employees are not encouraged as they create an unnecessary burden. Furthermore, while lead protects against x-radiation, it offers no protection against ELF radiation emissions.

## Vision

While visual fatigue may occur as a result of VDT work, the effects are not permanent and can be relieved by ergonomic considerations. Scientific and epidemiological data do not support claims of cataracts caused by radiation emitted by VDTs (or TVs, which are similar devices).

Eyes normally adjust to the viewing distance and the contrast lighting of the VDT screen display. However, visual discomfort may occur as a result of intensive VDT work, poor image quality, high lighting levels, glare, or insufficient humidity and ventilation.

Eyeglasses or contact lenses should not affect a worker's ability to use a VDT. However, wearers of bifocal or multifocal lenses should consult their eye specialist to determine the appropriate viewing distance and angle for VDT work or to obtain a suitable prescription.

## Musculo-skeletal problems

VDT users have reported muscular aches of the upper back, shoulders, arms, wrists and hands. These ailments, which are usually related to intense and prolonged repetitive activity, can lead to more chronic and debilitating problems known as repetitive strain injuries or cumulative trauma disorders. These include carpal tunnel syndrome (compression of the hand's median nerve), tendonitis (inflammation of the tendons) and tenosynovitis (inflammation of the tendons and surrounding sheath).

Lower back discomfort and circulatory problems of the legs are common complaints of sedentary workers. Jobs requiring static positions for lengthy periods have the potential to disturb the normal functioning of the circulatory and musculo-skeletal systems. Poorly designed seating that cannot adequately support the employee's size and shape contribute further to back and leg ailments.

Prolonged sedentary activity as well as repetitive activity may cause initial discomfort and may eventually lead to illness or injury.

## Stress

Stress, which is associated with all types of jobs, can adversely affect employee satisfaction and productivity. For employees using VDTs, the most commonly reported causes of stress after vision and musculo-skeletal factors are poor job design, lack of control over work requirements, under- or over-utilization of skills, high speed or repetitive work, electronic monitoring of productivity, and social isolation.

A key problem when dealing with work-related stress is the human perception of stress. People vary considerably in their shape, size, strength, aptitudes, and sensory and mental abilities. Similarly, what one employee may consider stressful, another may not. The workplace, the equipment and the job design should allow for these differences where possible.

### Addressing concerns

### Transfer of pregnant employees

Under the Transfer of Pregnant Employees Policy, a temporary transfer will be granted when requested by a pregnant employee working on a VDT, if other work is available. If other work is unavailable, the employee will be granted leave without pay upon request.

This policy is not an acknowledgement of a radiation hazard. Rather it is intended to relieve psychological stress caused, in part, by reports of VDT pregnancy hazards by the media and other sources.

### Ergonomics

All components of the work environment affect a person's well-being: lay-out, equipment, furniture, temperature, lighting, acoustics, and the like. Ergonomic considerations in the early design stages of the VDT work station as well as the tasks to be accomplished will minimize safety and health hazards as well as maximize employee productivity. Managers should think "ergonomics" when replacing equipment, setting up a new workplace, modifying tasks, or planning changes to the old ones. Employees may be able to suggest reasonable solutions to the difficult aspects of their work.

Considerations for the office workplace include:

1.     **Equipment** should be selected for ease of use. It should also be easy to install, operate and maintain. Shared equipment should be readily adaptable to different employees and to different tasks.

The placement of the VDT should allow the employee to look down at the middle of the screen, with the top of the screen being no higher than eye level. The screen should be at a comfortable viewing distance and slightly tilted backward.

There should be adequate space in front of the keyboard and the mouse to allow the employee's hands to rest.

2. **Visual displays** should be easy to read. Lay-out, viewing angle, size of characters, contrast, illumination level, reflections, glare and colour should all be considered. The shape and placement of controls, screen, keyboard, and document holder should also minimize employee effort. Non-reflective finishes on walls, office furniture and VDT equipment will reduce glare and reflections.

Employees can rest their eyes periodically by looking away from the screen and beyond the workstation. Eye exercises and periodic changes in activity also help to prevent eye strain.

3. **Furniture** should be arranged to allow for ease and efficiency of movement, taking into consideration the visual field, sequence and frequency of use, communication, reach requirements and storage facilities.

The worksurface should be at the appropriate height for the employee. The upper arms should hang from the shoulders and the forearms should be parallel to the floor with the hands in line with the forearms. A raised seat in combination with a footrest that is adjustable and large enough to allow movement, may be used as a substitute for a lower desk for short employees.

The worksurface should also be able to accommodate materials related to the task, and should allow for sufficient leg clearance underneath.

Chairs should be adjustable with attention given to: seat height, pan and angle; seat cushion; back and arm rests; and stability of the chair base.

The employee should be able to sit comfortably with the thighs parallel to the floor, the front of the seat not touching the backs of the knees, and the feet resting firmly on the floor or footrest.

The chair should allow the back to be upright and supported from below waist level to the shoulder blades. Armrests should also allow the forearms to rest at a right angle to the upper arms.

4. **Lighting levels for work at VDTs** should generally be lower than for paperwork. Consideration should also be given to contrast, daylight, glare, reflections, luminaires, ambient light, and lighting for paper tasks. VDTs should be placed away from windows or at a $90^0$ angle to the windows. Light from windows should be controlled with blinds or curtains.

5. To reduce **repetitive strain injuries** and add to employee satisfaction, repetitive work can be combined with other tasks requiring different skills. Physical exercise can play a role in prevention.

6. To reduce **lower back and leg discomfort**, sedentary work should allow for activities that enable periodic changes in posture. Work requiring static positions for lengthy periods can disturb normal circulation and muscle activity.

Employees can get up and stretch occasionally to relieve muscular tension. Physical exercise can also help to increase circulation and strengthen muscles.

7. To reduce **stress**, task demands can be fitted to employee characteristics through the design of the workplace and the task, through placement based on ability or through task-related training.

The involvement of employees in decisions that affect them, where practicable, helps to ensure a good match between the job and the employee.

**NOTE:** Additional information on this subject can be found in *Office Ergonomics: A National Standard of Canada*, published by the Canadian Standards Association. The standard has been adopted as a guide for the Public Service by the Treasury Board. The criteria for its implementation is found in the Treasury Board Information Technology Standard 13, Guideline on Office Ergonomics.

### VDT information sessions

The Occupational Health Nurses in departments are available to provide information sessions on the health aspects of VDTs. Departments may contact their occupational health nurse(s) or the nearest regional office of Health Canada.

### References

Treasury Board Information Technology Standard (TBITS) 13, Guideline on Office Ergonomics, 1991 (catalogue no. BT52-6/9), published in the *Treasury Board Manual*, Information and Technology Standards volume. Available from the Canada Communications Group — Publishing and departmental libraries.

*Office Ergonomics*: *A National Standard of Canada*, Canadian Standards Association (CSA), 1989, referenced in TBITS 13 on Office Ergonomics, 1991. Available from the CSA and departmental libraries.

Policy on transfer of pregnant employees, Chapter 13, Staff Relations Volume, *Treasury Board Manual*.

*Investigation of Radiation Emissions from Video Display Terminals* 83-EHD-91, Bureau of Radiation and Medical Devices, Health Canada.

Health Aspects of Work with Visual Display Terminals, *Journal of Occupational Medicine*, September 1988, Dr. Maria Stuchly and Dr. Ian A. Marriott, Bureau of Radiation and Medical Devices, Health Canada.

*Health and Ergonomic Aspects of Video Display Terminals*, Public Service Health Bulletin #7, Health Canada, available in departmental health units.

*Occupational Safety and Health Information Bulletin on Ergonomics, Safety*, Health and Employee Services, Treasury Board Secretariat, March 1993.

**Enquiries**

Enquiries should be directed to the responsible officers in departmental headquarters who, in turn, may seek interpretations from:

On general VDT Guide issues:

> Safety, Health and Employee Services Group
> Staff Relations Division
> Human Resources Policy Branch
> Treasury Board Secretariat

On the Treasury Board Information Technology Standard 13 on Office Ergonomics, published in the *Treasury Board Manual*:

> Office of Information Management, Systems and Technology
> Treasury Board Secretariat

On VDT radiation studies:

> Bureau of Radiation and Medical Devices
> Environmental Health Directorate
> Health Protection Branch
> Health Canada

On VDT Information Sessions:

> Departmental Occupational Health Unit or Regional Offices
> Occupational and Environmental Health Services Directorate
> Medical Services Branch
> Health Canada

## CHAPTER 6 - ADVISORY NOTICES

### Introduction

Occupational safety and health (OSH) advisory notices are special administrative instructions on important OSH issues for the immediate attention of departments. An advisory notice may be issued as a precursor to a new directive, as an adjunct to an existing directive, or simply as new instructions on a topical OSH issue for short-term or ongoing application.

## CHAPTER 6-1 - ADVISORY NOTICE - EMPLOYEES WORKING ALONE

Concerns have been expressed regarding the safety and well-being of employees who are required to work alone in situations where, in the event of serious injury or other emergency, the required assistance may not be readily available for the affected employee.

This advisory notice is to advise departments and agencies of the need to be aware of the precautionary and protective measures which may be required in respect of the safety of employees working alone, and to develop measures which are devised to provide assistance to such employees where and when required.

Local joint occupational safety and health committees should carefully review the various situations where employees are required to work alone, and make recommendations concerning the most reliable and practicable means of providing the required assistance to those employees, who may be subject to a significant risk or hazard.

In this regard, items that should be reviewed include:

(a)  a review of the work procedures and processes carried out in a "working alone" situation;
(b)  the adequacy of equipment and procedures used in the particular situation;
(c)  the adequacy of emergency support systems and monitoring mechanisms;
(d)  the adequacy of employee training and information respecting the use of survival equipment, protective equipment and clothing, and the use of communication or supervising systems.

A number of Public Service Occupational health and safety standards already incorporate specific references concerning the safety and health of persons working alone. These include such TB directives as: 2-3 (Electrical directive); 2-7 (Hazardous confined spaces directive); 2-10 (Materials handling directive); and 2-11 (Motor vehicle operations directive). The health and safety directives are now subject to cyclical review in the National Joint Council, and as each directive is consulted upon by the parties, particular consideration will be given to the incorporation, where required, of appropriate references to the safety and well-being of persons required to work alone.

**Enquiries**

Enquiries should be directed to the responsible officers in departmental headquarters who, in turn, may seek interpretation from the following:

> Safety, Health and Employee Services Group
> Staff Relations Division
> Human Resources Policy Branch
> Treasury Board Secretariat

**CHAPTER 6-2 - ADVISORY NOTICE - WEARING OF CONTACT LENSES IN LABORATORIES**

The purpose of this advisory notice is to provide general information concerning the precautions which should be taken in respect of persons who wear contact lenses while working in a laboratory environment.

As the type of work and the resultant risks involved in laboratory operations are variable, the precautions to be taken should be based upon the potential hazards identified in each laboratory environment.

Except for those situations where a significant risk of eye injury has been identified, the wearing of contact lenses in a laboratory is not hazardous. However, contact lenses should not be worn by those who are routinely exposed to irritating fumes, intense heat, liquid splashes, molten metals or other similar environments, and where the work requires the regular wearing of a respirator.

The wearing of contact lenses is not considered as a substitute for the required use of approved eye protection equipment and appliances, and those wearing contacts should use the same eye protection equipment as that required of other employees performing the same tasks.

Where there is doubt concerning the wearing of contact lenses in a laboratory, there should be consultation between the wearer, the supervisor and a medical adviser. Where there is doubt in a particular case, regular eye glasses should be worn.

Supervisors and those responsible for the administration of first aid should be aware of the identity of contact lens wearers.

Further information concerning this subject may be obtained through the appropriate Regional Office of the Medical Services Branch, Health Canada.

**Enquiries**

Enquiries should be directed to the responsible officers in departmental headquarters who, in turn, may seek interpretation from the following:

> Safety, Health and Employee Services Group
> Staff Relations Division
> Human Resources Policy Branch
> Treasury Board Secretariat

**CHAPTER 6-3 - ADVISORY NOTICE - THE EFFECTS OF EXTREME COLD**

The operational needs of the Public Service require some employees to work outdoors in conditions of extreme cold. The purpose of this advisory notice is to give information regarding the effects of cold, advice on how to prevent these effects from occurring, and how to deal with the effects before medical help can be reached.

*Frostbite* — This is the destruction of body tissues, usually in the face, hands or feet, by freezing. Circulation of blood in the tissues is slowed, then stopped; the skin appears a waxy white and becomes numb. Severe frostbite can lead to the destruction of tissue and even to the loss of fingers or toes (gelure).

*Hypothermia* — This is an actual lowering of body temperature due to the prolonged exposure to cold. Heat is lost from the body faster than it can be generated by metabolic activity. There is a gradual deterioration of body function leading eventually to loss of consciousness and death (hypothermie).

*Trench foot* — This is a form of local cooling of the feet, usually in cold, damp or even wet conditions, and can occur under conditions less cold than those producing frostbite (pieds gelés).

*Clothing* — The purpose of cold weather clothing is to maintain layers of warm air around the body. Clothing should be worn in a number of layers, so that one or more items of clothing can be discarded when doing heavy work. The outer layer should be windproof. Gloves should also be in layers, again with a windproof outer layer. Mitts are warmest, but a pair of gloves should be carried for more exacting work. "Mukluk" type footwear is best for the feet, provided that they will not get wet. Protection of the head is essential as this is where most heat loss occurs. Face masks may also be needed (porter des vêtements appropriés).

*Keeping dry* — Wet clothing will conduct heat away from the body. Sweating should be avoided by wearing only enough layers of clothing when working hard, and by opening clothing at the neck. Dry socks should be put on each morning. Snow must be brushed off all clothing before it is put on. Note that gasoline, when very cold, can cause severe skin burns. It must never be allowed to come in contact with the skin. The same applies to metal objects (rester au sec).

*Avoiding the wind* — Avoid exposure of the face to the wind. Build windbreaks or arrange to work on the lee side of buildings or trees (rester à l'abri du vent).

*Frostbite* — Use the buddy system, keeping a watch for the appearance of a white patch on another employee's face. If this happens the person affected should hold his/her bare hand over the area to warm it up. Fingers can be warmed by placing the hands close to the body, preferably under the armpits. If feet are frozen, the person affected should not walk, but should be transported to a warm place. Further treatment, in all frostbite cases, should be undertaken by a physician (gelure).

*Hypothermia* — Treatment is by slow re-warming. The victim should be protected from further heat loss by being wrapped in blankets and moved to a medical treatment centre or a hospital as soon as possible. Unconscious patients should be handled very gently. Hot water bottles or other warming devices, if available, should be no more than comfortably warm and must not be applied directly to the skin (hypothermie).

### Enquiries

Enquiries should be directed to the responsible officers in departmental headquarters who, in turn, may seek interpretation from the following:

> Safety, Health and Employee Services Group
> Staff Relations Division
> Human Resources Policy Branch
> Treasury Board Secretariat

## CHAPTER 6-4 - ADVISORY NOTICE - OCCUPATIONAL EXPOSURE TO BENZENE

Health Canada advises that exposure to benzene can be a serious health hazard.

The use of benzene should be avoided whenever possible. Substitutes can be used in almost all operations where benzene has been used as a solvent (see below). However, there may be a few operations where the use of benzene may be unavoidable (e.g., chemical synthesis or in certain analytical procedures). In these instances, meticulous care must be taken to prevent the escape of benzene vapour in order to reduce exposure of workers to the lowest practicable level. The high density of benzene vapour should be borne in mind and the workplace air should be tested periodically to ensure that precautions are adequate.

The most common use of benzene arises from its excellent solvent properties for a large number of materials. In substituting other solvents, it is recommended that the use of carbon tetrachloride and carbon disulphide be avoided. Preferred solvents include toluene, xylene, cyclohexane, aliphatic hydrocarbons, solvent naphthas and various alcohols, ketones, esters and some chlorinated derivatives of ethylene. However, it should be noted that adequate safety precautions are still required, depending upon the substitute.

Employees should not be exposed to benzene vapour in a concentration exceeding the time-weighted average recommended by the American Conference of Governmental Industrial Hygienists. Please refer to the Dangerous substances directive (chapter 2-2). Specific questions concerning exposure limits should be directed to the Medical Services Branch, Health Canada.

The following precautionary measures are also recommended:

(a) Employees should be informed of the hazards of exposure to benzene and advised of measures that should be taken to minimize such exposure.

(b) Operations and areas where benzene is used should be clearly identified and appropriate warnings displayed.

(c) Appropriate protective equipment should be used and operations involving benzene should be performed in enclosures provided with adequate ventilation to prevent the vapour from contaminating the work room. Information concerning protective equipment may be obtained from the Medical Services Branch, Health Canada.

(d) Periodic medical surveillance should be conducted for all workers employed in operations or areas where repeated exposure to benzene may occur, in accordance with the advice of the Medical Services Branch, Health Canada. The laboratory manager or person in charge should ensure that exposure or handling of benzene is appropriately indicated on the *General Physical Examination Report* (NHW 365).

(e)   Persons under 18, pregnant or nursing women, and persons with certain liver diseases may be at special risk and should, based upon the advice of Health Canada, be discouraged from working in areas where repeated exposure to benzene is likely to occur.

(f)   Whenever it is necessary to enter enclosures in which benzene has been recently used (e.g., for maintenance or repair), special health and safety precautions are required. See Hazardous confined spaces directive, chapter 2-7.

(g)   Benzene should be obtained in the smallest practicable quantities in order to minimize storage hazards.

Departments and agencies, particularly those involved in laboratory operations, should ensure that the use of benzene is avoided whenever possible, that exposures are kept to a minimum, and that the foregoing recommendations are implemented.

## Enquiries

Enquiries should be directed to the responsible officers in departmental headquarters who, in turn, may seek interpretation from the following:

> Safety, Health and Employee Services Group
> Staff Relations Division
> Human Resources Policy Branch
> Treasury Board Secretariat

## CHAPTER 6-5 - ADVISORY NOTICE - OCCUPATIONAL EXPOSURE TO SUNLIGHT

### Purpose

Concerns have been expressed about the effects of the sun on outdoor workers in the Public Service. These concerns have intensified as a result of recent announcements from the Government of Canada and other reputable sources on the depletion of the earth's ozone layer, the stratospheric shield that protects against hazardous ultraviolet (UV) rays.

This notice is to advise departments of their responsibility to minimize the hazard as far as possible and to inform their employees working outdoors of protective measures to be followed.

### Effects of Exposure

Health experts agree that excessive exposure to the sun, whether cumulative or in intense short periods, is hazardous to human health. Skin cancers (such as fatal melanomas), cataracts and suppression of the body's immune system are the most serious consequences of overexposure to the sun. Other effects include sunburn, increased aging of the skin, and heat-related conditions such as dehydration and related symptoms, rashes and heat stroke.

It should be noted that even in the shade, harmful effects may result from rays reflected off nearby water, snow, sand or cement. The sun's rays can also penetrate light cloud cover, fog and haze.

## Responsibilities

As an employer subject to the *Canada Labour Code*, Part II, the federal government has a responsibility to protect employees and inform them of potential occupational safety or health hazards.

## Protective Measures

Departments, in consultation with local occupational safety and health committees, should carefully review the various situations where employees are required to work outdoors and take all reasonably practicable measures to reduce employee exposure to the harmful effects of the sun.

Such measures could include avoidance of midday exposure (between 10:00 a.m. and 3:00 p.m.) when the sun's rays are most intense, where feasible, or providing shade in the form of tents or parasols when it is not otherwise available. Potable water must be provided as specified in the Treasury Board Sanitation standard to prevent dehydration and heat-related illnesses.

## Information/Education

It is also a departmental responsibility to inform employees working outdoors about sun-related health hazards and to advise them to protect themselves as follows:

–   wear suitable clothing, i.e. large-brimmed hats, long sleeves and long pants or skirts made of a tightly woven fabric that blocks the sun's rays yet allows perspiration to evaporate;
–   use sunglasses that protect against ultraviolet rays;
–   use effective sunscreens or sunblocks during periods of exposure.

Public Service occupational health nurses are prepared to help departments in providing educational sessions for employees. Information is also available from the Canadian Dermatology Association.

## Enquiries

Enquiries should be directed to the responsible officers in departmental headquarters who, in turn, may seek interpretation from the following:

Safety, Health and Employee Services Group
Staff Relations Division
Human Resources Policy Branch
Treasury Board Secretariat